Communicating Risk

Communicating in Professions and Organizations
Series Editor: **Jonathan Crichton**, University of South Australia

Titles include:

Glen Alessi and Geert Jacobs (*editors*)
THE INS AND OUTS OF BUSINESS AND PROFESSIONAL DISCOURSE RESEARCH
Reflections on Interacting with the Workplace

Christopher N. Candlin and Jonathan Crichton (*editors*)
DISCOURSES OF TRUST

Christopher N. Candlin and Jonathan Crichton (*editors*)
DISCOURSES OF DEFICIT

Jonathan Crichton
THE DISCOURSE OF COMMERCIALIZATION
A Multi-Perspectived Analysis

Jonathan Crichton, Christopher N. Candlin and Arthur S. Firkins (*editors*)
COMMUNICATING RISK

Cecilia E. Ford
WOMEN SPEAKING UP
Getting and Using Turns in Workplace Meetings

Sue Garton and Keith Richards (*editors*)
PROFESSIONAL ENCOUNTERS IN TESOL
Discourses of Teachers in Teaching

Rick Iedema (*editor*)
THE DISCOURSE OF HOSPITAL COMMUNICATION
Tracing Complexities in Contemporary Health Care Organizations

Louise Mullany
GENDERED DISCOURSE IN THE PROFESSIONAL WORKPLACE

Keith Richards
LANGUAGE AND PROFESSIONAL IDENTITY
Aspects of Collaborative Interaction

H. E. Sales
PROFESSIONAL COMMUNICATION IN ENGINEERING

Communicating in Professions and Organizations
Series Standing Order ISBN   978–0–230–50648–0
(*outside North America only*)

You can receive future titles in this series as they are published by placing a standing order. Please contact your bookseller or, in case of difficulty, write to us at the address below with your name and address, the title of the series and the ISBN quoted above.

Customer Services Department, Macmillan Distribution Ltd, Houndmills, Basingstoke, Hampshire RG21 6XS, England

Communicating Risk

Edited by

Jonathan Crichton
Senior Lecturer, University of South Australia

Christopher N. Candlin
Professor, Macquarie University, Australia

and

Arthur S. Firkins
Researcher, Macquarie University, Australia

Selection, chapter 1 and editorial content © Jonathan Crichton, Christopher N. Candlin and Arthur S. Firkins 2016
Individual chapters © Respective authors 2016

All rights reserved. No reproduction, copy or transmission of this publication may be made without written permission.

No portion of this publication may be reproduced, copied or transmitted save with written permission or in accordance with the provisions of the Copyright, Designs and Patents Act 1988, or under the terms of any licence permitting limited copying issued by the Copyright Licensing Agency, Saffron House, 6–10 Kirby Street, London EC1N 8TS.

Any person who does any unauthorized act in relation to this publication may be liable to criminal prosecution and civil claims for damages.

The authors have asserted their rights to be identified as the authors of this work in accordance with the Copyright, Designs and Patents Act 1988.

First published 2016 by
PALGRAVE MACMILLAN

Palgrave Macmillan in the UK is an imprint of Macmillan Publishers Limited, registered in England, company number 785998, of Houndmills, Basingstoke, Hampshire RG21 6XS.

Palgrave Macmillan in the US is a division of St Martin's Press LLC, 175 Fifth Avenue, New York, NY 10010.

Palgrave Macmillan is the global academic imprint of the above companies and has companies and representatives throughout the world.

Palgrave® and Macmillan® are registered trademarks in the United States, the United Kingdom, Europe and other countries.

ISBN 978–1–137–47877–1

This book is printed on paper suitable for recycling and made from fully managed and sustained forest sources. Logging, pulping and manufacturing processes are expected to conform to the environmental regulations of the country of origin.

A catalogue record for this book is available from the British Library.

Library of Congress Cataloging-in-Publication Data
Communicating risk / [edited by] Jonathan Crichton, Senior Lecturer, University of South Australia; Christopher N. Candlin, Professor, Macquarie University, Australia; Arthur S. Firkins, Researcher, Macquarie University, Australia.
    pages cm.— (Palgrave Studies in Professional and Organizational Discourse)
    ISBN   978–1–137–47877–1
    1. Risk communication.   I. Crichton, Jonathan, editor.   II. Candlin, Christopher, editor.   III. Firkins, Arthur S., 1963–
T10.68.C636   2016
361.001'4—dc23   2015023917

Typeset by MPS Limited, Chennai, India.

For Chris

Contents

List of Figures and Tables   x

Notes on Contributors   xii

1   Crucial Sites and Research Orientations: Exploring the Communication of Risk   1
    Christopher N. Candlin, Jonathan Crichton, and Arthur S. Firkins

Part I   Communicating Risk in Healthcare

2   Risk and Clinical Incident Disclosure: Navigating between Morality and Liability   17
    Rick Iedema, Donella Piper, Katja Beitat, Suellen Allen, Kate Bower and Su-yin Hor

3   'Being Diplomatic with the Truth': The Discursive Management of Risk in Accounts of People Leaving Forensic Psychiatric Settings   36
    Michael Coffey

4   Risk and Safety in Linguistic and Cultural Diversity: A Narrative Intervention in Residential Aged Care   51
    Jonathan Crichton and Fiona O'Neill

5   Choice, Risk, and Moral Judgment: Using Discourse Analysis to Identify the Moral Component of Midwives' Discourses   67
    Mandie Scamell and Andy Alaszewski

Part II   Communicating Risk in Legal Processes

6   Risk, Law, and Security   85
    Pat O'Malley

7   'Making a Raise' and 'Dusting the Feds': Contextualising Constructions of Risk and Youth Crime   103
    Joe Yates

Part III   Communicating Risk in Social Care

8   Communicating Risk in Youth Justice: A Numbers Game   121
    Stephen Case

9   Working with Risk in Child Welfare Settings   138
    Tony Stanley

Part IV Communicating Risk in Environmental Management and Biosecurity

10 Interpretive Environmental Risk Research: Affect, Discourses and Change — 155
Karen L. Henwood and Nick Pidgeon

11 The Communication and Management of Social Risks and Their Relevance to Corporate-Community Relationships — 171
Philippe Hanna, Frank Vanclay, and Jos Arts

12 How Structured Dialectical Discourse of Risk Eased Tension in North American LNG Siting Conflicts — 189
Susan Mello

13 Framing Risk and Uncertainty in Social Science Articles on Climate Change, 1995–2012 — 208
Christopher Shaw, Iina Hellsten, and Brigitte Nerlich

14 Between Two Absolutes Lies Risk: Risk Communication in Biosecurity Discourse — 229
Sue McKell and Paul De Barro

Part V Mediating Risk

15 Negotiating Risk in Chinese and Australian Print Media Hard News Reporting on Food Safety: A Corpus-based Study — 245
Changpeng Huan

16 The Uses of Biological Sciences to Justify the Risks of Children's Mental Health and Developmental Disorders in North American News Magazines: 1990–2012 — 267
Juanne N. Clarke and Donya Mosleh

17 'It's just statistics ... I'm kind of a glass half-full sort of guy': The Challenge of Differing Doctor-Patient Perspectives in the Context of Electronically Mediated Cardiovascular Risk Management — 285
Catherine O'Grady, Bindu Patel, Sally Candlin, Christopher N. Candlin, David Peiris and Tim Usherwood

Part VI Regulating Risk

18 Central Banking in Risk Discourses: 'Remaking' the Economy after Crisis — 307
Clea D. Bourne

19 Projecting a Definition of Risk Situation: Travel Advice and the Prudential Traveller — 323
Arthur S. Firkins and Christopher N. Candlin

20 Suicide Candy: Tracing the Discourse Itineraries of Food Risk 340
    Rodney H. Jones

Index 359

List of Figures and Tables

Figures

12.1	Graphical representation of social arena of risk in Québec – LNG debate	201
13.1	Distribution of frames across the 'risk' and 'uncertainty' categories	214
13.2	Distribution of papers by author country	214
13.3	Distribution of articles across the time period 1988–2012	215
13.4	Timeline of uncertainty frames, 1995–2012	216
13.5	Timeline of risk frames, 1995–2012	217
15.1	The framework of engagement	248
15.2	Examples of risk in relation to living creatures	255
15.3	Examples of risks in relation to babies	255
15.4	Examples of the semantic theme of government (G1.1) in the Australian corpus	256
15.5	Examples of the semantic theme of law and order (G2.1) in the Australian corpus	256
15.6	Percentage of key engagement features across Chinese and Australian corpora	257
15.7	Patterns of concordances of 'said' in the Australian food-safety corpus	258
15.8	Examples of collates of 'said' with 'he' in the Australian corpus	259
15.9	Distribution of attest sources in the Chinese food-safety corpus	261
15.10	Distribution of attest sources in the Australian food-safety corpus	262
20.1	'In the bin'	355

Tables

12.1	Alignment of Ryfe's (2003) factors for good public discourse and theoretical models of risk communication	193
13.1	Risk and uncertainty frames	212

15.1	Top 20 frequent lexical words in the Australian food-safety corpus	251
15.2	Top 20 frequent lexical words in Chinese food-safety corpus	252
15.3	Top 20 key words in the food-safety corpus against the BNC written sampler	253
15.4	Top 20 frequent semantic tags in the Australian food-safety corpus against the BNC written	254
15.5	Distribution of attest voices across Chinese and Australian corpora	260
16.1	Summary of magazine readerships	272
16.2	Answers to the three guiding questions	279
A17.1	Transcription conventions	302

Notes on Contributors

Andy Alaszewski is an emeritus professor at the University of Kent, Canterbury, the editor of *Health, Risk and Society*, and Director of the NHS-funded Research and Development Support Unit which provides support for research capacity development in Kent and Surrey. Since 1972 he has been the grant holder or co-grant holder of over 50 grants valued at over £3.2 million and his research has resulted in nearly 200 publications.

Suellen Allen is Director of Clinical Communication and Mental Health at the Australian Commission on Safety and Quality in Health Care. Her doctoral research focused on safety culture in maternity services, and she has published widely in the area of health communication.

Jos Arts is a special professor in the Department of Planning, Faculty of Spatial Sciences at the University of Groningen, the Netherlands. He is also strategic environment advisor with Rijkswaterstaat, Ministry of Infrastructure and Environment and a specialist in environmental impact assessment and environmental planning procedures.

Katja Beitat holds a master's in Communications and Media Science from the University of Leipzig, and a master's in International Business and Law from the University of Sydney. She works for the NSW Health Care Complaints Commission where she is responsible for outreach programmes. Her doctoral research focused on the relationship between medical practitioners and patients before, during, and after incidents in the patient's healthcare. Her thesis was published in 2015.

Clea D. Bourne is a lecturer at Goldsmiths, University of London. Her research focuses on the role played by communication professionals in deploying, interceding in, and influencing global discourses, and the ways in which these discourses privilege and marginalise their subjects. She has written several articles and chapters on financial communication. Her forthcoming book, *Trust, Power and Public Relations in Financial Markets*, will be published in 2016.

Kate Bower is a postdoctoral research fellow in the Faculty of Health at the University of Tasmania, Sydney campus. She is researching communication in healthcare incident disclosure funded by an ARC Discovery Grant. She has previously worked on the 100 Patients Project on incident disclosure for the Australian Commission on Safety and Quality in Health Care.

Christopher N. Candlin was Senior Research Professor Emeritus in the Department of Linguistics at Macquarie University, Sydney, and Fellow of the UK Academy of the Social Sciences. His research and publications lie in the critical analysis of professional/institutional discourses, particularly in the domains of healthcare and law. He published widely, edited/co-edited several international book series, and was a member of the editorial boards of key international journals. For two terms he was President of the International Association of Applied Linguistics (AILA). He co-edited, with Srikant Sarangi, the *Journal of Applied Linguistics and Professional Practice*. Recent publications include *Discourses of Deficit* (2011) and *Discourses of Trust* (2013) both co-edited with Jonathan Crichton.

Sally Candlin is an honorary senior research fellow in the Department of Linguistics at Macquarie University, Sydney. She has published widely in international journals and is the invited author of numerous book chapters. She is the author of *Therapeutic Communication: A Lifespan Approach* (2008) and the co-author with Peter Roger of *Communication and Professional Relationships in Healthcare Practice* (2013). She holds a master's degree in public health from the University of Hawaii, an honours degree in linguistics and psychology and a PhD in Linguistics from Lancaster University, UK. She is a registered nurse, registered midwife, and health visitor.

Stephen Case is an associate professor at the Centre for Criminal Justice and Criminology at Swansea University. He co-authored *Understanding Youth Offending: Risk Factor Research, Policy and Practice* (with Prof. Kevin Haines) and has published numerous academic articles in international, multi-disciplinary journals such as *Youth Justice*, the *Howard Journal*, *Children and Society*, the *Journal of Substance Use* and the *British Journal of Social Work*. His 2007 article 'Questioning the evidence of risk' won the British Society of Criminology Brian Williams Prize for original scholarship.

Juanne N. Clarke is a medical sociologist at Wilfrid Laurier University in Waterloo, Ontario, Canada. She is interested in critical analyses of medicalisation and biomedicalisation in the context of increasingly 'risky' post-modernity. The ways that illnesses are constructed to serve the interests of large social, economic, and political interests is of particular concern. In recent years much of her work has focused on the 'mediation' of, or the portrayal of, health and illness in mass media such as popular high-circulating magazines, newspapers, and blogs.

Michael Coffey is an associate professor at Swansea University. He specialises in researching the delivery and organisation of community mental healthcare with a specific interest in exploring multiple perspectives. He has published research on aspects of transition and identity work in forensic mental healthcare, experiences of auditory hallucinations and also occupational stress. He is investigating recovery outcomes of care co-ordination in mental health and social care services.

Jonathan Crichton is Senior Lecturer in Applied Linguistics and Member of the Research Centre for Languages and Cultures at the University of South Australia. His research focuses on the role of language in professional–lay interactions. He has written interdisciplinary papers in a wide range of international journals and edited collections, and is the author of *The Discourse of Commercialization* (2010), and co-editor, with Christopher N. Candlin, of *Discourses of Deficit* (2011) and *Discourses of Trust* (2013).

Paul De Barro is a senior principal research scientist and Leader of the Reducing Likelihood Theme in the CSIRO Biosecurity Flagship. He is internationally recognised for his research on various aspects of biosecurity. He holds a PhD from the University of Adelaide in ecology and has published over 100 ISI cited papers.

Arthur S. Firkins is a researcher in the Department of Linguistics at Macquarie University, Sydney. He has had a professional engagement with risk analysis and risk communication throughout his professional career and risk remains the focus of his research. His thesis titled 'Discourse and the framing of risk' is focused on the communication of risk in organisations and professions. Firkins has co-authored several papers on risk with Christopher N. Candlin, most recently contributing a chapter to *Discourses of Deficit*.

Philippe Hanna is a PhD student in the Department of Cultural Geography at the University of Groningen, the Netherlands. He has an MBA in project management and has previously been an anthropologist working in a variety of settings with indigenous peoples in Brazil.

Iina Hellsten is an associate professor at the VU University Amsterdam, Department of Organisation Sciences and affiliated to the Network Institute. Her research has focused on the politics of metaphors in the public debates on genetics, genomics and climate change. She is interested in the dynamics of communication networks, and the development of new methods for the analysis of web-based texts.

Karen L. Henwood is a professor at the Cardiff School of Social Sciences. She has a psychological background, and a long-standing research interest in understanding the dynamics of environmental and sociocultural change, with particular reference to discourse, knowledge, relationships and subjectivity. In her empirical work she conducts in-depth longitudinal and community case studies, and uses interpretive, qualitative methods to engage with local communities on issues of risk, environmental controversy, and identity.

Su-yin Hor is a postdoctoral research fellow at the School of Health Sciences, University of Tasmania. She studies patient safety using ethnographic and video-reflexive ethnographic methods, exploring how healthcare workers create safety amidst the complexity of their everyday work.

Changpeng Huan is pursuing a PhD in the Department of Linguistics at Macquarie University. His current research focuses on the contrastive analysis of journalists' stance in Chinese and Australian hard news print media in relation to discourses of risk. His publications appear in *The Journal of Pragmatics*, *The Journal of Contemporary Linguistics* (Chinese) and *Discourse & Society*. His research has been presented at international conferences and symposiums.

Rick Iedema manages the Research Portfolio at the NSW Ministry of Health – Agency of Clinical Innovation. He is Professor of Healthcare Innovation at the University of Tasmania's Faculty of Health. Recent books include *Visualising Healthcare Improvement* (2013) and *Communicating Safety and Quality in Health Care* (2015).

Rodney H. Jones is an associate professor in the Department of English at City University of Hong Kong. His research interests include professional communication (especially health communication), computer-mediated discourse, and language and sexuality. He is co-author (with Christoph Hafner) of *Understanding Digital Literacies: A Practical Introduction* (2012), and author of *Health and Risk Communication: An Applied Linguistic Perspective* (2013).

Sue McKell is Deputy Director (Innovation) of the Institute for Social Science Research at the University of Queensland where she leads business processes for research development and translation. She has held a number of strategic communication and technology transfer positions for national research organisations, and has lectured at UQ's Business School. She holds a master degree in discourse analysis, a Bachelor of Commerce, and Honours in Arts from UQ.

Susan Mello is Assistant Professor of Communication at the Ohio State University. Her research interests lie at the intersection of risk perception, communication, and the environment. Mello was awarded a 2007–2008 US Fulbright Fellowship to l'Université Laval in Québec City, Canada, where she examined the impact of risk communication on regional energy development. She received her master and PhD in communication from the University of Pennsylvania.

Donya Mosleh is an MA candidate at Wilfrid Laurier University of Waterloo, Ontario. Her academic interests concern critical perspectives on health and illness. Mosleh's research focuses on the everyday lived materiality of post-modern health and the severe restrictions this imposes for certain groups and individuals.

Brigitte Nerlich is Professor of Science, Language, and Society at the University of Nottingham. She has a background in linguistics and philosophy and now works mainly on issues relating to the framing of scientific controversies in the media and in policy. She is a fellow of the Academy of Social Sciences.

Catherine O'Grady is an honorary associate of the Department of Linguistics at Macquarie University, and a teacher and researcher with an abiding interest in the application of discourse analytical findings to clinical contexts. Awarded the Vice Chancellor's Commendation for her PhD thesis, a discourse analytical study of communicative expertise required for the general practice of medicine, her recent publications examine empathy and trust in primary care and surgical settings.

Pat O'Malley is an honorary professor at the University of Sydney, and previously was Canada Research Chair in Criminology and Criminal Justice in Ottawa. Most of his work has focused on issues of risk and security, especially in relation to crime prevention, drug harm minimisation, insurance and insurance law. He has been awarded various professional honours including the American Society of Criminology's 2000 Sheldon and Eleanor Gleuck Award for contributions to international criminology, and distinguished lectureships at Amherst College, the University of Toronto, Leipzig University, New York University and the Victoria University of Wellington. In 2012 he was elected as a Fellow of the Academy of the Social Sciences in Australia (FASSA).

Fiona O'Neill is a researcher in the Research Centre for Languages and Cultures at the University of South Australia. Her doctoral research focused on the intercultural experience of multilingual professionals. Her publications include 'From classroom to clinical context: the role of language and culture in communication for nurses using English as a second language' (*International Journal of Nursing Studies*, 2011) and 'Making sense of being between languages and cultures: a performance narrative inquiry approach' (*Journal of Language and Intercultural Communication*, 2013).

Bindu Patel is a PhD student at the University of Sydney and Project Manager at the George Institute for Global Health. Her research focuses on the process evaluation of the impact of a computerised multifaceted quality improvement intervention to improve CVD management in primary healthcare. Patel has a particular interest in improving health systems and outcomes.

David Peiris is Head of Primary Health Care Research at the George Institute for Global Health and an associate professor, Sydney Medical School, University of Sydney. He is a Sydney GP and his research interests are in the field of implementation of science which looks at strategies to translate evidence into routine healthcare.

Nick Pidgeon is Professor of Psychology at Cardiff University, where he directs the Understanding Risk Research Group (www.understanding-risk.org). His research looks at risk perception, risk communication, and public engagement with environmental and technological risks. He was co-editor with Roger Kasperson and Paul Slovic of *The Social Amplification of Risk*, 2003 and with the late Barry Turner of the second edition of *Man Made Disasters*, 1997. He is a fellow of the Society for Risk Analysis and an honorary fellow of the British Science Association.

Donella Piper consults to a number of health service providers, universities and policy bodies on medico-legal, regulatory, and safety and quality issues. She is a social scientist with a legal background. Piper completed her PhD at the Centre for Health Communication at the University of Technology, Sydney, in 2010.

Mandie Scamell is a medical anthropologist and midwife specialising in risk and the maternity services in the UK. She joined City University London in 2013 having previously been part of the Florence Nightingale School of Nursing and Midwifery at King's College London. Her main area of work has been on midwifery care in the UK, with particular interests in clinical governance and institutionalised risk management technologies and in the culture and organisation of maternity care.

Christopher Shaw is a research fellow at the Environmental Change Institute, University of Oxford where his research explores the mediating role of language in climate politics. Shaw's work has a particular focus on the interdisciplinary study of risk communication. He has written extensively on the two-degree limit and the politics of defining an acceptable level of climate risk.

Tony Stanley is the principal child and family social worker for Children's Social Care at the London Borough of Tower Hamlets. In this role, and as part of the senior management team, he is involved in strategically reforming Tower Hamlets' child welfare system, and he provides social work leadership across the local authority. Holding postgraduate awards in sociology and social work, he has published in both, with a particular focus on risk and child welfare. Stanley has a practice background in statutory child welfare.

Tim Usherwood is a professor and the head of the Department of General Practice, University of Sydney, Westmead. His clinical practice is in a community-controlled aboriginal medical service, and his research focuses on the evaluation of interventions to improve health outcomes in chronic disease and primary care.

Frank Vanclay is Professor and Head of Cultural Geography in the Faculty of Spatial Sciences at the University of Groningen, the Netherlands. Previously located in Australia, Vanclay specialises in the areas of social performance and social impact assessment, social aspects of place, community engagement, and social aspects of agriculture, farming, and natural resource management

Joe Yates is Director of the School of Humanities and Social Science in the Faculty of Arts, Professional and Social Studies at Liverpool John Moores University. His current research interests and publications revolve around youth justice and policy responses to marginalised children involved in crime and antisocial behaviour.

1
Crucial Sites and Research Orientations: Exploring the Communication of Risk

Christopher N. Candlin, Jonathan Crichton, and Arthur S. Firkins

Issues of risk are foundational to people's lives in contemporary societies, a fact sharply highlighted by the recent history of practices associated with the financial markets, science and technology, workplace health and safety, environmental policy and biosecurity, law enforcement and criminal justice. Exploring such issues is central to our understanding of how professional practice impacts on human relationships in contemporary social life.

Drawing on invited and original contributions from key practitioners and researchers, this book explores how people routinely and across professional domains and sites are discursively engaged in the assessment, management, and communication of risk in ways that materially affect human lives. The book thus recognises that risk, as a major theme in contemporary social and professional life, is both an overarching theoretical construct and one which is constructed in communication among people across diverse sites of practice according their particular expertise and circumstances. It is the argument of the book that if we are to understand the significance of risk in contemporary life, both constructions of risk – the macro and the micro – need to be brought into play, explored, and engaged with each other through a research process that involves interdisciplinary dialogue between professionals, participants, and researchers.

The need for this agenda is pressing. We live in a world described by Beck (1992, 1998, 1999) and by Giddens (1991, 1998) as a 'risk society' in which risk is a 'systematic way of dealing with hazards and insecurities induced and introduced by modernization itself' (Beck, 1992, p. 21) and 'the concept of risk becomes fundamental to the ways that lay actors and technical specialists organize the social world' (Giddens, 1991, p. 3). The public and private spheres increasingly turn on the management of a portfolio of disparate risks, while the assessment of risk has become a focal activity of government, organisations, and the professions, where the communication of risk occurs as a crucial component of daily work. In essence, as Beck argued, risk is the defining macro construct of the modern age. At the same time, it is increasingly imperative to understand how different 'societal members' define, analyse and communicate risk to a range of increasingly diverse audiences, and for what purposes (Horlick-Jones, 2005). That is to say that risk communication has increasingly become a rhetorical activity and the accomplishment of such activity across a wide range of professional fields is fundamentally embedded in discourse and interaction. These include the fields of health (Alaszewski, 2005a, 2005b; Alaszewski & Horlick-Jones, 2003; Hoffman, Linell, Lindh-Astrand, & Kjellgren; Linell, Adelswärd, Sachs, Bredmore, & Lindstedt, 2002; Moore, Candlin, & Plum, 2003), with specific foci such as genetic counselling (Sarangi, Bennert, Howell, & Clarke, 2003; Sarangi & Clarke, 2002; Wood, Prior, & Gray, 2003) and health policy (Bancroft & Wilson, 2007); as well as social work (Firkins & Candlin, 2006, 2011; Hall, Slembrouck, & Sarangi, 2006; Hall & Slembrouck, 2009); international security (Jore & Kain, 2010); and science and technology (Grundmann & Krishnamurthy, 2010; Petersen, 2005).

Despite the emphasis placed by many theorists on the rational and relational nature of risk, what appears to be absent in studies that have sought to define, categorise, and appraise risk, especially within organisations, is a significant focus on how such risk is *communicatively* and jointly accomplished through interpersonal interaction employing various modalities and across diverse contexts of use. Those studies which have done so have focused on few and single domains and sites, and not explicitly sought the inter-domain and inter-site perspective provided in this collection. Exceptions include journal Special Issues devoted to risk discourse (for example, Candlin & Candlin, 2002; Sarangi & Candlin, 2003b; Zinn, 2010) and a sustained focus in the journal *Health, Risk & Society* (see, for example, Alaszewski, 2005c; Horlick-Jones, 2003; Sarangi & Candlin, 2003a).

Taking this orientation to risk research, this book brings together macro and micro perspectives on the communication of risk within and across professional domains and, most consequentially for participants, 'crucial sites' (Candlin, 2002b) within which:

> occur what I have called critical moments, where the communicative competence of the participants is at a premium and at its

greatest moment of challenge. This may be due to the heightened significance of the subject matter, for personal, professional, or ideological reasons. These moments may be defined generically across topics and conditions, such as the breaking of bad news, or individually sited within particular conditions in particular contexts, such as the issue of disclosure of sexual and HIV+ status ... What then becomes interesting is to map the critical moments on to the crucial sites and to calibrate these against the participant perspectives of those involved. (p. 10)

This focus immediately raises the question of who a site is crucial for and how such critical moments could be identified, described, interpreted and explained within a programme of research that is accountable to participants. For the researcher this raises the question of how to conceptualise the 'site' of such research, and – more specifically – what social/theoretical understanding we can bring in doing so, and how we can warrant our answers to this question.

No single methodology will be able to match the demands of such a programme. Rather, it requires the engagement of researcher, practitioner, and participant expertise, brought to bear on the integration of multiple methodologies in seeking to make visible and connect the different perspectives that may be relevant (Candlin, 1997, 2006). Such a 'multi-perspectival' approach (Candlin & Crichton, 2011; Crichton, 2010) is not limited to particular theoretical positions or methodologies but open to and able to bring into play multiple theoretical and methodological perspectives on the communication of risk depending on relevant and emergent understandings of the research site under scrutiny. These understandings will depend on collaborative interpretation among researchers and participants, raising what Sarangi has identified as 'the analyst's paradox' (Sarangi, 2007): the problem of how the researcher can align her analysis with the perceptions of participants without either having to become a faux participant or, if not, being irrelevant to their world. Achieving this 'mutuality of perspective' (Sarangi & Candlin, 2001) among researchers and participants is a particular challenge because:

For the participants, then, workplace discourse is a process; for the analyst it is inevitably a product, and, so achieving a reciprocity of perspectives is not only a matter of mutualising view and stance, it is also a matter of (re)vitalising what is necessarily an ecology. (Candlin, 2002a, p. 5)

Key to meeting this challenge is an orientation to research that 'starts with the site'; in other words, that acknowledges from the outset Cicourel's (1992, 2007) call for 'ecological validity':

> Validity in the non-experimental social sciences refers to the extent to which complex organisational activities represented by aggregated data from public and private sources and demographic and sample surveys can be linked to the collection, integration and assessment of temporal samples of observable (and when possible recordable) activities in daily life settings. Fragments of discourse materials always are shaped and constrained by the larger organizational settings in which they emerge and simultaneously influenced by cognitive/emotional processes despite the convenience of only focusing on extracted fragments independently of the organizational and cognitive/emotional complexity of daily life settings ... the challenge remains how daily life activities simultaneously constrain and shape more complex organizational structures. (1992, p. 736)

Cicourel here underscored the need for sensitivity to the different participant perceptions and 'interpenetrating contexts' that localise and situate any particular instance of communication: that is, as he (1992, p. 294) explained:

> Verbal interaction is related to the task in hand. Language and other social practices are interdependent. Knowing something about the ethnographic setting, the perception of, and characteristics attributed to, others, and broader and local organisational conditions becomes imperative for an understanding of linguistic and non-linguistic aspects of communicative events.

Acknowledging the institutionally situated, locally accomplished nature of risk communication makes visible as a focus of research sites involving professional communication that is institutionally located. Particularly relevant here is Sarangi and Roberts' (1999) account of how such sites are characterised by the intersection of both professional and institutional discourses that may be more or less commensurate, serving different or competing purposes, and creating the potential for shifting or realigned constructions of professional, institutional, and personal experience (Roberts & Sarangi, 1999). For example, drawing on ethnographic data Yates (Chapter 7) examines the institutional context of the contemporary policy trajectory in youth justice and the risk research paradigm which underpins it, arguing that the voices of young people provide insight into the complexity of risk in their worlds – a complexity which is easily obfuscated by this risk factor paradigm. In the context of health, O'Grady et al. (Chapter 17) use a combination of discourse analysis and ethnography to highlight the disparate interpretations that patients and doctors might bring to risk calculations by examining the use of software designed to assist general practitioners to engage patients in considering their cardiovascular disease risk. And drawing on critical discourse analysis McKell and De Barro (Chapter 14) reveal how risk activities described in public messages about biosecurity negotiate, and make trade-offs between, institutional, professional and personal knowledge and responsibility.

Together the chapters in the book elaborate, and the book as whole models, this inter-domain and inter-site 'mapping' of crucial sites explored through the thematic focus on risk. The communication of risk is revealed, not as restricted to particular disciplinary formulations or theoretical orientations, but as inherently and multiply interpretable, depending on the particular locations, participants, professional, institutional and research orientations and modes of collaboration between and among participants, and between them and researchers. From this inter-domain and inter-site perspective it becomes clear that risk is not simply concerned with the identification of 'hazards', in its negative projection and 'opportunities' in its positive projection. Risk is also and crucially implicated in communicative issues of power (essentially, who defines risk and who challenges them); categorisation (how risk are categorised and given priorities); distribution (how risks are distributed through a community and how such distribution is controlled); and the cross-cutting issues of regulation (how systems of governance are applied to regulate risks); negotiation (how social and cultural interests and values affect the framing, interpretation, and presentation of risks); and mediation (how the communication of risks is mediated through, for example, regulatory frameworks, methods of assessment and modelling, new technologies, media organisations, public relations, marketing and social networks).

By taking this orientation, the book as a whole seeks to enrich and explore the potential of risk as an overarching and motivating theme (Candlin & Crichton, 2011) informing applied linguistic, sociological, professional and communication research. The volume thus centrally positions risk as part of a broader research orientation premised on communication in interaction, highlighting the following key ways of positioning the construct of risk:

- Risk as based on intention and choice, and socially and contextually located
- Risk as not an event or state but a process
- Risk as relational, interpersonal and intersubjective
- Risk as a foundational socio-cultural category
- Risk as a strategic accomplishment involving risk makers, risk takers, risk perceivers and risk receivers in relation to the objects, processes and outcomes of risk
- Risk as diverse in its accomplishment, in terms of people, domains, sites and foci of risk concern

This orientation to risk underscores the need for such research to be practically relevant (Roberts & Sarangi, 1999) to specific sites and participants, and foregrounds the question of how this relevance is to be accomplished within particular projects. The orientation includes, but is not restricted to, risk as associated with particular interpretive repertoires, in particular social, organisational, and professional settings. It naturally implicates, as Luhmann (1979) emphasised, associated and personally, professionally, and organisationally relevant themes such as trust, accountability, blame, stigma, or confidence, as well as research oriented constructs such as identity, capacity, and agency, depending on the particular site of engagement (Candlin & Crichton, 2013). For example, Coffey (Chapter 3) draws on discourse analysis of talk from people leaving forensic settings and returning to live in the community to argue that the successful handling of concerns about mental illness and risk clears space for participants to deploy emergent identities. And in the context of clinical incident disclosure, Iedema et al. (Chapter 2) draw on more than 300 in-depth interviews conducted with clinicians and healthcare consumers to examine how errors in healthcare are disclosed to the harmed patient and/or their family, arguing that disclosure conversations harbour the risk of reducing complex circumstances to simplistic causation and reductive attributions of guilt and blame.

The orientation taken here seeks to distinguish risk from semantically allied terms such as danger or hazard and encompasses both external and manufactured risk (Giddens, 1998), risk perception (Slovic, 1987), risk as understood from particular cultural perspectives (Rayner, 1992) and risk as understood from particular institutional memberships (Wynne, 1992, 1996, 2002). This orientation is brought out in sharp relief by Scamell and Alaszewski (Chapter 5) who draw on ethnographic discourse analysis to explore the ways in which midwives' discourse on risk and place of birth take place at the intersection of two discrete imperatives: to provide pregnant women with choice over where and how they give birth; and to protect mothers and babies from harm.

More broadly, this orientation acknowledges that the majority of fields of social inquiry have been redefined in terms of the identifying and communicative managing of risk. Quite generally, the circumstances, responsibilities and entailments of risk have been shifted to individuals, leaving decision-making process to fall onto the client, the patient or the customer. For example, in the context of food health risks, Jones (Chapter 20) argues that decisions regarding risk take place at a complex nexus where different people, texts, objects, and practices, each with their own histories, come together. In such contexts, professional workers, be they a lawyer, nurse, engineer, economist, or social worker, must engage in the activities of risk assessment and risk communication; it falls to them to define how risk is to be constructed in relation to their field. This point is taken up, in the context of social work, by Stanley (Chapter 9) who explores how risk assessment has taken a discursive hold in the work of child protection, affecting and shaping practice and decision making; similarly, but now in the context of residential aged care, Crichton and O'Neill (Chapter 4) report on a collaborative study that drew on a narrative intervention to reframe and enhance the communication of risk and safety among staff and residents in aged care. This expansion of risk assessment into an onus on *all* professionals brings its own risks, for as Candlin and Candlin (2002, p. 130) have explained:

> Risk talk can mean both talking about risks and generating risks within talk, and can be initiated by both parties to the interaction. Risk poses a dilemma for professionals (Adelswärd & Sachs, 1998); to talk about risks may exacerbate tensions concerning risk, yet to avoid talk about risk may also lead to anxiety. Risk talk is, accordingly, a risky business for all participants.

This in turn presents the challenge, explored throughout the chapters in this book, of how assessments of risk are to be communicated in ways that make sense for and to those who may be affected by what is at stake but do not share the expertise to interpret the assessment. The challenge is compounded when the bases and calculation of risk are institutionally and discursively distanced from those who need to work with them and are affected by them. Here, for example, O'Malley (Chapter 6) examines how the development of crime prevention, risk-based policing, the redefinition of offences in risk terms, and the use of risk factors in sentencing and parole have transformed much of the field of law enforcement and criminal justice. In the context of liquefied natural gas projects, Mello (Chapter 12) combines in-depth stakeholder interviews with analysis of secondary sources to explore the results of divergent risk-communication strategies. Also in the area of corporate-community relations, but now in relation to cross-cultural communication, Hanna et al. (Chapter 11) consider the consequences of the corporate culture of risk assessment on Indigenous communities. These chapters exemplify how the constructions of risk that are in play in contemporary risk communication may be both anonymous and authoritative, a feature of risk summarised by Beck:

> We no longer choose to take risks, we have them thrust upon us. We are living on a ledge – in a random risk society, from which nobody can escape. Our society has become riddled with random risks. Calculating and managing risks which nobody really knows has become one of our main preoccupations ... The basic question here is: how can we make decisions about a risk we know nothing about? (1998, p. 12)

Allied to risk as an unknowable imposition, the institutionalisation of risk creates powerful – because taken for granted and therefore unnoticed – ways in which potentially unaccountable constructions of risk underlie the categories by which issues are framed and people are grouped for institutional purposes (Sarangi & Candlin, 2003a). This point is exemplified in the context of financial regulation by Bourne (Chapter 18) who examines the construction of risk during a 2008 Congressional hearing into the global financial crisis and the role of US regulators. Similarly, but in the context of youth justice, Case (Chapter 8) explores how risk is constructed within risk assessment and risk-focused intervention practices. In the context of climate change, Shaw, Hellsten, and Nerlich (Chapter 13) trace the framing of risk and uncertainty through the recent history of scholarly articles in that domain; and, in the context of popular media, Clarke and Mosleh (Chapter 16) analyse the construction of risks associated with children's mental health issues in articles in high-circulating English-language news magazines. The focus on institutional and cultural variability in the construction of risk is taken up by Huan (Chapter 15), who draws on concordance analysis to examine how the risk of food safety is negotiated in Chinese and Australian print media. And in the context of international security and safety, Firkins and Candlin (Chapter 19) examine how a 'definition of risk situation' as framed by a governmental travel authority is projected to a specific audience, 'the traveller'.

By foregrounding the inter-domain and inter-site complexity of communicative challenges around risk, the book recognises that *all* professionals are in fact risk communicators and that risk communication is not the domain of a narrow field of expert risk communicators. Hence, in part, why the book in recognising this generality seeks to explain risk communication across a diverse number of professional sites, while at the same time arguing how *across* such sites the essential characteristics of risk communication can be defined. The point is illustrated by the chapters as a whole, and exemplified within the context of research on environmental risk and socio-technical change by Henwood and Pidgeon (Chapter 10) who draw on case studies of living with nuclear risk and gender and risk to argue for the value of interpretive risk research across such domains.

Accordingly, the book models an agenda for risk research that brings together professionals with expertise in diverse sites of practice with those researchers and practitioners from a commensurate range of theoretical and analytical orientations who are engaged in the communication of risk. The individual chapters and the book as a whole, show how this engagement of expertise in communication and risk within and across diverse sites can offer fresh insights into the overarching concept of risk, and how this is realised communicatively across different sites and professional domains, and as a consequence can provide mutually and practically relevant insights for both professional and analytical practice.

The overall objective is for the book to clarify the twin themes of communication and risk by engaging each with the other, within, and across the sites. Consequently, it is the overarching argument of the book that:

1) Risk is informed by tacit models of professional understandings, discursively realised, which invoke positive and negative relationships in terms of which risk comes to be presumed as a theme against which actions and their proponents are judged, their identities co-constructed, and the institutions which they represent measured, at particular sites.
2) A research agenda is necessary which focuses on the interdisciplinary identification and analysis of risk as a situated accomplishment – involving 'joint problematisation' (Roberts & Sarangi, 1999) – among

participants and researchers, offering ways of elucidating underlying issues, and holding out the promise of explaining how risk has come to be a criterial theme in terms of which relationships, individuals, and groups are identified and appraised, and the effectiveness of actions, pro and contra, evaluated.

3) It is by reference to the particular modes of inter-professional engagement among researchers and their interactions with participants at these sites that the meanings of communication and risk may be constructed, extended, and interpreted in such research.

These sites may involve inter- or intra-professional encounters in particular professions and organisations, whether or not engaging laypersons, exemplified by those domains and sites represented among the chapters in the book. Each selected domain is represented by one or more chapters in which authors report on how they have explored the central theme of communicating risk. Each chapter illustrates, and the chapters taken as a whole, present and explore, distinctive perspectives arising from these different inter-relationships. These include:

- Approaches to defining communication in professional practice: including interaction, communication, construction, discursive strategy, reflexivity, interdiscursivity, and the need to adopt both historical and contemporary perspectives
- Approaches to defining risk: including locating the origins, appraisal, management, and mitigation of risk in relation to particular professions sites, events, participants: issues of positioning and subjectivisation; and differential understandings and social-theoretical orientations
- Approaches to researching the communication of risk: ways of interconnecting different methodological perspectives in research in relation to specific sites; how such perspectives are grounded in different motivational relevancies (Sarangi & Candlin, 2001); inter-professional relationships of researchers and joint problematisation with participants, and challenges of achieving practical relevance (Roberts & Sarangi, 1999)
- Approaches to locating sites of the communication of risk: domains, sites, participants, data and foci within particular sites
- Approaches to deriving practical action consequent upon research into the communication of risk: including systemic interventions within organisations, education initiatives, mentoring and professional development

In developing this argument, the book addresses the methodological question in general, and at each site of engagement, of how one might go about such a research agenda, and takes up the interrelated question of how we are to understand the ways in which methodology can be 'commensurate' with such a focus. In so doing, it provides not only a response to the question of what constitutes 'communicating risk' but also how manifestations and interpretations of risk may be defined and researched.

For that reason, the chapters in the book construct the theme of risk in distinctive ways, displaying how risk can be illuminated by exploring diverse and crucial sites of communicative engagement. At the same time, such engagements parallel themselves across domains, revealing quite general cross-cutting critical moments of risk-accountability, perception and mitigation including moments of risk-anticipation, management, exacerbation, and reduction. In doing so, the book elaborates 'communicating risk' as an overarching 'focal theme' (Roberts & Sarangi, 2005) that can inform research and practice across diverse sites of professional, institutional, and disciplinary engagement and practice. Elucidated here are underlying issues linked to the unlocking and making available of expertise relevant to risk discovery, risk perception, risk analysis and risk communication among persons and organisations. At the same time, and now of more general relevance, the chapters of the book argue that it is by reference to the particular modes of inter- and intra-professional engagement among researchers, and their interactions with participants at these sites, that the meanings of communication and risk can be constructed, interpreted, and explained.

In developing the argument of the book, the chapters are organised from the perspectives of domains, sites, and the overarching processes that shape risk and communication across them. Accordingly, the chapters are organised into six sections. The first four sections foreground particular domains and illustrative sites that may be associated within and across them, bringing together studies of diverse sites within the broad domains of Healthcare, Legal Processes, Social Care, and Environmental Management and Biosecurity. The final two sections move from the perspectives of particular domains to processes of Mediating and Regulating Risk that overarch such domains and sites: this allows the book to introduce further domains and sites, namely: Mass Media Representations, Technology Mediated Health Care, Financial Regulation, International Risk Assessments, and Product Safety and Regulation. Organised in this way, the book elaborates and guides readers through the multiple and dynamic, macro and micro, inter-domain and inter-site interrelations that shape processes of communicating risk.

References

Adelswärd, V., & Sachs, L. (1998). Risk discourse: Recontextualisation of numerical values in clinical practice. *Text, 18,* 191–211.
Alaszewski, A. (2005a). A person-centred approach to communicating risks. *PloS Medicine, 2,* 93–95.
Alaszewski, A. (2005b). Risk communication: Identifying the importance of social context. *Health, Risk & Society, 7*(2), 101–105.
Alaszewski, A. (2005c). Risk, safety and organizational change in health care? *Health, Risk & Society, 7*(4), 315–318.
Alaszewski, A., & Horlick-Jones, T. (2003). How can doctors communicate information about risk more effectively? *British Medical Journal, 327,* 728–731.
Bancroft, A., & Wilson, S. (2007). The 'risk gradient' in policy on children of drug and alcohol users: Framing young people as risky. *Health, Risk & Society, 9*(3), 311–322.
Beck, U. (1992). *Risk society: Towards a new modernity.* London: Sage.
Beck, U. (1998). Politics of risk society. In J. Franklin (Ed.), *The politics of risk society* (pp. 9–22). Cambridge Polity.
Beck, U. (1999). *World risk society.* Cambridge: Polity Press.
Candlin, C. N. (1997). General editor's preface. In B.-L. Gunnarsson, P. Linell & B. Nordberg (Eds.), *The construction of professional discourse* (pp. viii–xiv). London: Longman.
Candlin, C. N. (2002a). Introduction. In C. N. Candlin (Ed.), *Research and practice in professional discourse* (pp. 1–36). Hong Kong: City University of Hong Kong Press.
Candlin, C. N. (2002b). *Reinventing the patient/client: New challenges to healthcare communication.* Cardiff University: Healthcare Communication Research Centre.
Candlin, C. N. (2006). Accounting for interdiscursivity: Challenges to professional expertise. In M. Gotti & D. Giannone (Eds.), *New trends in specialized discourse analysis* (pp. 21–45). Bern: Peter Lang.
Candlin, C. N., & Candlin, S. (2002). Editorial: Discourse, expertise, and the management of risk in health care settings. *Special Issue of Research on Language and Social Interaction, 35*(2), 115–137.
Candlin, C. N., & Crichton, J. (2011). Introduction. In C. N. Candlin & J. Crichton (Eds.), *Discourses of deficit* (pp. 1–22). Basingstoke: Palgrave Macmillan.
Candlin, C. N., & Crichton, J. (2013). From ontology to methodology: Exploring the discursive landscape of trust. In C. N. Candlin & J. Crichton (Eds.), *Discourses of trust* (pp. 1–18). Basingstoke: Palgrave Macmillan.
Cicourel, A. V. (1992). The interpenetration of communicative contexts: Examples from medical encounters. In A. Duranti & C. Goodwin (Eds.), *Rethinking context: Language as an interactive phenomenon* (pp. 291–310). Cambridge: Cambridge University Press.
Cicourel, A. V. (2007). A personal, retrospective view of ecological validity. *Text & Talk, 27*(5/6), 735–759.
Crichton, J. (2010). *The discourse of commercialization.* Basingstoke: Palgrave Macmillan.
Firkins, A., & Candlin, C. N. (2006). Framing the child at risk. *Health, Risk and Society, 8*(3), 273–291.

Firkins, A., & Candlin, C. N. (2011). 'She is not coping': Risk assessment and claims of deficit in social work In C. N. Candlin & J. Crichton (Eds.), *Discourses of deficit* (pp. 81–98). Basingstoke: Palgrave Macmillan.

Giddens, A. (1991). *Modernity and self identity*. Cambridge: Polity Press.

Giddens, A. (1998). Risk society: The context of British politics. In J. Franklin (Ed.), *The politics of risk society* (pp. 23–34). Cambridge: Polity.

Grundmann, R., & Krishnamurthy, R. (2010). The discourse of climate change: A corpus-based approach. *Critical Approaches to Discourse Analysis across Disciplines*, 4(2), 125–146.

Hall, C., Slembrouck, S., & Sarangi, S. (2006). *Language practices in social work: Categorisation and accountability in child welfare*. London: Routledge.

Hall, C. J., & Slembrouck, S. (2009). Professional categorization, risk management and interagency communication in public inquiries into disastrous outcomes. *British Journal of Social Work*, 39(2), 280–298.

Hoffman, M., Linell, P., Lindh-Astrand, L., & Kjellgren, K. I. (2003). Risk talk: Rhetorical strategies in consultations on hormone replacement therapy. *Health, Risk & Society*, 5(2), 139–154.

Horlick-Jones, T. (2003). Managing risk and contingency: Interaction and accounting behavior. *Health, Risk & Society*, 5(2), 221–228.

Horlick-Jones, T. (2005). On 'risk-work': Professional discourse, accountability, and everyday action. *Health, Risk & Society*, 7(3), 293–307.

Jore, S. J., & Kain, D. J. (2010). Risk of terrorism: A scientifically valid phenomenon or a wild guess? The impact of different approaches to risk assessment *Critical Approaches to Discourse Analysis across Disciplines*, 4(2), 197–216.

Linell, P., Adelswärd, V., Sachs, L., Bredmore, M., & Lindstedt, U. (2002). Expert talk in medical contexts: explicit orientation to risks. *Research on Language & Social Interaction*, 35(2), 115–137.

Luhmann, N. (1979). *Trust and power*. Chichester: John Wiley.

Moore, A., Candlin, C. N., & Plum, G. (2003). Making sense of viral load: One expert or two? *Culture, Health & Sexuality*, 3(429–450).

Petersen, A. (2005). The metaphors of risk: Biotechnology in the news. *Health, Risk & Society*, 7(3), 203–208.

Rayner, S. (1992). Cultural theory and risk analysis. In S. Krimsky & D. Golding (Eds.), *Social theoris of risk* (pp. 83–115). Wesport: Praeger.

Roberts, C., & Sarangi, S. (1999). Hybridity in gatekeeping discourse: Issues of practical relevance for the researcher. In S. Sarangi & C. Roberts (Eds.), *Talk, work and institutional order: Discourse in medical, mediation and management settings* (pp. 473–504). Germany, Berlin: Mouton de Gruyter.

Roberts, C., & Sarangi, S. (2005). Theme-oriented analysis of medical encounters. *Medical Education*, 39, 632–640.

Sarangi, S. (2007). Editorial. The anatomy of interpretation: Coming to terms with the analyst's paradox in professional discourse studies. *Text & Talk*, 27(5/6), 567–584.

Sarangi, S., Bennert, K., Howell, L., & Clarke, A. (2003). Relatively speaking: Relativisation of genetic risk in counselling for predictive testing. *Health, Risk & Society*, 5(2), 155–170.

Sarangi, S., & Candlin, C. N. (2001). 'Motivational relevancies': Some methodological reflections on social theoretical and sociolinguistic practice. In

N. Coupland, S. Sarangi & C. N. Candlin (Eds.), *Sociolinguistics and Social Theory* (pp. 350–387). Harlow, London: Longman.

Sarangi, S., & Candlin, C. N. (2003a). Editorial: Categorization and explanation of risk: A discourse analytical perspective. *Special Issue of Health, Risk and Society, 5*(2), 115–124.

Sarangi, S., & Candlin, C. N. (2003b). Editorial: Trading between reflexivity and relevance: New challenges for applied linguistics. *Special Issue of Applied linguistics, 24*(3), 271–285.

Sarangi, S., & Clarke, A. (2002). Zones of expertise and the management of uncertainty in genetics risk communication. *Research on Language and Social Interaction, 35*(2), 139–172.

Sarangi, S., & Roberts, C. (1999). The dynamics of interactional and institutional orders in work-related settings. In S. Sarangi & C. Roberts (Eds.), *Talk, work and institutional order: Discourse in medical, mediation and management settings* (pp. 1–60). Germany, Berlin: Mouton de Gruyter.

Slovic, P. (1987). Perception of risk. *Science, 236*(4799), 280–286.

Wood, F., Prior, L., & Gray, J. (2003). Translations of risks: Decision making in a cancer genetics service. *Health, Risk & Society, 5*(2), 185–198.

Wynne, B. (1992). Risk and social learning: Reification to engagement. In S. Krimsky & D. Golding (Eds.), *Social theories of risk* (pp. 275–297). London: Praeger.

Wynne, B. (1996). May the sheep safely graze? A reflexive view of the expert-lay knowledge divide. In S. Lash, B. Szersynski & B. Wynne (Eds.), *Risk, environment & modernity* (pp. 44–83). London: Sage.

Wynne, B. (2002). Risk and environments as legitimatory discourses of technology: Reflexivity inside out? *Current Sociology, 50*(3), 459–477.

Zinn, J. O. (2010). Editorial. Risk as discourse: Interdisciplinary perspectives. *Special Issue of Critical Approaches to Discourse Analysis across Disciplines 4*(2), 106–124.

Part I
Communicating Risk in Healthcare

2
Risk and Clinical Incident Disclosure: Navigating between Morality and Liability

Rick Iedema, Donella Piper, Katja Beitat, Suellen Allen, Kate Bower and Su-yin Hor

Introduction

In recent years, healthcare incident disclosure has gained increased attention from policy makers, academics, insurers, clinical professionals, patients and consumer groups and lawyers (Australian Commission on Safety & Quality in Health Care, 2013; Clinton & Obama, 2006; Lamo, 2011; Levinson & Pizzo, 2011; Sage et al., 2014; Studdert & Richardson, 2010; Wojcieszak, Banja, & Houk, 2006). Variously described as a form of restorative justice (Berlinger, 2005), a feasible financial risk reduction strategy (Kachalia et al., 2010) and a service responsiveness philosophy (Iedema & Allen, 2012), incident disclosure appears sufficiently flexible to accommodate stakeholders' different and often competing interests. The institutional and personal benefits of incident disclosure have now been widely reported (Boothman, Blackwell, Campbell, Commiskey, & Anderson, 2009; Kachalia et al., 2010).

To facilitate and support incident disclosure, there has been much progress in legal reform (Mastroianni, Mello, Sommer, Hardy, & Gallagher, 2010), policy development (Australian Commission on Safety & Quality in Health Care, 2013; Canadian Patient Safety Institute, 2011; U.K. National Patient Safety Agency, 2009), and incident disclosure research (O'Connor, Coates, Yardley, & Wu, 2010) internationally. There have also been consistent efforts to develop targeted training (Iedema, Jorm, Wakefield, Ryan, & Dunn, 2009), procedures (Australian Commission on Safety & Quality in Health Care, 2013), models (Boothman et al., 2009), and detailed advice for policy and law makers about how to further strengthen the practice of disclosure (Sage et al., 2014). The needs and wishes of patients who have suffered harm from incidents have been studied as well, revealing a tension between regulatory constraints and patients' expectation of openness (Iedema, Allen, Britton, & Gallagher, 2011; Iedema, Allen, Britton, Grbich, et al., 2011).

The locus where professionals', institutions', patients' and family members'[1] interests are arbitrated is during the incident disclosure discussion (Gallagher et al., 2013). Here, the main challenge arises from what should and can be said (Iedema, Allen, Sorensen, & Gallagher, 2011; Lipsky, 1980). Undeniably, patients consider the proceduralisation of resolution programs and formalised communication strategies as important (Iedema, Mallock, Sorensen, Manias, Tuckett, Williams, Perrott, Brownhill, Piper, Hor, Hegney, & Scheeres, 2008). But there is one thing which they see as still more basic: discussion about people's and/or services' responsibility for what happened. It is precisely this expectation that most exposes the legal, ethical, reputational and psychological risks that are inherent in incident disclosure. The present article draws on more than 300 interviews with clinicians, patients and family members to explore these risks.

Background

Reluctance on the part of services and professionals to discuss responsibility for something that has gone wrong is common (Gallagher, 2009; Iedema, Mallock, Sorensen, Manias, Tuckett, Williams, Perrott, Brownhill, Piper, Hor, Hegney, Scheeres, et al., 2008). To some extent this is understandable. First there are professionals' *reputational* fears about falling out with their colleagues over how an incident is recounted, being reported to a professional registry, or – even when they do elect to communicate openly – patients' emotions preventing the disclosure being heard as open and just, or the information being misinterpreted (Iedema, Allen, Sorensen, et al., 2011).

Another reason for professionals' reluctance may be that responsibility is often not readily determinable immediately after the incident. This may be because not enough is known about what went wrong, or, if enough is known, responsibility may be dispersed to different degrees across different actors, equipment, and processes, rendering simple identification and clear attribution of a causal relationship difficult. Yet another reason for reluctance to disclose is that while the source of responsibility may seem clear to one person, it may not be so for others, and responsibility claims may be contested and denied. Confounding matters still further, people's view on or the facts about responsibility may also shift as time goes on and as more becomes known. Here, professionals face a *legal* risk, given the significance that is given to incident explanations and the role they play in subsequent dealings between the harmed patient and the health service.

Finally, professionals' expressing uncertainty about what went wrong may, in some cases, intensify patients' insistence on being 'told the truth' (Iedema, Mallock, Sorensen, Manias, Tuckett, Williams, Perrott, Brownhill, Piper, Hor, Hegney, & Scheeres, 2008). Patients may be unable to reconcile care givers' original claim to clinical authority with their subsequent acknowledgement of uncertainty. Patients may well respond to such acknowledgement with suspicion, questioning professionals' stated ignorance about what went wrong. Patients may also demand that someone accept personal responsibility for the incident even when this is unwarranted (Iedema, Allen, Britton, Grbich, et al., 2011; Iedema, Sorensen, et al., 2008), threatening professionals' psychological wellbeing (Nash et al., 2007). The policy and research literatures treat the question as to whether to discuss individual professionals' responsibility for what went wrong as a problem that has a pre-scribable solution (Mello et al., 2014; Sage et al., 2014). Clearly, however, here we have a 'wicked problem' (Rittel & Webber, 1973) that professionals are faced with having to solve each time anew.

What severely complicates incident disclosure is that attributing responsibility is not just a technical accomplishment that involves locating the origin of cause, or a legal-cum-risk managerial strategy centring on identifying negligence risk. Responsibility attribution involves making retrospective value judgments about what should have happened, interpreting and navigating around different people's claims, expectations and interpretations about what happened, and trying to re-calculate how things might have unfolded with the benefit of hindsight.

In its aim to assist services in determining the nature and extent of responsibility for a harm, the patient safety literature has sought to map out professionals' actions when things go wrong. One such categorisation distinguishes 'unintentional error' from 'risky action', 'reckless action', 'malicious action' and 'impaired judgment' (Leonard & Frankel, 2010). While such distinctions may seem useful for separating forgivable actions from unforgiveable ones, they operate on the basis of the legalistic assumption that individuals' actions and intentions can be safely categorised as wholly and only theirs, and as therefore comprehensively explained by classifying them on a scale ranging from benevolent to malevolent. Here, risk is inherent in and arises from single people's actions.

Aside from the problem of never *really* knowing individuals' intentions, judgments about the risk inherent in an action are however hard pushed to account comprehensively and exhaustively for clinicians' actions and decisions as they occur *in situ*. This is because these actions and decisions are often too complex and entwined with other people's to allow a precise calculus of individuals' roles in an incident and yield a final judgment (S. Dekker, 2008). Clinicians', patients' and family members' own stories attest that incidents are rarely simple (Iedema, Allen, Britton, & Gallagher, 2011; Iedema, Jorm, & Lum, 2009). Indeed, no *post hoc* account may ever be sufficiently explanatory to afford definite conclusions about what happened and who was responsible, and how specific actions converted risk into disaster, even when informed by extensive investigation and analysis (Iedema & Allen, 2012).

In all, the disclosure literature tends to eschew the ambiguity and uncertainty that are often at the heart of things going wrong. To limit the risk of conveying inaccurate accounts and encouraging the harmed patient to take legal action, this literature warns healthcare professionals not to stray beyond discussing 'the facts of the incident' with the patient harmed (Australian Commission on Safety & Quality in Health Care, 2013; Disclosure Working Group, 2011; U.K. National Patient Safety Agency, 2009). This literature also advises professionals to avoid exploratory or 'speculative' dialogue about what went wrong, and they are certainly not to 'parse responsibility' for the incident (Truog, Browning, Johnson, & Gallagher, 2011).

Effectively, these proscriptions harbour two problematic consequences. One, if the communication is limited to only known 'facts' and everything else is considered as matters we cannot talk about, the incident disclosure discussion will remain unnaturally 'cramped' (Iedema, Allen, Britton, Grbich, et al., 2011). Indeed, dichotomising the factual and the 'unknown' is tantamount to giving the unreasonable impression that incidents can always be satisfactorily discussed only when sufficient information becomes available (S. W. A. Dekker, 2005).

Two, the patient tends to be excluded from discussions in which 'factual' but perhaps rather more 'emotional' matters are addressed; think of patients' impressions, uncertainties, intuitions, opinions, judgments and doubts. In turn, the exclusion of what we could term 'factual sentiments' from incident disclosure amounts to an *a priori* denial of the value of patients' 'stories' for understanding and learning from the incident (Iedema, Allen, Britton, & Gallagher, 2012).

To explore how these delicate matters are negotiated *in practice*, we probed our data with the following question: how is the risk inherent in acknowledging responsibility for what went wrong negotiated by those engaged in incident disclosure conversations?

Methods

Context

Our study was done in Australia where patients need to pursue compensation for their hospital-caused harm through the legal system. Except for New South Wales and the Australian Capital Territory, where any apology utterance is legally protected from being used to support liability claims, Australian states have 'partial apology' ('I'm sorry this happened') protection laws; that is, only apologies acknowledging personal fault ('I'm sorry I did the wrong thing') are admissible in court. Since 2003, Australia has had an 'Open Disclosure' Standard, and its new Open Disclosure Framework (2013) is now part of the National Quality and Safety Standards against which public healthcare services are accredited. Australia does not have a formalised disclose-and-offer programme.

The interviews took place over some time (2006–2012), mapping states', services', and professionals' adaptation to the national 'Open Disclosure' policy first published in 2003 and revised (in part in response to our patient-oriented research) in 2013.

Study samples

Over a period of six years, as part of three parallel studies, a total of 302 semi-structured, open-ended interviews were conducted with health professionals and patients (and family members) regarding their incidents and incident disclosure, as summarised as follows. The interview schedule was consistent across studies. The first study (2006–2008) evaluated the Australian Open Disclosure Pilot that involved 41 hospitals (by and with funding from the Australian Commission on Safety and Quality in Health Care or ACSQHC). Ethics approval for this pilot evaluation study was obtained for only 21 of the total of 41 health service sites that had agreed to participate in the ACSQHC pilot (Iedema, Mallock, Sorensen, Manias, Tuckett, Williams, Perrott, Brownhill, Piper, Hor, Hegney, & Scheeres, 2008).

The second study (2009–2011) was a follow-up evaluation after Australian State Health Ministers formally endorsed the ACSQHC's Open Disclosure Policy (2008). This study was granted ethics approval by nine of the 18 health services that were approached for the study (Iedema, Allen, Sorensen, et al., 2011).

For these first and second studies, the project team invited the participating sites to provide contact details of practitioners who had volunteered to become involved in Open Disclosure; 131 practitioner interviews were conducted in Study 1, 16 in Study 2. Interviewees came from nursing (20), medicine (49), and hospital management and administration (80). Information about patients who had suffered incidents was sought from participating healthcare organisations, through the national print media, and with the help of internet marketing companies, yielding 23 patient interviews in Study 1 and 100 interviews in Study 2, for a total of 44 patients and 88 relatives, with some interviews involving more than one person.

The third study (funded by the Australian Research Council), begun in 2012 and currently ongoing, targets clinicians and patients involved in the same incident. Designed along the lines of the previous two studies and with the purpose of evaluating whether and how incidents are disclosed and discussed, this study has ethics approval from the administering university (the University of Tasmania). Four clinicians and 12 patients have been interviewed to date.

Across the three studies, 50% of the interviews were (and have been) conducted face-to-face, and the remainder over the telephone. The interviews were in-depth and semi-structured (open-ended yet guided by a general interview protocol), and were of between 45 minutes and 2½ hours in duration. No interviewees who participated in Study 1 were reinterviewed for Study 2, and none of those interviewees was reinterviewed for Study 3. The interviews were transcribed verbatim from the digital audio or video sound files. For the purpose of clarity, we defined an incident as an unexpected event that registers with the patient as psychological and/or bio-physiological harm.

Data analysis

Three coders participated in the coding. Open coding served to identify the 'discursive themes' (Martin & Rose, 2004) to which the interviewees devoted particular attention in their interviews. Themes were imported into and reconciled in QSR NVivo® (Qualitative Solutions and Research Pty. Ltd. Version 9), a code-and-retrieve computer software package for compilation of the original transcript segments (quotes) and grouping of the themes. Coding discrepancies were few, and importation into NVivo provided an additional opportunity for coders to check consensus. Each theme abstracted from the data revealed a frequency or 'intensity' (for example, the theme 'angry' encompassed a higher number of transcript data quotes than the theme 'satisfied'). The resulting NVivo theme network or 'node tree' enabled identification and specification of the main thematic clusters (e.g. 'emotions', 'causation'). Quantification of the responses proved not possible because of the often multiple and overlapping codings of data (interview) segments. For the purposes of this chapter, we focus on the risk inherent in offering the harmed patient an explanation for the incident that acknowledges 'responsibility for what went wrong'.

Results

The results revealed the thematic node 'responsibility for what went wrong' to encompass four sub-themes: (1) practitioners refusing to attribute responsibility for what went wrong, (2) practitioners attributing responsibility for what went wrong to others, (3) practitioners admitting to being unable to determine responsibility for what went wrong, and (4) practitioners being open to consider and accept responsibility for what went wrong. This analysis is presented below, and draws on both clinicians' and patients' interview responses for exemplification.

Clinicians refusing to attribute responsibility for what went wrong

The following quotes demonstrate the difficulty of separating attributions of responsibility from explanations of 'what went wrong' when discussing incidents with patients. This first quote is an example of a healthcare professional acknowledging this tension:

> You can apologise and say ... we are sorry that this has happened to you, but we cannot turn around and say, yes we can offer you [an explanation of how and why it happened and who was responsible] ... and that is some of the anger, because they keep coming back through the course of the meetings and say 'Why don't you just say that you stuffed [screwed] up?' – Nursing clinician

This nurse interviewee acknowledged that their stance is often hard to accept for those harmed, and may induce patients' anger. Indeed, patients may experience clinicians' unwillingness to address the issue of responsibility as an incident in its own right, as expressed in the extract below:

> There was one admission [of responsibility], if you like, because I mentioned the lack of Clexane. And he said, 'Yes, that protocol was

not observed, and I am totally at a loss to understand why it was not done.' In other words, it's still not *his* [the doctor's] fault: 'somebody else should have organised it'. That [doctor's unwillingness to accept responsibility for the consequences of unsatisfactory care] is so much deeper and wider and higher than the effects of chopping a toe off, or a hand, or something. – Relative of patient

Here, the relative of the harmed patient regarded the doctor's use of the passive voice ('was not observed') as depersonalising the problem of the protocol not having been observed, and thereby as signalling that he was not prepared to enter into the issue of who was responsible for having caused the incident. The interviewee responded to the doctor's unwillingness to 'parse responsibility' with blame: 'It's still not his fault.' This set the patient and the doctor on divergent courses, the one seeking to understand whose actions produced the misadministration, and the other, for whatever reasons, seeking to avoid specifying for the patient who was responsible.

In cases where the service or its clinicians refused to discuss retrospective responsibility, the clinicians were also heard by interviewees as refusing them access to discussions about how the service would take *proactive, future-oriented* responsibility for ensuring that similar incidents would not happen again, as occurs in the quote below. This family member interviewee expressed their frustration about being confronted with what seemed like an excessively cavalier attitude towards the risk that led to the harm in the first place. Potentially because of this, the family member's response hardened into a demand that they be told 'what happened':

> I wish there was some way you could have full disclosure, without the thought that people are going to sue other people. If they could fully disclose, 'Okay, this is what's happened', and it's not a blame game. 'This is what's happened, okay, and how can we make sure this doesn't happen again', rather than, 'I'm terrified you're going to sue me.' You know, because nothing's getting changed. I mean, I can tell you, if someone like [patient's name] went down to [major hospital] tomorrow, the same thing could still happen. – Relative of patient

The problem of rising tension and of intensifying demand for honest identification of 'who was responsible' in response to health professionals' reluctance to discuss responsibility is also strong in the next extract. The family member quoted below recounted how they got an acknowledgement of responsibility that remained clouded by professionals 'covering their backside':

> It was implying that there was quite a big fault on the part of the hospital and I think she has her backside to cover too. And she doesn't want all the details out there. She's happy to let us know what we need to know, to help us get through, but somehow somebody has pulled this investigation and so there's a lot gone on, even to this day, that [name patient] and I don't know about and as I said there's been no follow-up with us, so – and for somebody to say, you know, this is a catastrophe, this should never have happened ... you know, it ... I know my mother's got to die of something, but I don't want her dying from the fault of some stupid person who didn't do their best. She put a lot of faith in the system. – Patient family member

We summarise the above analysis of risk in incident disclosure communication as follows: patients who are denied open discussion about what went wrong and not offered answers to all aspects of the incident that they wish to have clarified may feel strengthened in their determination to obtain not just a comprehensive explanation, but also a full acknowledgement of personalised responsibility. That is, clinicians' or services' unwillingness to openly discuss responsibility for what went wrong may exacerbate the risk of patients demanding a full acknowledgement of fault, and lessen the chance of their being willing to accept uncertainties and ambiguities as part and parcel of things going wrong.

Clinicians attributing responsibility to others

The data further revealed instances where the occurrence of an incident was acknowledged but where the acknowledgement did not address the needs of the patient and their family by deflecting responsibility (sub-theme b). The patient's mother quoted below recounted how her doctor offered news of 'some nurses even quit[ting] their jobs' over the incident. The mother explained her dissatisfaction with this disclosure by suggesting that the doctor wrongly interpreted her request for a full explanation as a need for retribution that he thought would be satisfied by telling the mother the news of the nurses' quitting, and implying the nurses' responsibility for what had gone wrong. Yet, as the interviewee explained, the news of the nurses' quitting remained unsatisfactory as it did not contain an explanation for how the overdose had come about:

> No [I did not get an explanation for the overdose]. All I could get out of the doctor was, 'The nurses feel very bad. Some nurses even quit their jobs over this', you know, 'when they make a mistake'. And, well, I don't really care. I don't really care if they quit their jobs or whatever. – Patient's mother

Allegations of responsibility by clinicians directed at colleagues also have to be taken into consideration by those who are charged with enacting disclosure. The safety coordinator quoted below for instance actively shielded family members from such attributions of blame between clinicians:

> I mean the death of the baby – one that I've talked about, but you know there was a midwife involved, there was two registrars [residents] involved, there were two consultants involved. So I did the open disclosure on that one ... because I was worried the consultant would say something [negative] about the midwives ... that was very, very delicate. – Safety coordinator

On our interpretation, the phrase 'saying something' in the extract above means 'attributing responsibility' or 'casting blame'. The problem of 'the consultant [potentially saying something negative] about the midwives' was a risk factor that the safety coordinator knew needed to be carefully managed. The safety coordinator sought to avoid discussion about these matters to occur in front of the mother of the dead infant on the view that she should not be privy to inappropriate attributions levelled among clinicians.

The safety coordinator's comments above further suggest that the mother was not going to be informed about the inter-professional tensions affecting the team who had cared for her infant. The quote also implies that the wish to locate responsibility for actions and reach closure is not only common among patients, but also among the clinicians, although underlying motivations can vary. While in one sense understandable, denying the mother information about inter-professional tensions institutes deep rifts between what the patient wants to know, what the patient is told, the need to tackle team members' dysfunctional approach to negotiating what went wrong, and the ethics of the service's approach to risk, incident disclosure, and practice improvement. Leaving these rifts unresolved means not just that the mother was denied important answers about how the incident came about, but also that team members were likely to continue to avoid discussing how what happened led to the detrimental outcome and what they now were to do to avoid further incidents.

Clinicians acknowledging they are unable to determine responsibility

A third sub-theme (c) centred on clinicians' acknowledging that they were not able to pinpoint the exact reasons and causes for the incident. Given the increasing complexity of healthcare, it is not surprising that clinician interviewees reported situations in which they faced the challenging task of persuading the patient that the harm had resulted from complex circumstances, and, concomitantly, that both the risk of harm and responsibility for the harm were therefore quite *dispersed*. The interviewee quoted below relates how he or she impressed on the patient that not a single individual would have been responsible, but a whole team:

> And because there was no primary person involved in the care, I didn't think we'd get the value of, you know, the person honestly expressing regret from the mistake they made. There was no one person that was [responsible] ... it was a team. And so in that one, a very delicate team approach, I chose to do the open discussion. – Safety coordinator

Delving yet more deeply into the matter of dispersed risk and responsibility, the medical director quoted below acknowledged that several individuals may have a hand in what goes wrong, because there may be long chains of responsibility, with each link playing a role in the mitigation of risk. At the same time, this interviewee admitted that some people more than others may have responsibility for mitigating specific risks: 'it will be her manager, not her manager's manager's manager':

> So it might not be the nurse that administered the wrong drug, or the wrong dose, but it will be her manager, not her manager's manager's manager's manager [who is answerable] ... clearly most incidents aren't blameworthy acts [but] there's still a professional responsibility to not continually administer the wrong dose to patients. – Medical director

For this interviewee, professional responsibility pertained to identifying and obviating risks, as in 'not continually administering the wrong dose to patients'. The notion that responsibility for particular acts can extend to one's manager, but maybe not one's manager's manager's manager, at once captures the intricate ways in which responsibility for mitigating specific risks may implicate some but not others, and the notion that such responsibility may apply to people in different degrees (Wachter, 2012).

This last extract reveals at once the importance of acknowledging and explaining the likelihood that responsibility for a wrong is dispersed and calibrated across different actors and past decisions. It also reveals the importance of being prepared to explore the complex remit of the original risk, and of the wrong that led to the harm. Lacking insight into the full complexity of practice and its risks, we suggest, patients and healthcare professionals alike may make the mistake of working with simplistic attributions of responsibility, and thereby unfairly blaming particular individuals.

Clinicians being willing to discuss responsibility for what went wrong

The data revealed few cases where responsibility for failing to obviate a risk and for occasioning a harm was not simply acknowledged but – to some extent – explored (sub-theme d). As one of these exceptions, the interviewee quoted below made herself available to the family to discuss what went wrong. In this case, she acknowledged the service might not have done the right thing, and intimated, without however openly agreeing, that the family was correct in directing their blame at the nurse ('you're right') who 'was supposed to be looking after [the patient]':

> ... the family are livid, they're really, really angry. ... so I said ok, well we need to ... meet with them and talk through this. ... it took quite a while for them to just get rid of their anger. ... I didn't try and interrupt them or stop them from talking. I didn't try and deflect the blame because they were wanting to blame somebody, they were wanting to know the name of the nurse that was supposed to be looking after [the patient] ... for 25 minutes, they ranted and shouted and were very scathing of, of the service. I guess at the end of that ... they were just getting tired from being so angry [and] I ... really just apologized and said, you're right. – Nursing manager

This interviewee described how her 'active listening' enabled the family to 'vent' their anger. Disclosure training has emphasised the importance of 'active listening' for the purpose of allowing patients to articulate their views and emotions (Queensland Health, 2006). Practising 'active listening', the nursing manager allowed the issues of risk and responsibility to be put on the table, and, while not entering into discussion about the name of the nurse whom the family understood had been looking after the patient, in the end signalled that the family's assessment was correct, saying 'you're right'.

In saying 'you're right', the nurse manager acknowledges that the nurse failed to manage the original risk and allowed the patient to come to harm. In saying this, the nurse manager may be seen to risk the family taking her admission to a lawyer with the intent to sue, or to risk her own credibility if in future it transpires that the responsibility for the incident was with another provider and has to correct her acknowledgement. However, in this case, she felt that the family 'were getting so tired from being angry', and that having vented their anger, they may have been less inclined to sue.

Besides partaking in discussions about particular other individuals' responsibility, clinicians also at times proactively *assumed* responsibility themselves for what went wrong without as yet fully clarifying their reasons or exploring their uncertainties:

> And it was just the most powerful thing I've ever seen, this guy [clinician] sort of saying, 'I really don't know what happened. I really can't explain what happened, but it shouldn't have happened, and I have to take the responsibility for that. I was the one that had the responsibility for it'. You could see he was gutted and the family [of the patient] responded to that. – Support personnel

Here, the doctor personally assumed responsibility for the incident, even if he was not in fact 'the last pair of hands' that brought it about. His taking the risk of assuming full responsibility had a good outcome in this instance, despite the absence of a detailed explanation or an honest exploration of what happened. This may have been due to the doctor's sincerity and willingness to shoulder full responsibility *in principle* for the actions of members of his team. In identifying with the incident in this way, the doctor not only took responsibility for what happened but also intimated that it was he who would ensure that similar risks should be avoided in future.

Discussion

The results and analysis presented above showed that disclosure conversations are in the first instance framed by how professionals position themselves in relating to clinical risk: do they position themselves as carrying responsibility for ensuring that that 'primary' clinical risk not convert into patient harm? Or do they position themselves as *not* responsible for obviating that risk? These two stances have important implications for how the disclosure conversation proceeds. We saw above that, first, professionals may choose to constrain the disclosure conversation by refusing to discuss responsibility or attribute it to non-present others (sub-themes a and b). While attributing responsibility to non-present others seems inherently problematic (even if technically justified), avoiding to acknowledge responsibility for what went wrong may be justified by pointing to the need to first establish 'what are the facts', to 'avoid speculation', or to 'not blame anyone for what happened'.

The secondary risk that arises here is that harmed patients and their family members may not understand professionals' or services' reasons for not immediately clarifying the source and cause of the harm. Indeed, as heard above, harmed patients may regard professionals' unwillingness to discuss what went wrong as an incident in its own right. This risk becomes all the more pronounced when, in the patient's eyes, the harm is very easily explained. Above we heard a family member blaming 'a nurse that was supposed to be looking after [the patient]'. This secondary risk may convert into serious adverse consequences, such as a formal complaint or even litigation (Iedema, Allen, Britton, Grbich, et al., 2011). An unwillingness to discuss what happened may thus escalate, particularly when in the patient's eyes matters are crystal clear: 'the nurse should have looked after me with greater diligence'.

This now leads us to conceptualise a tertiary risk: that of reducing care and risk complexity to simplistic explanations, guilt and blame. Without denying that at times guilt and blame are justified, we propose this tertiary risk conception on the ground that contemporary care is immensely complex, and therefore the causes of patient harm are likely to be complex as well. Consider the following: many patients present with co-morbidities (diseases additional to their primary diagnosis) and medically unexplainable symptoms, and services face constant disruption due to staff churn (due to rotations, shifts, and attrition), changing technologies (IT as well as drugs and diagnostic techniques), strained resource allocations, and changing regulations (new evidence necessitating updated policies and guidelines). Given this pervasive complexity, we could indeed ask why care does not go wrong more frequently in ways that are hard if not impossible to fully explain. A real and tertiary risk that presents itself here, then, is that those involved in a harm reduce an incident to a single individual's actions (usually 'the last pair of hands'). But this inappropriately converts a complex event with an extended genealogy into a single 'wrong' action deserving of a simplistic judgment of guilt and blame.

The results and analysis above also revealed responses from professionals who were inclined to acknowledge patients' and families' right to be angry about what happened and discuss their views on causes and responsibilities (sub-themes c and d). Here, by showing themselves willing to discuss patients' and families' understanding of what went wrong without ruling such discussion out of court, professionals are able to obviate the secondary risk: that of the harmed patient and their family feeling that their right to an immediate and open discussion is not dismissed. Evident in the analysis however was that the conversations tend to centre around the patient's views and understandings, enabling them to 'vent their anger'.

From this second set of responses it appears that circumventing secondary risk (patients regarding professionals' unwillingness to discuss what went wrong as an incident in its own right) may also avoid tertiary risk. That is, if patients are trusted to conduct disclosure conversations in a tenor and register of their choosing, they may prove more willing and ready to consider and take seriously contributory causes, ancillary people's roles, as well as less easily explainable dimensions of the incident. Here, openness need not equate with full provision of facts and explanations ('hard information'), but may also pertain to professionals' and services' willingness to discuss harmed patients' views, emotions, and understandings ('soft information'). On this latter view, disclosure pertains not principally to fact provision, exhaustive explanation and logical causation, but is significantly contingent on a respectful sharing of sentiments and views with the aim of achieving agreement about how the original risk, the incident, and the resulting harm are to be understood. In this latter formulation, disclosure principally provides opportunity for interpersonal rapprochement, as well as setting out the causes for what happened in a discourse with which all stakeholders can identify.

Conclusion

Our analysis shows that in practice incident disclosure may relieve professionals of having to 'deny and defend' their responsibility for a wrong or a harm. Professionals tend to limit discussion about why care went wrong and whose responsibility it was, ostensibly to satisfy legal and investigative prioritisation of the 'facts'. Effectively making it possible for professionals to 'collude in anonymity', this prioritisation may risk contravening patients' desire to delve into the complexities, uncertainties, and emotionality of incidents.

Where patients perceive that their need to understand not only what exactly happened, but also how it could have happened, is explored openly by the provider, they may be more inclined to accept that there may not be a singular responsibility for an incident. In other words, they may be more open to accept the realities of complex factors that may have contributed to the incident rather than looking for attribution of responsibility to one person. Hence, identifying causation and responsibility for what went wrong need not act as endpoint.

The analysis revealed that, in practice, openness may trump the cramped approaches to incident disclosure that result from subjecting it to legalistic and risk managerial constraints, rather than framing it in the first instance as a restorative and deliberative process. Even when no 'facts' are forthcoming, openness may be experienced by patients as 'good disclosure' because it respects their wish to delve into the unfolding of an incident. Granting patients this wish is important also for another reason. If patients are indeed as central to patient safety as they are advocated to be (viz. 'patient-centred care'), no regulatory, legal, or policy constraints should limit their right to openness.

Finally, delving into the complexities of care in general, of what went wrong in specific instances of care, and of people's various interpretations and uncertainties, should also enable *clinicians* to become more effective learners themselves, positioning themselves better to improve care practices. Dialogue among professionals and patients about the realities, complexities, compromises, uncertainties, and ambiguities intrinsic to healthcare is a hallmark of relation-centred professionalism, as advocated by Beach and Inui (Beach et al., 2006). This approach does not ignore legal risk and liability. It does prioritise open dialogue however as the means to obviating secondary and tertiary risks. Simply put, this means professionals and services show willingness to discuss the fraught nature of care gone wrong and to restore their relationship with those harmed on their terms.

Acknowledgement

We thank all the patients and relatives who participated in our study.

Note

1. From here on, and for brevity's sake, the term *patient* refers to the person harmed and/or their family members.

References

Australian Commission on Safety & Quality in Health Care. (2013). *The Australian open disclosure framework*. Sydney: Australian Commission on Safety & Quality in Health Care.

Beach, M. C., Inui, T., Frankel, R., Hall, J., Haidet, P., Roter, D., ... et al. (2006). Relationship-centered care: A constructive reframing. *Journal of General Internal Medicine, 21*(SUPPL. 1), S3–S8.

Berlinger, N. (2005). *After harm: Medical error and the ethics of forgivenness*. Baltimore: Johns Hopkins Press.

Boothman, R., Blackwell, A. C., Campbell, D. A., Commiskey, E., & Anderson, S. (2009). A better approach to medical malpractice claims? The University of Michigan experience. *Journal of Health and Life Sciences Law, 2*(2), 125–159.

Canadian Patient Safety Institute. (2011). *Canadian disclosure Guidelines: Being open with patients*. Edmonton: Canadian Patient Safety Institute.

Clinton, H. R., & Obama, B. (2006). Making patient safety the centerpiece of medical liability reform. *New England Journal of Medicine, 354*(2006), 2205–2208.

Dekker, S. (2008). *Just culture*. London: Ashgate.

Dekker, S. W. A. (2005). *Ten questions about human error: A new view of human factors and system safety*. Mahwah, NJ: Lawrence Erlbaum.

Disclosure Working Group. (2011). *Canadian disclosure Guidelines: Being open and honest with patients and families*. Edmonton: Canadian Patient Safety Institute.

Gallagher, T. H. (2009). *Disclosing harmful errors to patients: Recent developments and future directions*. Paper presented at the Open Disclosure Seminar, University of Technology Sydney.

Gallagher, T. H., Mello, M. M., Levinson, W., Wynia, M. K., Sachdeva, A. K., Snyder Sulmasy, L., ... Arnold, R. (2013). Talking with patients about other clinicians' errors. *New England Journal of Medicine, 369*(18), 1752–1757. doi: 10.1056/NEJMsb1303119

Iedema, R., & Allen, S. (2012). Anatomy of an incident disclosure: On the importance of dialogue. *US Joint Commission Journal of Quality and Patient Safety, 38*(10), 435–442.

Iedema, R., Allen, S., Britton, K., & Gallagher, T. (2012). What do patients and relatives know about problems and failures in care? *BMJ Quality and Safety, 21*(3), 198–205 doi: doi:10.1136/bmjqs-2011–000100

Iedema, R., Allen, S., Britton, K., Grbich, C., Piper, D., Baker, A., ... Gallagher, T. (2011). Patients' and family members' views on how clinicians enact and how they should enact Open disclosure – The '100 Patient Stories' qualitative study. *British Medical Journal, 343*. doi: doi: 10.1136/bmj.d4423

Iedema, R., Allen, S., Sorensen, R., & Gallagher, T. H. (2011). What prevents the disclosure of clinical incidents and what can be done to promote it? *US Joint Commission on Quality & Patient Safety, 37*(9), 409–417.

Iedema, R., Jorm, C., & Lum, M. (2009). Affect is central to patient safety: The horror stories of young anaesthetists. *Social Science & Medicine, 69*(12), 1750–1756.

Iedema, R., Jorm, C., Wakefield, J., Ryan, C., & Dunn, S. (2009). Practising open disclosure: Clinical incident communication and systems improvement. *Sociology of Health & Illness, 31*(2), 262–277. doi: 10.1111/j.1467-9566.2008.01131.x

Iedema, R., Mallock, N., Sorensen, R., Manias, E., Tuckett, A., Williams, A., ... Scheeres, H. (2008). *Final report: Evaluation of the national open disclosure pilot program.* Sydney: The Australian Commission on Safety and Quality in Health Care.

Iedema, R., Mallock, N., Sorensen, R., Manias, E., Tuckett, A., Williams, A., ... Jorm, C. (2008). The national open disclosure pilot: Evaluation of a policy implementation initiative. *Medical Journal of Australia, 188*(2008), 397–400.

Iedema, R., Sorensen, R., Manias, E., Tuckett, A., Piper, D., Mallock, N., ... Jorm, C. (2008). Patients' and family members' experiences of open disclosure following adverse events. *International Journal for Quality in Health Care, 20*(6), 421–432. doi: 10.1093/intqhc/mzn043

Kachalia, A., Kaufman, S. R., Boothman, R., Anderson, S., Welch, K., Saint, S., & Rogers, M. (2010). Liability claims and costs before and after implementation of a medical error disclosure program. *Annals of Internal Medicine, 153*(2010), 213–221.

Lamo, N. (2011). Disclosure of medical errors: The right thing to do, but what is the cost? *Lockton Companies Newsletter, 2011*(Winter), 1–7.

Leonard, M., & Frankel, A. (2010). The path to safe and reliable health care. *Patient Education and Counselling, 80*(2010), 288–292.

Levinson, W., & Pizzo, P. A. (2011). Patient-physician communication: It's about time. *JAMA: The Journal of the American Medical Association, 305*(17), 1802–1803. doi: 10.1001/jama.2011.556

Lipsky, M. (1980). *Street-level bureaucracy: Dilemmas of the individual in public services.* New York: Russell Sage Foundation.

Martin, J. R., & Rose, D. (2004). *Working with discourse: Meaning beyond the clause.* London: Continuum.

Mastroianni, A. C., Mello, M. M., Sommer, S., Hardy, M., & Gallagher, T. H. (2010). The flaws in state 'Apology' and 'Disclosure' laws dilute their intended impact on malpractice suits. *Health Affairs, 29*(9 (September)), 1611–1619.

Mello, M. M., Boothman, R. C., McDonald, T., Driver, J., Lembitz, A., Bouwmeester, D., ... Gallagher, T. H. (2014). Communication-and-resolution programs: The challenges and lessons learned from six early adopters. *Health Affairs, 33*(1), 20–29.

Nash, L., Daly, M., Johnson, M., Walter, G., Walton, M., Willcock, S., ... Tennant, C. (2007). Psychological morbidity in Australian doctors who have and have not experienced a medico-legal matter: Cross-sectional survey. *Australian and New Zealand Journal of Psychiatry, 41*(2007), 917–925.

O'Connor, E., Coates, H. M., Yardley, I. E., & Wu, A. (2010). Disclosure of patient safety incidents: A comprehensive review. *International Journal for Quality in Health Care,* August (2010), 1–9.

Queensland Health. (2006). *Queensland health open disclosure peer support training.* Brisbane: Queensland Health.

Rittel, H., & Webber, M. (1973). Dilemmas in a general theory of planning. *Policy Sciences, 4*(1973), 155–169.

Sage, W., Gallagher, T. H., Armstrong, S., Cohn, J. S., McDonald, T., Gale, J., ... Mello, M. (2014). How policy makers can smooth the way for communication-and-resolution programs. *Health Affairs, 33*(1), 11–19.

Studdert, D. M., & Richardson, D. M. (2010). Legal aspects of open disclosure: A review of Australian law *Medical Journal of Australia, 193*(3), 273–276.

Truog, R., Browning, D., Johnson, J., & Gallagher, T. H. (2011). *Talking with patients and families about medical error: A guide for education and practice.* Baltimore: The Johns Hopkins University Press.

U.K. National Patient Safety Agency. (2009). *Being open: Communicating patient safety incidents with patients, families and carers.* London: National Patient Safety Agency.

Wachter, R. (2012). Personal accountability in healthcare: searching for the right balance. *BMJ Quality and Safety.* doi: doi:10.1136/bmjqs-2012-001227

Wojcieszak, D., Banja, J., & Houk, C. (2006). The sorry works! Coalition: Making the case for full disclosure. *Journal on Quality and Patient Safety, 32*(6), 344–350.

3
'Being Diplomatic with the Truth': The Discursive Management of Risk in Accounts of People Leaving Forensic Psychiatric Settings

Michael Coffey

Introduction

Mental ill-health can be a very distressing experience for those with the condition, their families, and the wider community. Unpredictable and strange behaviours can lead to strains in the social fabric that make up the support network surrounding the individual. Many of these behaviours will have caused embarrassment to the individual and their families. Perhaps most significantly the response of the wider community to such behaviours (and confirmation of contact with mental health services) is often to reject the person or refrain from further contact. It is perhaps understandable then that many people with mental health problems are cautious about revealing information about their condition. In circumstances where the individual has engaged in serious criminal offences, which have the effect of confirming public misconceptions about mental illness, it is likely that the maintenance of privacy and decisions about disclosure are constant challenges. The social identity implications of stigma, discrimination and exclusion make the maintenance of privacy an area of particular sensitivity. In this chapter I will explore how privacy and disclosure were handled in the talk of people returning to community living following detention in forensic mental health facilities. Discursive practice among patients and workers is examined to show how risk is constructed and dealt with in accounts. I suggest that the successful handling of concerns about mental illness and risk in talk clears space for participants to deploy emergent identities that in itself attempts the work of moderating ideas of risky behaviours. The balancing of privacy and disclosure decisions can be seen to be one part of the transition towards full reintegration in community settings.

Goffman (1963) noted that individuals may show visible signs in their presentation suggesting *discredited* identities or if they are able to conceal their condition in some way they may be regarded as possessing a *discreditable* identity should this information later come to light. Discreditable social identities such as mental illness and/or serious criminal offending have the potential to be regarded as highly socially sensitive (Brannen, 1988). Many people using mental health services do not fall neatly into a discredited/discreditable dichotomous divide. Mental distress and the effects of the treatment for it such as medication effects and intrusive supervision by workers, can work to unmask what otherwise would be concealed (Coffey, 2012a). Keeping hidden or at the very least managing the disclosure of these socially sensitive elements of identity is one way of easing social participation and allowing space for generating new identities. My analysis will show that service-user participants' accounts do the work of accomplishing emergent identities by dealing with the ever-present threat of their status being revealed to the wider community, which itself is a source of risk to the individual. Discredited social identities are handled in talk by creating distance and difference between current and previous identities (Coffey, 2012b). Distance is signalled in both time and space through talk which positions dangerous behaviours as temporally and geographically located (Coffey, 2013). Difference was accomplished by reference to individual moral characteristics in opposition to the more serious (other) cases that populate the forensic mental health system.

Identity-relevant information which is liable to attract negative labelling is associated with discrimination, stigma, reduced social capital, and exclusion (Thornicroft, 2006). Managing these identity threats is a task which is achieved in the talk of individuals in social situations. One way this is accomplished is to keep some information private. Westin (1967: 7) has suggested that privacy is the assertion by individuals to determine,

> when, how and to what extent information about them is communicated to others.

Privacy management is seen as a dynamic process by Westin in that it is regulated through the control of information for the purposes of securing or servicing role requirements (Margulis, 2003). It has been argued that privacy regulation is a task for individuals in situations where for instance disclosure of normally private information is being considered (Altman et al., 1981). Interest in self-disclosure has led to concerns about variable definitions of the concept which Fisher (1984: 278) argues is,

> verbal behaviour through which individuals truthfully, sincerely and intentionally communicate novel, ordinarily private information about themselves.

Decisions to reveal private information appear to be influenced, at least in part, by identity performance and the handling of threats which may result from socially sensitive information. Sensitivities surrounding certain types of information have implications for the individual in that they imply risk or threat to identity (Sieber and Stanley, 1988). As such, concealing or keeping hidden those aspects germane to identity in mental illness and related dangerous behaviours is a tactical move by actors which may be seen to help with social bonding. Altman and Taylor's (1973) initial theorising on social penetration held that people revealed superficial non-intimate details of themselves in early exchanges with the assumption that this was a cumulative and directional process. That is, social bonds deepened over time in part oiled by the incremental and reciprocal disclosure of more intimate information from social actors. This is a view that was subsequently revised to recognise that social actors may engage in cyclical, reversible, and non-linear exchanges in relationships (Altman et al., 1981). Recognising that social interaction in which disclosure is achieved has implications for identity, Foddy and Finnigan (1980: 6) argued that privacy was,

> the possession by an individual of control over information that could interfere with acceptance of his claims for an identity within a specified role relationship.

This suggests that privacy management and disclosure of socially-sensitive information achieved through talk is a dynamic process and fundamental to establishing emergent identities.

I suggest here that privacy management is something that is actively accomplished in the talk of people leaving forensic mental health settings in at least two ways. The first way in which privacy management is actively accomplished is in relation to the how and what of revealing identity sensitive risk information in research interviews. Occasioned interview talk functions to elaborate identity work in interaction for the purposes of laying claim to, or resisting, available labels which are identity-relevant. Everyday reciprocal rules for information disclosure appear not to apply, although a measure of trust in the interviewee/researcher relationship may be necessary. The second way in which privacy management is actively accomplished and which I focus on in this chapter is in re-telling of events where disclosures to the wider community were warranted. Rhetorical devices are employed in the talk of participants to establish plausible, credible, and convincing stories of how the day-to-day management of emergent identities are handled.

The study

This research was conducted with discharged patients and staff contacted through two National Health Service settings in the UK. These settings provided forensic mental health inpatient and aftercare services in a large geographical region of the UK consisting of urban locales and sparsely populated rural areas. The settings provided for both the preparatory stages of discharge and aftercare monitoring and supervision of individuals. Service-user participants were all subject to legally mandated aftercare supervision, monitoring, and follow-up referred to as 'conditional discharge'. These conditional discharge orders are reserved as a legal outcome in criminal convictions for serious offences involving attempted or completed homicides and where a mental illness is implicated. Individuals in this study had spent a minimum of two years in a forensic inpatient facility and there is no maximum period for such orders. Leaving hospital is a highly managed affair in such situations and once achieved individuals are keen to maintain their new found status but must do so under the threat of being recalled to hospital should they demonstrate risky actions deemed indicative of active mental illness.

The approach in this study was to examine everyday understandings in accounts offered by discharged patients and workers involved in aftercare monitoring. A multiple perspectives approach in which both worker and service-user accounts are analysed provides the opportunity to examine the differing views and separate stances of a vulnerable group and those working with them. In treating talk as action-oriented, and therefore functional, this study approached verbal communication as constitutive of and as a means for accomplishing socially relevant actions. The study adopted an ethnomethodological stance of seeking to explore people's own displays of their understandings of life events, as produced in talk (Garfinkel, 1967). The context for this interview talk was one of discharge and return to community living. In-depth interviews were used to facilitate participants' accounts. The intention was to focus not only on what 'really happened' but also on the situated talk that illustrates the social organisation of everyday life as evident in the utterances of actors (Edwards and Potter, 1992: 57). All participant names used here are pseudonyms.

Approach to data analysis

The study consisted of a total of 59 research interviews with service-users, their social supervisors, and community psychiatric nurses (CPNs). Social supervisors were social workers with responsibility for reporting to the Ministry of Justice on the progress of the patient in the community. The approach was to allow participants to prioritise what they wished to in their accounts of discharge and follow-up aftercare. Interviews were recorded for later transcription and took place in the person's home or for workers, in their places of work.

The analysis considered the sequential organisation of talk and the discursive actions which accounts worked to achieve (Edwards and Potter, 1992). This approach is concerned with examining people's practices such as communication, interaction, and argument. Narratives provided by participants in research interviews allow actors to account for both their view of themselves, and the social world around them (Scott and Lyman, 1968). This approach recognises that accounts function to achieve particular rhetorical ends in that they persuade, argue, convince and show awareness of competing versions (Radley and Billig, 1996). Edwards and Potter (1992) argue that a discursive action approach positions language as representation subordinate to language as action. The focus of analysis was on what purposes talk was put to by speakers themselves. Additionally a focus on multiple perspectives may highlight discrepancies between worker and patient accounts and signal problems in agreements about treatment goals (Anderson et al., 1989).

Findings

In the analysis presented here I show how interview talk often involves an active process of negotiating the boundary between private and public, determining what identity-relevant information is disclosed and when. For example, during the course of this study some participants disclosed highly personal information such as talking about past risky behaviours. Others however chose not to reveal this information. Access to this information however was handled by participants so that what was revealed about their risk status and the positioning of this information within the research interview was functional or action-oriented. In almost all cases where participants spoke about the risk behaviours they did so only after the interview was well established. This usually followed turns where alternative versions of identity claims appeared within stories establishing illness claims as credible and warranted. For example I have shown previously (Coffey, 2012b), how one participant (Maeve) goes to some lengths to establish the severity and persistence of her mental distress. The criminal offending behaviour was positioned sequentially to the description of ill-health. This action orientation is one way of managing the identity-relevant elements of socially sensitive information. Lee's (1993) concept of sensitivity as threat can be extended therefore to consider identity threats involved in interaction (at least within the interview setting but also perhaps more broadly) which are then handled in the talk of participants.

Returning to live in the community following lengthy detention in a forensic mental health facility can present challenges in managing social interactions. Establishing relationships along with the potential for rejection or isolation by other social actors is one area of interaction which service-users referred to in their talk. Service-users participating in this study had been detained in forensic hospitals for long periods of time ranging from three to 25 years. In most cases these facilities were distant from their home communities making continued contact with friends and family more difficult. Although there were exceptions, the criminal offence, conviction, and subsequent detention in a secure mental institution had in many instances led to the person being rejected by family and friends. In some instances, close members of the person's social network were victims of the offence. Risk behaviours included dangerous and often fatal behaviours towards others. In circumstances such as these there are obvious difficulties in maintaining social networks. The spectre of risk is implicated in every interaction whether stated or otherwise. For workers and perhaps society more widely, risk is embodied in the forensic patient who has manifested widely held fears about mental illness and those afflicted by it. Patient's accounts show an awareness of these concerns but also an anxiety associated with deviance labels. Their accounts can be seen to do the work of managing these concerns discursively.

Dave: What you've done in the past sticks with you all the time, you know um, I'm still classed as a dangerous person cause if I wasn't dangerous I wouldn't have anybody with me would I? So it just shows along that way that people still don't trust me like.

I: do you see yourself as being dangerous?
Dave: No I'm not dangerous but like I said though, it's like talking to the brick wall isn't it; I can't get no sense into anybody.

In this extract Dave highlighted his concern that deviance is still a current concern for workers as evidenced by the terms of his discharge order. Dave was required to be in the presence of a worker at all times when he went outside his supported accommodation. Dave's account displayed his understanding that this implied a concern on the part of workers that he was not to be trusted. The phrasing 'I'm still classed ...' worked to indicate that an external authority was responsible for this designation and it was one that Dave rejected in his next turn. Externalising the designation to a faceless or even omnipotent source did the moral work of placing Dave as the 'victim' of the system of care and control. 'I'm still classed' also works to show that this designation is one version only. It brings to the fore the frequently contested nature of mental illness designations and the lack of predictive power of contemporary risk assessment tools (Swanson, 2008). Dave skilfully introduces doubt and highlights the uncertainty of these judgements. By doing so he opens the way for alternative understandings of his situation which he ably addresses by indicating that workers may be impervious to reasoned challenge from patients, 'it's like talking to a brick wall'.

The concerns expressed by service-users with regard to handling deviance labels may be related to social sensitivity (Lee, 1993). Sensitivities surrounding certain types of information have implications for the individual in that they imply risk or threat to identity (Sieber and Stanley, 1988). Hiding the relevance of particular social identities from others can be difficult in the face of intensive aftercare by the clinical team. In the following extract Tony expressed this concern. Tony was concerned with preserving his anonymity and avoiding the label 'murderer'. He articulated concerns that background information about him would become known to others.

Tony: ... But I wonder is it worth? Well I wonder what these people think me living here on my own sort of thing.
I: your neighbours?
Tony: Yeah, well they see people coming back and forth like [CPN] and (2). I haven't got a problem with it mind, they can think what they like really. Mind if they do know I can't do anything, I can't change it, (2) I can't change it. (3) ... Unless they know the actual story, the whole story, they could sit in judgement and say murderer innit.

Tony showed background expectancies of how others might respond to him. He wished to preserve his anonymity as a means of easing his way into a new social setting. Tony's account indicated that both he and his neighbours accessed everyday knowledge about people who require regular visits from statutory services. In other words, people who are visited at home by workers must be in need of special supervision and are therefore liable to be categorised as someone to be cautious about. Tony accessed this culturally available knowledge as a way of explaining concerns with establishing social bonds in his community. This account however also does the work of shifting the locus of concern from the danger that Tony may be seen to represent and onto the danger that he may be subject to as a discharged patient at the whim of public opinion. These are real anxieties to be sure but in structuring his account in this way Tony discursively manages concerns about his risk by bringing to the fore the risk presented by the community in which he resides. Tony uses the social relevance of his identity as a previously dangerous offender to make relevant the danger that he will be identified and subject to negative public reaction. Handling of information about oneself and choosing when, who, and how much to disclose are routine decisions for many people in the same situation as Tony.

It was also the case that for many respondents there was a high-profile element to their offending histories. Participants reported that events surrounding the criminal offence had featured in local and national newspaper reports, suggesting particular problems for reintegration. For example, some spoke of their notoriety which resulted in visits from the police whenever an offence similar to the one they had committed was reported in their area. This was not only anxiety-provoking but led to strains in their relationships with neighbours and seemed to contribute to their notoriety. This was often precisely the opposite of what many participants sought to achieve on discharge.

'Being diplomatic with the truth': managing disclosure in social settings

One way in which awareness of social sensitivity was evident in interview talk was related to how service-users reported their handling of disclosure issues with the wider community in relation to their previous offences and mental ill-health. Service-users reported that when called to explain themselves in social situations they did so with recourse to illness rather than criminal offending labels. In most cases, participants reported active decisions to keep hidden identity-relevant information in social situations unless it was impossible to avoid. The exception to this was when encountering officials in formal settings for instance when claiming social security benefits or registering with a new general practitioner where participants reported disclosing both their mental illness and criminal offending.

In the extracts presented below both Tim and Martin discuss adopting strategic approaches to manage disclosure of their history. Tim and Martin had been convicted of serious assaults and located these as past events that needed to be managed in the present. Their talk indicated that they were very much aware of available identity labels and reported adopting tactical approaches to making new friends because of concerns about how past events may be viewed by others.

Tim: They don't realise it's a sentence you are under you know it's a sentence. I mean for instance I'm obviously very well stable and ah, I go out socialising and I have come to the stage that um, I'm talking to people socialising, keeping, being diplomatic with the truth. Because obviously I don't want to mention my past you know most of, a lot of or some of it um but you know because what's ongoing that's not so much of a problem because if they get to know me then I can sort of you know divulge that information. But you know when I can't even bring them back and I have social workers coming on if I stay overnight which is very difficult if you've met someone in the pub ... how can you do that when you've got to let the Home Office know? You can't do it you know so it's very hard even trying to live the sort of life you know um, get um, you know um, doing normal things even you know.

I: Are there recent experiences you have had of telling somebody about yourself and it not working out the way you wanted it to?

Martin: I always try to find some sort of neutral ground between people like if they can't get their minds round it like, cause if it's difficult for me it's difficult for everyone I suppose. I try to make friends with everybody like, who needs enemies isn't it? So I try to make peace with everybody and get along as long as people like me you know I am happy like you know. (14) There's some people you can't please isn't there [little laugh]

In this extract Tim's account opened with a complaint about the incremental staging of his aftercare monitoring which he argues was unnecessary, having already proved himself capable of independent living in an interim placement. He had previously outlined the stages he had already completed indicating how he believed he had satisfactorily demonstrated his social and domestic skills. These opening lines show the positive directionality and time-oriented nature of conditional discharge narratives (Coffey, 2013). The story was one of achieving incremental improvements which are benchmarks on the path to eventual discharge (Roth, 1963). He likened his continuing aftercare and the placement in supported accommodation as 'a sentence you are under you know it's a sentence'. Tim was using the language of the penal system to establish a complaint of unreasonableness in relation to his placement in supported accommodation. The sequential placing of this complaint following on from Tim's statement of his life skill achievements works to accomplish an interrupted recovery trajectory set up in the previous lines.

Being 'very well' was used to introduce the talk about going out socialising. Taken-for-granted notions of being well were accessed here through the use of the directional phrase 'I have come to the stage', to establish the basis for continued claims of independence which were realised in seeking friendships outside the mental health system. Going out socialising was constructed as part of an incremental process in which Tim had now 'come to the stage' of talking to people external to the mental health system and getting to know them. In such a process he indicated that sensitive identity-relevant information about his past may be understood to lead to premature termination of new friendships. In demonstrating his achievement of rational expectations of progress Tim puts space between his past risky behaviour and his current (and potentially future) self.

For some respondents it was the ability to control information about their past that was a pre-requisite for new identity work. Both Tim and Martin suggested that they managed information about themselves in a titrated need-to-know basis. Their talk suggested an awareness that revealing too much too soon about their history may prove socially detrimental. Tim had earlier signalled that when he had told people about his mental illness, 'then they look at you in a different light' while Martin tries to find 'some sort of neutral ground'. Negative identity labels resulting from divulging a mental illness in social situations were active within these accounts 'if it's difficult for me, it's difficult for everyone I suppose' (Martin). These labels were understood to have consequential effects for those labelled. In such circumstances effort may then be taken in social settings to avoid repeated exposure to these experiences.

These extracts indicated that both speakers were oriented to the notion that concerns about their past (and for workers perhaps their ever-present) riskiness had to be managed in what they say about themselves to others. Indeed a new risk is implied here, the risk of failed social connectedness. Both Tim and Martin accomplished in their talk a sense of what was required to establish social bonding and integration as a discharged forensic mental health service-user. These accounts show awareness of likely negative reaction to disclosure of mental illness and how it was managed in interaction in social situations. However the research interview itself was also a social situation and Tim's account was oriented towards managing the interview interaction by establishing claims for what were to be understood as valued categories. Being 'diplomatic with the truth' or 'finding some sort of neutral ground' were tactics for managing disclosure issues in the world outside the interview.

These accounts also achieve something far more subtle in their telling. They establish through talk identity claims as people no longer in need of continued aftercare monitoring and supervision, in essence as no longer to be considered risky. This was achieved in Tim's account by foregrounding what was reported as his ability to live and manage his own life independently. Martin handled this differently; perhaps not needing to assert himself so forthrightly given his long established community tenure. Martin's account was populated with his interests as a collector of various pop culture paraphernalia, his routine of attending specialist markets and his quest for a partner at numerous singles clubs. For Martin his independence from the health and social care services is displayed through his ordinariness.

Tim categorised himself as an 'independent-minded person' which he elaborated as a category of service-user who wished to live without the interference of health and social care services. Limited in his ability to secure full independence by virtue of the conditional discharge order, Tim constructed the category of 'independent-minded person' as one who seeks or desires this state rather than someone who has already achieved it. This identity-oriented category was an important one for Tim as he contrasted this with another category at other stages of the research interview that is, those having poor illness control, those who were unable to care for themselves, and those who were dependent upon others. Tim's account therefore was concerned with the problem of establishing the category of independent-mindedness for the purposes of the interview. This was contrasted with the need for continued intrusive support and monitoring by the aftercare team and housing workers, which he constructed as synonymous with long-term institutionalised mental illness categories.

Institutionalised categories may include those who lack competence to live independently outside of hospital, see themselves as sick, are passive, helpless, and have lost daily living skills (Townsend, 1976). This category development functioned to establish Tim as capable, rational, and therefore able to determine and manage disclosure of private matters in social relationships.

Having established identity claims to the category of independence, Tim and Martin address the dissonance encountered in social situations in which they were required to actively establish emerging identities in social interaction. Dissonance in this sense refers to managing social interactions in which maintenance of privacy was required to ensure initial engagement with other social actors while being fully aware that this was unlikely to be a successful strategy for establishing sustained social bonds. 'being diplomatic with the truth' and 'some sort of neutral ground' involved managing the flow of information about the past. This type of information is highly identity-relevant (Estroff, 1989). Participants made use of common understandings of societal reactions to mental illness to support decisions to keep hidden identity-relevant aspects of his past. Having already established through talk his independent-mindedness, Tim used this category to warrant an additional claim, 'I'm obviously very well, stable'. Martin established his independence through the everyday hobby of being a collector of memorabilia. Thus the category of independence was presented as verifiable and credible. This becomes increasingly important for individuals when the status of their mental health is directly implicated in risk behaviours. Establishing independence and sustained mental health is crucial to ward off concerns about risk although ultimately these concerns are highly durable and unlikely to be so easily dispensed with.

Tim's claim to the obviousness of his wellness was in part due to his established independent-mindedness but also perhaps as a result of an unspoken assessment that may be perceived to have occurred in the research interview. Tim's claim to a wellness category was necessary to reinforce his claims to the category of independent-mindedness. This normative category orientation is required in social situations when illness labels are still current and available. The continued supervision and monitoring of Tim by workers made available and warranted the use of the label of mental illness. In challenging and resisting this label Tim was required to do the work in his talk of establishing an alternative identity which was realised in the category of independent-mindedness. The statement that his wellness was obvious functioned to shore-up and protect the category from challenge. If something is claimed to be obvious to one participant it may be difficult for the listener to question this. This may be particularly so in an interaction where the speaker is expected to hold the floor such as in providing narrative accounts in research interviews. Tim shows that he is highly attuned to the rules of social engagement. Premature disclosure of information would threaten social relations given previous risky behaviours. Tim's 'being diplomatic with the truth' appears to find a corollary in Martin's 'I always try to find some sort of neutral ground'. Tim shifts the locus of concern onto the risk to him of failing to meet his obligations for supervision and residence as required by his restriction order. Martin's account works to construct his vulnerability to 'attacks'. Risk in these accounts is therefore placed as emanating elsewhere. The concern with disclosure of past offending and mental illness histories is mobilised for the purposes of managing implications of risk and shifting the locus to the community rather than individual.

However for many people the conditions of their discharge may place them under obligations to be fully open, for instance in situations involving prospective employers or new intimate partners.

Colum: ... (2) What I find now, now I am out is the restrictions on me if I go for a job and things like that they want me to disclose my offence and things like that which I find puts the employer off and (1) you know.

It appeared that for Colum agreements made in secure settings about future community living and without full knowledge of the likely outcomes had presented unexpected challenges to the work of fitting-in. Colum provided his understanding that disclosure of past events ostensibly for risk management reasons had negative consequences for his prospects of securing employment. The insistence of workers to require disclosure may run the parallel risk of prolonged unemployment and continued exclusion for Colum and others in the same situation. This may be 'a risk worth taking' for those providing aftercare but it remains a problem for patients themselves. Interestingly Colum's social worker constructed the need for disclosure as an opportunity for him to learn and gain positive outcomes.

Colum's social worker: .... if we're right about this we should be encouraging him to do this for himself, if he can do that for himself, go in, deal with the disclosure, have his own interview, he'll get more self-esteem from that, he'll feel good about himself.

For workers disclosure was seen as risk management tool, a means to ensure communication of past behaviours to those who may need to know this information in the present. Workers show that while they are aware of the intrusive nature of their monitoring they remained convinced that this was the best way of managing potential risk behaviours and avoiding blame for untoward events (Douglas, 1992). The belief that this intensive care supervision provides sufficient leverage over individuals to reduce risk behaviours in the absence of attention to social and structural factors may be unfounded (Swanson, 2008) but in the absence of technologies that provide more certainty they may be all that is available.

Conclusion

The path towards establishing new and emerging identities is not as straightforward as simply articulating new versions of the self. Within the context of serious and enduring mental ill-health and criminal offending risk then articulating new risk-free versions is perhaps more difficult still. Instead these new versions are realised and recast within talk where identity claims are at stake. This interactive element of identity formation is ongoing and in some ways may mirror or parallel transitions from hospital to the community, from detention towards liberty and from mentally ill offender towards community citizen, from risky individual to ordinary person. It has been suggested that the deployment of social identities always relates to some matter in hand in talk (Antaki et al., 1996). In interactions with others, Antaki and colleagues suggest, this deployment may be briefly over and done with but can have a cumulative effect. This is a process of assembling, rehearsing, and refining identities in talk and is a necessary and crucial element in the move towards achieving successful community return and full involvement in social life.

References

Altman, I., & Taylor, D. A. (1973). *Social Penetration: The Development of Interpersonal Relationships*. New York: Holt.

Altman, I., Vinsel, A., & Brown, B. R. (1981). Dialectic conceptions in social psychology: An application to social penetration and privacy regulation. *Advances in Experimental Social Psychology*, 14, 107–160.

Anderson, J. M., Elfert, H., & Lai, M. (1989). Ideology in the clinical context: Chronic illness, ethnicity and the discourse on normalisation. *Sociology of Health and Illness*, 11(3), 253–278.

Antaki, C., Condor, S., & Levine, M. (1996). Social identities in talk: Speakers own orientations. *British Journal of Social Psychology*, 35(4), 473–492.

Brannen, J. (1988). The study of sensitive subjects. *Sociological Review*, 36, 552–563.

Coffey, M. (2013). Time and its uses in accounts of conditional discharge in forensic psychiatry. *Sociology of Health and Illness*, 35(8), 1181–1195.

Coffey, M. (2012a). A risk worth taking? Value differences and alternative risk constructions in accounts given by patients and their community workers following conditional discharge from forensic mental health services. *Health Risk and Society*, 14(5), 465–482.

Coffey, M. (2012b). Negotiating identity transition when leaving forensic hospitals. *Health: An Interdisciplinary Journal for the Social Study of Health, Illness and Medicine*, 16, 489–506.

Douglas, M. (1992). *Risk and Blame: Essays in Cultural Theory*. London: Routledge.

Edwards, D., & Potter, J. (1992). *Discursive Psychology*. London: Sage Publications.

Estroff, S. E. (1989). Self, identity, and subjective experiences of schizophrenia: In search of the subject. *Schizophrenia Bulletin*, 15(2), 189–196.

Fisher, D. V. (1984). A conceptual analysis of self-disclosure. *Journal for the Theory of Social Behaviour*, 14(3), 277–296.

Foddy, W. H., & Finnigan, W. R. (1980). The concept of privacy from a symbolic interaction perspective. *Journal for the Theory of Social Behaviour*, 10, 1–17.

Garfinkel, H. (1967). *Studies in Ethnomethodology*. Englewood Cliffs, New Jersey: Prentice-Hall.

Goffman, E. (1963). *Stigma: Notes on the Management of Spoiled Identity*. New Jersey: Prentice-Hall Inc.

Lee, R. M. (1993). *Doing Research on Sensitive Topics*. London: Sage Publications.

Margulis, S. T. (2003). On the status and contribution of Westin's and Altman's theories of privacy. *Journal of Social Issues*, 59(2), 411–429.

Radley, A., & Billig, M. (1996). Accounts of health and illness: Dilemmas and representations. *Sociology of Health and Illness*, 18(2), 220–240.

Roth, J. A. (1963). *Timetables: Structuring the Passage of Time in Hospital Treatment and Other Careers*. Indianapolis: Bobbs-Merrill.

Scott, M. B., & Lyman, S. M. (1968). Accounts. *American Sociological Review*, 33(1), 46–62.

Sieber, J. E., & Stanley, B. (1988). Ethical and professional dimensions of socially sensitive research. *American Psychologist*, 43(1), 49–55.

Swanson, J. (2008). Preventing the unpredicted: Managing violence risk in mental health care. *Psychiatric Services*, 59(2), 191–193.

Thornicroft, G. (2006). *Shunned: Discrimination against People with Mental Illness*. Oxford: Oxford University Press.

Townsend, J. M. (1976). Self-concept and the institutionalization of mental patients: An overview and critique. *Journal of Health and Social Behavior*, 17(3), 263–271.

Westin, A. F. (1967). *Privacy and Freedom*. New York: Anthenaeum.

4
Risk and Safety in Linguistic and Cultural Diversity: A Narrative Intervention in Residential Aged Care

Jonathan Crichton and Fiona O'Neill

Introduction

This chapter reports on an intervention that sought to enhance how safety is understood and communicated in residential aged care. Workplace safety in aged care is a growing concern internationally because the combination of older clients with increasingly complex health profiles and growing linguistic and cultural diversity among healthcare workers and their clients is raising the physical and psychosocial risks for both groups (Pearson et al., 2007).

Current approaches to safety communication in the sector and more generally emphasise the provision and auditing of information about safety procedures and safe work practices. This 'transactional' approach (Clarke, 2013) is limited by not being sensitive to diverse understandings of safety among staff or to the complex social, linguistic and cultural environment in which they understand and interact with each other and with residents. Moreover, the fact that these groups are from different linguistic and cultural backgrounds further risks the mutual understanding on which meaningful communication depends (Fryer, Mackintosh, Stanley, & Crichton, 2013).

The study (Scarino, O'Keeffe, Crichton, O'Neill, & Dollard, 2014) aimed to address this limitation by collaborating with staff, residents, and managers to develop interventions that foster shared understanding and communication about safety among staff and residents in the context of linguistic and cultural diversity.

We explore here the potential that narrative offers for such interventions. We start by outlining the broader context of change in residential aged care before explaining the study and the narrative approach taken to the intervention. Using illustrative examples of data and drawing in particular on the work of Garfinkel and Ricoeur we argue that narrative offers a way of promoting an 'interactional' approach to safety communication: focusing on how narratives were elicited during the study, used to intervene in training and reflected on by staff in light of their subsequent practice.

The context of residential aged care

The aged care sector in Australia exemplifies the global phenomenon of increasing and more complex, shifting and dynamic connections between people of diverse linguistic and cultural backgrounds (Blommaert & Rampton, 2011). These changes are combining with economic, technological and demographic trends to transform contemporary life, reshape societies and challenge the categories by which they have been understood (Blommaert, 2013). The demographics of aging are one such trend. Australia like many other Western countries is facing a 'crisis in aged care' (Hugo, 2007) due to a dramatic increase in the number of people requiring aged care and a shortfall of workers to meet this need, which is increasingly being met by recent migrants of diverse linguistic and cultural backgrounds (Fine & Mitchell, 2007). Currently, 35% of residential aged care staff in Australia were born overseas, and a third of these come from countries where English is not the primary language (King et al., 2013). At the same time, the proportion of people in care for whom English is not the primary language is also increasing as previous generations of migrants – from different linguistic and cultural backgrounds to more recent migrants – enter aged care, with a shift predicted from the post-war European migrants to an increasing number of aging Asian migrants (Productivity, 2011).

The implications of this increasing and increasingly complex diversity for communication in aged care is not well understood but an emerging body of literature has identified concerns that centre on risks associated with care delivery and workplace safety (Diallo, 2004; Fine & Mitchell, 2007; Hugo, 2009), and emphasising the physical and psychological wellbeing of both those who deliver and those who receive care (Myers, Silverstein, & Nelson, 2002). In a national survey (King & Martin, 2008), communication was identified by 70% of aged care facility managers as the most pressing issue for workers for whom English is not their primary language, and linguistic and cultural difference have been identified as particular concerns where shift handovers, patient assessment and interactions with residents and colleagues are just some of the potential communicative demands to be navigated (Crawford & Candlin, 2012).

Risks associated with communication in culturally and linguistically diverse healthcare settings are both psychosocial and physical, include misunderstandings which affect care delivery, the challenge of learning medical and lay language, and the risk of negative evaluations, stereotyping, and professional marginalisation (Bosher & Smalkoski, 2002; Dreachslin, Hunt, & Sprainer, 2000; Johnstone & Kanitsaki, 2007; Olson, 2012). A key message is that, as in other areas of work, assumptions and judgements of linguistic and cultural proficiency are inseparable from evaluations of professional competence, and that these evaluations can become expectations against which people's expertise is stereotyped, further putting at risk the mutual understanding and communication on which workplace safety depends. Relevant here is Ulrey and Amason's (2001) point that it is important not to focus solely on people as 'other' in the host culture, but to foster awareness that intercultural communication is something that everyone, not just those who may not have full proficiency in the dominant language and culture, is increasingly called to do.

Pearson et al. (2007) drew on a systematic review of 659 papers to explore how cultural competence can be promoted among staff. Defining cultural competence as 'the knowledge, skills, attitudes and behaviours required of a practitioner to provide optimal health care services to persons from a wide range of cultural and ethnic backgrounds' (2007, p. 59), the review underscores the complexity created by increasing linguistic and cultural diversity in healthcare. A key finding was the urgent need for intercultural awareness education and training for such professionals. This is relevant internationally and certainly in the Australian context, in which healthcare settings are still trying to navigate the complexity of communication arising from linguistic and cultural diversity, despite 30 years of multicultural policy (Johnstone & Kanitsaki, 2007).

The intervention

Some background on the study will be helpful before getting to the account of the intervention. The study involved researchers in applied linguistics and in occupational health and safety, was funded by SafeWork South Australia and conducted in collaboration with a major aged care organisation which provides home care services, retirement living, and residential care homes to over 5,000 clients in metropolitan and regional areas, with residential care in eight sites.

The study sought to understand how and why participants communicate, understand and accomplish safety within linguistic and cultural diversity while ensuring the 'practical relevance' (Roberts & Sarangi, 1999) of the study to the sector. This emphasis on developing a mutual understanding required 'thick participation' (Sarangi, 2005), which

> constitutes a form of socialisation and it should not be equated with becoming a professional expert. There is more to expertise than a familiarisation with experience from the periphery. What I have in mind here is more of an acquisition of professional/organisational literacy that would provide a threshold for interpretive understanding. (p. 377)

Seeking this 'threshold' meant close and extended immersion by the researchers within the everyday life of the organisation; in particular, it required ongoing collaboration and dialogue among the researchers and the participants, within the organisation more broadly and among the different disciplines represented in the research team.

Data was gathered at two of the organisation's residential aged care facilities over six months and combined ethnographic approaches with ongoing interviews. At these sites residents self-identified with seven linguistic groups and staff with 12 and there was no overlap between them. In all 67 shifts were observed, and 43 nurses and carers and 22 residents interviewed. The resulting observational notes and interview transcripts were analysed thematically from the perspectives of the research team and in collaboration with participants (Creswell, 2007; Riessman, 2008). For this chapter, the importance of the findings lie in how they informed the intervention that followed.

In sum, the findings emphasised that in contexts of linguistic and cultural diversity, psychosocial and physical risks increase for all participants because of the complexity of interactions and understandings between people. This complexity is not, as it were, one sided; with the dominant linguistic and cultural groups stable and homogeneous and 'diversity' a characteristic only of those groups who are not identified with the dominant language and culture. Such categorisation is inevitably imposed by the dominant group in a way that positions non-dominant groups as different, identifying a particular category of people as 'a problem' and implying normative expectations for how that group will behave in contrast to others (Sacks, 1992). Rather the findings emphasised that all individuals are examples of diversity and therefore mutually involved in the complexity that results. In dealing with this complexity, evidence was found that carers and residents developed 'local methods' (Garfinkel,

1967) to develop shared understandings of safe practices and to interact in ways that attended to safety considerations that were not part of documented procedures. In describing their experiences participants did not cite safety regulations – as if their actions were to be explained as following rules. This perhaps should not be surprising for, as Garfinkel (1967) explained, every rule ends in an implicit 'et cetera': in other words, no rule can determine in advance how it should be used because no rule can specify all the conditions of its own application. Rather, the 'application of a rule' is tied to the local circumstances in which people make sense of the rule for practical purposes. It is only then – in other words retrospectively – that actions may be interpreted to have been in 'accordance' or at 'variance' with the rule. Very much in keeping with this emphasis on the meaning of rules as necessarily interpreted against local methods participants in the study recounted their actions as inextricably tied to, and revealing of the circumstances and relationships in which they are produced i.e. as 'accounts' in Garfinkel's (1967) sense: accounts which together constituted the larger narratives through which participants recounted and made sense of their working lives (Riessman, 2008).

Examples included the following account by a nurse of how she explored with staff 'what it is they are going to do, rather than ask, "Do you understand?"':

> Sometimes when there's a misunderstanding I'm not sure if it's a language issue or a cultural difference. I've found it's more helpful to ask questions to clarify or to get the carer to explain what it is they are going to do, rather than ask, 'Do you understand?', because not many people answer 'No I don't understand', they just go off and try to figure it out for themselves. This is really important when lifting or transferring, because that's one of those moments when everyone needs to know what's happening so no-one gets injured.

This emphasis on accomplishing understanding through mutual involvement in the task at hand rather than through checking comprehension of instructions or antecedent procedures was paralleled in interviews with residents. In the following example, in the context of showering, a situation involving potential psychosocial and physical risks, a resident recounts her use of metaphor to frame the activity so as to reduce the potential stress for both participants:

> It's a bit shattering at first especially when you get to the stage where you've got to be showered by other people ... especially if a great big

> six foot ten Nigerian comes in to shower you and you've never met him before ... I make him laugh and I say to him 'Look just pretend you're washing a car ... scrub my back' I said 'You're doing the bodywork on the car now your arms and legs they're the four wheels so give them all a good scrub ... now.' I said 'Just give me the sponge I'll be doing my face while you clean the windscreens. 'Well, they finish up laughing and enjoying themselves instead of being embarrassed as I was you see ... they're more embarrassed than you, especially when they start ... after they've done me a few times, we just have a good laugh.

This local, mutual, in vivo accomplishment of safety in interactions among individuals in accordance with how they understood each other in the interaction at hand stood in contrast to the reliance in the sector on such methods of distributing safety information as procedures, checklists and the like. Moreover, while these 'transactional' methods readily offer desktop compliance with relevant regulatory requirements they do not acknowledge the local 'interactional' methods on which the day-to-day and usually taken-for-granted accomplishment of safety may depend. While these ways of managing workplace risks were routine for participants experienced in this environment, the fact that they were 'hidden in plain sight' made them especially daunting to acquire for people new to the sector and less familiar with the dominant language and culture. This challenge was summed up by one participant as follows:

> Nothing ... you can't do anything about it ... it's this way I will explain ... there is a highway ... everyone else is going over the speed of 100k per hour and you're just trying to enter this highway and your speed is probably 45 what do you do you can't take your car there and let them smash into you so what do you do? You just drive carefully around the corner and try to get to the speed 100k and then when you have 100k you manipulate yourself you can't fight with everyone can you?

A key implication of the study's findings was that training in this area needs to recognise and promote safety as an ongoing, in large part locally accomplished, project of organisational learning and change. The challenge in designing the intervention was how to accomplish this and in doing so to trial an appropriate model of training that attended to the policies and local understandings and practices of the collaborating organisation and encouraged each person in the organisation to understand herself as example of linguistic and cultural diversity.

The intervention drew on an approach to narrative intervention developed by Crichton and Koch (2007, 2011, 2007) in which narratives of experience gathered in collaboration with participants were used to understand, reframe and enhance practices involved in the transition into residential aged care for a person living with dementia. The approach owes a considerable debt to Mattingly's (1998) study in the context of occupational therapy in which she showed how shared narratives are co-constructed by professionals as 'narrative contexts' within which they can interpret as meaningful the behaviours of clients and of each other. More generally, these studies draw on the work of Ricoeur who emphasised that narrative is not only a means by which people represent their experiences to themselves and each other but that narrative makes human experience itself meaningful by constituting people's retrospective and ongoing interpretations of their perceptions of themselves and others in time. He stated that:

> time becomes human to the extent that it is articulated through a narrative mode, and narrative attains its full meaning when it becomes a condition of temporal existence. (Ricoeur, 1985, p. 52)

In other words, people's understanding of each other and their capacity for narrative are mutually dependent: we interpret who we are and what we do as meaningful to the extent that we are able to bring a narrative structuring to our perceptions of ourselves and others in an ongoing, interwoven, ever expanding 'cloth woven of stories told' (Ricoeur, 1988, p. 246). Through this process of 'emplotting' (Ricoeur, 1991), narrative enables a shared moral world because how people figure in such narratives – their 'narrative identity' (Ricoeur, 1988) – reflexively informs how they are understood and judged as a people, including questions of ethical and professional behaviour, accountability, and competence:

> the category of character is therefore a narrative category as well, and its role in the narrative involves the same narrative understanding as the plot itself. (Ricoeur, 1992, p. 143)

Drawing on this feature of narrative as the primary medium through which people construct, understand and evaluate themselves and others as part of a shared moral world, the intervention sought to 'play back' into the ongoing narratives of those working in the organisation selected narrative 'vignettes' from the experiences of those who had participated in the study. Specifically, extracts from the interview data were included in an online training module on safety procedures in 'manual handling', which includes all aspects of work in which staff may be involved in moving people or objects. The aim was to reframe how the module could be read in light of the local methods of safety communication that had been recounted in the interviews, thereby enabling the training materials not only to transmit information about safety in the form of procedural sequences and check lists but also to invite readers to draw on, include, share and extend these narratives in understanding themselves among others in accomplishing safety. In other words to provide a narrative context within which the local methods of others could be shared and understood, not as rules to be applied but as invitations to participate in their lives. As an intervention, the inclusion of the vignettes thus sought to raise awareness of how people take up and act upon safety information, facilitating reflection on individual safety practices (Iedema, Mesman, & Carroll, 2013), and sensitising them to how (in)attention to linguistic and cultural difference contributes to how they can together communicate, interpret, and accomplish safety.

In line with the aim of achieving 'practical relevance' (Roberts & Sarangi, 1999) to those who would be participating in it, the design of the intervention involved collaboration between the researchers, trainers, staff, and managers working together in three stages: (1) development of the intervention; (2) online training module undertaken by ten staff members; and (3) follow up journal completed by staff to give feedback on their perceptions of the effect of the intervention on their work practices.

Together, over a series of meetings the group chose extracts from the interviews along with photographs of organisational staff and residents that would be included at the beginning of each section in the training module. These include the extract quoted earlier in the chapter which was chosen to introduce the module as a whole. While discussing how the vignettes and images would be inserted and the pages reformatted to foreground them, the group decided that the original text that had been used in the module would need to be edited to reference their understanding of the conjunction of narrative vignette and image. This recognition by the designers of a need to alter the original text can perhaps be seen as the vignette, image, page formatting and original text combining with the intervention agenda to 'resemioticize' (Iedema, 2003) how their reading of training materials, reflecting:

> a multimodal appreciation of meaning making [that] centres around two issues: first, the de-centring of language as favoured meaning

making; and second, the re-visiting and blurring of the traditional boundaries between and roles allocated to language, image, page layout, document design, and so on. (Iedema, 2003, p. 33)

These revisions to the original text occurred through all sections of the module in response to the inclusion of vignettes and images. The sentence 'Safety is about caring for each and communicating with everyone' was added as a 'principle' that might be inferred from the experience recounted in the narrative, against which the subsequent information could be read.

The intervention sought to make available and invite a new reading of this text, one that could include the person pictured and the reader together within the narrative context implied by the vignette and photograph, in effect seeking to 'revoice' (Bakhtin, 1981) the text as part of a shared narrative.

In order to understand how the redesigned training materials were interpreted by the staff who undertook the training module, a reflective journal was used. Ten staff members undertook the revised training module: an RN, four enrolled nurses, and five carers. Four of the nurses and one of the carers were from an Anglo-Australian background; one of the nurses and four of the carers identified with other linguistic and cultural backgrounds. Completed within two weeks over five shifts, the journal included five prompts for reflection, one for each shift, and each with two vignettes from the e-learning programme: the introductory quote and a second from a later section of the module.

After each shift the participants were asked and reflect on their experience of the communication of safety in relation to the experience recounted in the vignette, as shown in the following example:

During today's shift, please consider these quotes and the question below:

> Sometimes when there's a misunderstanding I'm not sure if it's a language issue or a cultural difference. I've found it's more helpful to ask questions to clarify or to get the carer to explain what it is they are going to do, rather than ask, 'Do you understand?', because not many people answer 'No I don't understand', they just go off and try to figure it out for themselves. This is really important when lifting or transferring, because that's one of those moments when everyone needs to know what's happening so no-one gets injured. (Nurse)
>
> I could see the carer was embarrassed about helping me in the shower, but I needed him to stay so I didn't fall, so I told him jokes.

> In the end this made us both feel better about it, and I certainly felt safer because he stayed and helped me. (Resident)

Reflecting on these experiences, how would you change the way you care for yourself, your colleagues and residents?

Reflect on your experience of communicating safety with others on the shift and write a paragraph or two comparing your experience with the experiences described in these quotes. For example, were there any similarities or differences in the way you cared for colleagues and clients?

The final section of the journal invited participants to reflect more broadly on any ways in which the vignettes, prompts, and questions from the training module might have affected their understanding. The aim of these further questions, and of the journals as a whole, was to provide examples of experience that could be included in and further improve safety training.

Participants' reflections

Eight nurses and carers completed the redesigned training module and reflective journal. Five identified as Anglo Australian and three with other linguistic and cultural backgrounds. All the participants commented on the value of understanding safety as interactionally accomplished. This interactional account of safety communication can be seen in the experience of the participants in their feedback after the intervention in the following examples:

> I worked with a carer, she was a new employee and it was her third shift. I told her she is going to shower one of the residents, when I looked at her face, I found it looked like she was not sure how she was going to do it. I thought it's better the first time to be with her and explain to her what to do and make sure she is doing the right manual handling and safe working. I told her how she needs to do things, we did them together and at the end she told me how she was confident for the next resident. That was a good experience for the shift. She learned she needs to ask if she is not sure, before making mistakes. (Middle Eastern carer)

> I've had this happen, a new carer had started and was struggling with how to do things. After sitting down and discussing what they were struggling with, the nurse and myself then showed her. After a few doubles (two staff working together) she got the hang of it.

I encouraged her to ask questions if she was unsure. After the shift I asked them if they wanted some clarification. I answered questions for her. (Australian carer)

The quotes exemplify the ways in seeking to develop a shared understanding of safe practice participants attended in the first instance not to the relevant regulation by telling and checking understanding of this but to their understanding of each other together in the task at hand. In elaborating this recognition of the interactional nature of safety, a second theme emerged: that of mediating languages and cultures. Here the journal entries recounted how participants sought to anticipate preconceptions and managed potential misunderstandings:

This [narrative vignette] reminded me that assuming familiarity with terms can be a mistake. After taking a resident's blood pressure as I'd requested, the carer told me the result which was on the low side, I automatically queried 'Lying?' (as in lying/standing position) and her horrified response was 'No I'm telling the truth'. This occurred in front of the other carers and after brief embarrassment and laughter a discussion with the relieved carer about understanding English language with all its sayings, slang and colloquialisms. I often feel I may be patronising in querying understanding if language is an issue and have occasionally sensed resentment if I reiterate or oversimplify. I have used this anecdote and other examples to ensure colleagues understand that I do not mean to demean them in so doing. (Australian nurse)

The quote illustrates the reflective and reflexive character of local methods, in which here the nurse draws on her understanding of the situation to interpret her action as it is interpreted by the carer as falling short because it threatens the face of the carer, then recovers mutual understanding by adjusting her action and understanding of herself and the carer in the task at hand, and infers implications for her own ongoing practice. In doing so she bridges a gap that arises in vivo from linguistic and cultural differences, potentially compounded by different levels of status and experience. The nurse's stance on her own presence and actions as being subject to diverse interpretations by others points to the third way in which interactional nature safety was elaborated in the journals, by understanding oneself as an example of difference, illustrated in the following extracts:

A lot of carers in this industry are much younger and even if I were not senior to them would possibly feel disrespectful in asking me to repeat myself. I also work quickly and this can suggest inferior skills and impatience which could erode confidence so I try to be encouraging and positive. I also use my hearing deficit to identify with staff from different language and culture backgrounds, as I often have to ask people to repeat themselves and have sensed irritation at times. I certainly know I have assumed that I have heard correctly and not asked for repetition for fear of seeming 'dense'. I do have the confidence to ask for clarification where safety may be compromised and have explained my hearing deficit and the speaker accent combined are a problem. I find this deflects the tension and clears the air for much safer communication. Clients can be influential in persuading staff to perform unsafe manual handling. We need to give each other strategies to refuse without causing offense or tension. (Australian nurse)

Obviously our backgrounds have a lot to do with how we live our lives every day and how we care for people. For example the way I talk to residents is a bit different. Most of the time I speak to them in a low voice and this has something to do with my background and my personality, as I see all of them as my grandparents and in my culture you are never allowed to yell at elderly ones and maintaining eye contact with them is considered being rude and disrespectful. So I take my time and I'm extra careful while communicating with my colleagues and residents because it's always good to treat people the same way you will love to be treated. (African carer)

Each of these participants draws on their autobiographical narratives to interpret and present themselves as examples of difference, thereby invoking and contributing to the ongoing accomplishment of a shared narrative context. This shared narrative depends on the participants understanding themselves as examples of difference, underscoring the point noted above that as long as linguistic and cultural diversity is understood only as a characteristic of 'others', those who identify with the dominant language and culture will not understand themselves as embodying cultural assumptions and linguistic categories that underlie the judgements and associated narratives that they may be predisposed to construct about these others. This leads to the fourth way in which participants elaborated the interactional nature of safety communication, that of enabling people to speak up for themselves and work together, illustrated in the following quotes:

When 12 years ago I started working as a carer I was very shy, but I remember one sentence in my book, 'Before you make a mistake always ask' That's my experience and I always tell to new staff 'Please it's not embarrassing to ask, it's embarrassing to make a mistake not asking'. We are from different cultures and backgrounds, but it doesn't mean we can't learn from each other and ask questions, nothing to do with culture and background you have to be confident in your job otherwise you're not going to be a good carer or other job. I love to get to know new staff and found it's what's in common rather than 'I'm better than you', which makes them to hate me. (Middle Eastern carer)

Other staff members have approached me and said they were not sure that she could understand them, and that they had difficulty understanding her. When I asked them why they said she spoke very quietly and when they asked her something she would respond with very short answers or nod. When I worked alongside her I found I would have to initiate conversation, ask her understanding of what tasks we were about to do so as to see clarification she understood. Once I offered her some prompts I found she would offer more in her conversation on that one-on-one basis. I asked other carers to discuss more with this carer what tasks they were doing, assist her to understand, ask her to speak up a bit more so clients could hear her and engage her in other conversations so she felt more included in the team. (Australian nurse)

These examples show not only how high the stakes are but also the sophistication of communicative expertise that is needed to create and maintain the mutual trust on which participants' professional practice depends (Crichton, 2013). The point here is not that regulation is irrelevant but, as Garfinkel (1967) explained, that no rule can determine or anticipate how people make sense of their immediate situation; rather, the relevance of a rule can only emerge out of how people understand each other in interaction within this situation.

Conclusion

In this chapter we have argued for the value of understanding the communication of safety not only as transactional – as a matter of transmitting, learning, and complying with information about safety processes and requirements – but also as interactional: specifically as local methods that reflect the myriad ways in which people accomplish care for each other by communicating and interpreting safety and risk in any given situation. As Candlin and Candlin (2002, p. 103) state, such 'expertise in the management of risk is not solely – or even primarily – a matter of knowledge but one of discursive negotiation among participant values and experiences'. This understanding of the relationship between regulation and local methods in the workplace becomes all the more important in contexts of linguistic and cultural diversity. Here the psychosocial and physical risks are heightened and the intercultural communicative expertise of all participants is at a premium, requiring ongoing alertness to oneself as an example of difference, the potential for unexpected understandings from others, and ongoing communication about safety that is not based on preconceptions about the way others understand them. This expertise is dependent on the narratives by which people interpret themselves and each other. A narrative approach can, we have argued, promote the accomplishment of a shared narrative context within which such local methods of safety communication may be acknowledged and enhanced. We suggest that this approach provides a possible model for reframing current transactional approaches to safety communication.

References

Bakhtin, M., M. (1981). *The dialogic imagination: Four essays by M. M. Bakhtin* (C. Emerson & M. Holquist, Trans.). Austin: University of Texas Press.

Blommaert, J. M. E. (2013). Citizenship, language, and superdiversity: Towards complexity. *Journal of Language, Identity & Education and Society, 12*(3), 193–196.

Blommaert, J. M. E., & Rampton, B. (2011). Language and superdiversity. *Diversities, 13*(2), 1–21.

Bosher, S., & Smalkoski, K. (2002). From needs analysis to curriculum development: Designing a course in health-care communication for immigrant students in the USA. *English for Specific Purposes, 21*, 59–79.

Candlin, C., & Candlin, S. (2002). Discourse, expertise, and the management of risk in health care settings. *Research on Language & Social Interaction, 35*(2), 115–137.

Clarke, S. (2013). Safety leadership: A meta-analytic review of transformational and transactional leadership styles as antecedents of safety behaviours. *Journal of Occupational and Organizational Psychology, 86*(1), 22–49. doi: 10.1111/j.2044-8325.2012.02064.x

Crawford, T., & Candlin, S. (2012). Investigating the language needs of culturally and linguistically diverse nursing students to assist their completion of the bachelor of nursing programme to become safe and effective practitioners. *Nurse Education Today*. doi: http://dx.doi.org/10.1016/j.nedt.2012.03.005.

Creswell, J. W. (2007). *Qualitative inquiry and research design: Choosing among five approaches*. Thousand Oaks, CA: Sage.

Crichton, J. (2013). 'Will there be flowers shoved at me?': A study in organisational trust, moral order and professional integrity. In C. N. Candlin & J. Crichton (Eds.), *Discourses of trust* (pp. 119–132). Basingstoke: Palgrave Macmillan.

Crichton, J., & Koch, T. (2007). Living with dementia: Curating self identity. *Dementia, 6*(3), 365–381.

Crichton, J., & Koch, T. (2011). Narrative, identity and care: Joint problematization in a study of people living with dementia. In C. N. Candlin & J. Crichton (Eds.), *Discourses of deficit* (pp. 101–118). Basingstoke: Palgrave Macmillan.

Diallo, K. (2004). Data on the migration of health-care workers: Sources, uses and challenges. *Bulletin of the World Health Organisation, 82*(8), 601–607.

Dreachslin, J., Hunt, P., & Sprainer, E. (2000). Workforce diversity: Implications for the effectiveness of healthcare delivery teams. *Social Sciences and Medicine 50*, 1403–1414.

Fine, M. D., & Mitchell, A. (2007). Immigration and the aged care workforce in Australia: Meeting the deficit. *Australasian Journal in Ageing, 26*(4), 157–161.

Fryer, C. E., Mackintosh, S. F., Stanley, M. J., & Crichton, J. (2013). 'I understand all the major things': How older people with limited English proficiency decide their need for a professional interpreter during health care after stroke. *Ethnicity & Health*, 1–16. doi: 10.1080/13557858.2013.828830

Garfinkel, H. (1967). *Studies in ethnomethodology*. Englewood Cliffs, CA: Prentice-Hall.

Hugo, G. (2007). Contextualising the 'crisis in aged care': A demographic perspective, Australian Journal of Social Issues *Australian Journal of Social Issues, 42*(2), 169–182.

Hugo, G. (2009). Care worker migration, Australia and development. *Population, Space and Place, 15*, 189–203. doi: 101 1002/psp

Iedema, R. (2003). Multimodality, resemiotization: Extending the analysis of discourse as multi-semiotic practice. *Visual Communication, 2*(1), 29–57. doi: 10.1177/1470357203002001751

Iedema, R., Mesman, J., & Carroll, K. (2013). *Visualising health care practice improvement: innovation from within*. London: Radcliffe.

Johnstone, M.-j., & Kanitsaki, O. (2007). An exploration of the notion and nature of the construct of cultural safety and its applicability to the Australian health care context. *Journal of Transcultural Nursing, 18*(3), 247–256.

King, C., & Martin, B. (2008). *Who cares for older Australians? A picture of the residential and community based aged care workforce, 2007*. Flinders University: National Institute of Labour Studies.

King, D., Mavromaras, K., Wei, Z., He, B., Healy, J., Macaitis, K.,... Smith, L. (2013). *The Aged Care Workforce Final Report 2012*. Flinders University.

Koch, T., & Crichton, J. (2007). Living with dementia: Innovative methodological approaches toward understanding. In M. Nolan, E. Hanson, G. Grant & J. Keady (Eds.), *User participation in health and social care research: Voices, values and evaluation* (pp. 89–103). Maidenhead: Open University Press.

Mattingly, C. (1998). *Healing dramas and clinical plots: The narrative structure of experience*. New York: Cambridge University Press.

Myers, D., Silverstein, B., & Nelson, N. A. (2002). Predictors of shoulder and back injuries in nursing home workers: A prospective study. *American Journal of Industrial Medicine, 41*, 466–476.

Olson, M. A. (2012). English as a second language (ESL) nursing student success: A critical review of the literature. *Journal of Cultural Diversity, 19*(1), 26–32.

Pearson, A., Srivastava, R., Craig, D., Tucker, D., Grinspun, D., Bajnok, I., ... Gi, A. A. (2007). Systematic review on embracing cultural diversity for developing and sustaining a healthy work environment in healthcare. *International Journal of Evidence Based Healthcare, 5*, 54–91. doi: 10.1111/j.1479-6988.2007.00058.x

Productivity, C. (2011). *Caring for older Australians*. Canberra.

Ricoeur, P. (1985). *Time and narrative* (K. Mclaughlin & D. Pellauer, Trans. Vol. 2). Chicago: University of Chicago Press.

Ricoeur, P. (1988). *Time and narrative* (K. Mclaughlin & D. Pellauer, Trans. Vol. 3). Chicago: Chicago University Press.

Ricoeur, P. (1991). Life in quest of narrative. In D. Wood (Ed.), *On Paul Ricoeur: Narrative and interpretation* (pp. 20–33). London: Routledge.

Ricoeur, P. (1992). *Onself as another* (K. Blamey, Trans.). Chicago: Chicago University Press.

Riessman, C. K. (2008). *Narrative methods for the human sciences*. Thousand Oaks, CA: Sage.

Roberts, C., & Sarangi, S. (1999). Hybridity in gatekeeping discourse: Issues of practical relevance for the researcher. In S. Sarangi & C. Roberts (Eds.), *Talk, work and institutional order: Discourse in medical, mediation and management settings* (pp. 473–504). Germany, Berlin: Mouton de Gruyter.

Sacks, H. (1992). *Lectures on conversation, volumes I and II*. Oxford: Blackwell.

Sarangi, S. (2005). The conditions and consequences of professional discourse studies. *Journal of Applied Linguistics, 2*(3), 371–394.

Scarino, S., O'Keeffe, V., Crichton, J., O'Neill, F., & Dollard, M. (2014). *Communicating work, health and safety in the context of linguistic and cultural diversity in aged care*. Adelaide: Safework, SA.

Ulrey, K. L., & Amason, P. (2001). Intercultural communication between patients and health care providers: An exploration of intercultural effectiveness, cultural sensitivity, stress and anxiety. *Health Communication, 13*(4), 449–463.

5
Choice, Risk, and Moral Judgment: Using Discourse Analysis to Identify the Moral Component of Midwives' Discourses

Mandie Scamell and Andy Alaszewski

Introduction

In this chapter we examine midwives' discourses in relationship to risk and place of birth. We analyse the ways in which these discourses take place at the intersection of two discrete imperatives: to provide pregnant women with choice over where and how they give birth; and to protect mothers and babies from harm. When midwives' assessment of risk of harm during birth is aligned with their assessment of the riskiness of a woman's preferred place of birth then there is little need or purpose in scrutinising this choice. However where there is a misalignment then midwives feel obliged to interrogate the choice, especially when midwives categorise a mother as high risk and they want to restrict the range of choices. In this chapter we focus on the discursive methods that midwives use to shape mothers' decisions when pregnant women are unwilling to accept midwives' risk categorisation and/or the recommended place and method of birth. We examine the ways in which implicit moral judgments underpin and are evident in such discourses.

Using ethnography for discourse analysis

Discourse analysis at the micro level, examines texts (written, spoken, and/or visual) to examine the ways in which language both creates meaning and constitutes relations of power. Through detailed analysis of text (using a wide and in some cases disparate range of approaches) the intentional, and arguably more superficial, process of communication can be penetrated to expose discourse as an instrument of power (Fairclough, 1992; Weedon, 2004). Detailed discourse analysis seeks to move beyond the overt and obvious meaning of the texts, the words, and utterance, to the underlying socio-political purpose of the text. Thus discourse analysts are interested in what lies behind the text such as 'participants' role-relationships and their motives/accountability as well as wider institutional/professional and socio-political underpinnings' (Sarangi and Candlin, 2003, p. 116).

Such methodological approaches focus on the internal structures of language, but can be criticised for their inward and dislocated focus. Through the detailed analysis of text there can be a tendency to overlook the importance of the context from which the texts emerge, especially the conditions under which they are produced and how this shapes their meaning (Sarangi and Candlin, 2003, p. 116). Analysis of such contextual elements is often limited with little consideration of the context and the ways in which these contribute to purpose and function of the text. Ethnographic discourse analysis seeks to combine the interest in discourse as a form of social action with an analytical sensitivity to the social context from which utterances emerge. From this perspective the finer details of language can be examined not as a dislocated and isolated text but an embedded process of meaning making (Sarangi and Roberts, 1999).

The texts we use in this chapter are derived from an ethnographic discourse analysis of midwifery and childbirth that Mandie Scamell undertook in four clinical settings in England in 2009 and 2010. This approach combines a textual analysis of policy documents with the analysis of the ways in which midwives talk and act in their everyday practice. The texts include national policy documents, local clinical practice protocols, interview transcripts and ethnographic field notes and memos. The four clinical settings accessed in the study represent the major settings for birthing and midwifery practice in the UK: doctor-led obstetric units with all the medical facilities for high-risk births; midwife-led units, located in a hospital with access to back-up medical facilities if and when a birth shifted from low- to high-risk category and free standing units where the reclassification of a birth into the high-risk category involves an ambulance transfer journey; and the woman's own home.

In this chapter we focus on the texts that specifically relate to midwives' interaction with mothers in the context of choice and safety. We focus on the social context and use textual material from the ethnographic discourse analysis to explore not only the ways in which midwives made sense and defined their work but also the ways in which moral judgment permeated and was expressed through texts. We examine the use of texts as a way of defining the situation and exercising power and explore how midwives sought to impose their own definitions of the situation. We show how this exercise of power involved them in remaining relatively silent and neutral in some situations but assertive and judgmental in others.

The national discourse: empowering pregnant women through choice while ensuring they are safe

Midwives' discourses are shaped by national discourses on the nature of childbirth and the role and rights of pregnant women. As in most areas of healthcare, these discourses have been shaped by shifting notions of power, especially the shift from the paternalist notion of the individual as the passive recipient of healthcare to a more enlightened approach in which the individual is respected as an active agent exercising power and control through informed consent though as we will show in this section this is tempered by a concern with minimising harm to the pregnant women and her unborn foetus.

The dominant choice discourse

In maternal health, *Changing Childbirth* (Department of Health Report of Expert Maternity Group, 1993), identified the ideal of service user autonomy and informed choice as the key element of maternity policy.

Subsequent government policy statements endorsed and provided more substance to the principle of choice. At the 2001 Royal College of Midwives Conference the then Secretary of State for Health, Alan Milburn, pledged £100 million for maternity services to 'ensure that pregnant women have more choice and access to improved maternity services' (House of Commons Health Committee, 2003 p. 4); while in the 2007 *Maternity Matters* White Paper the word 'choice' dominates, appearing no less than seven times in the short preamble address written by the then Secretary of State for Health, Patricia Hewitt. In the White Paper the government gave 'choice guarantee', the Department of Health promising that by 2009 all women were to be offered a choice of birth settings.

This commitment to 'choice' permeates the midwifery discourse on birth and in the professional literature midwifery is positioned as a mechanism for empowering women by providing them with choice. In this literature midwives are described as politically and ethically aligning themselves with the concept of informed choice and woman-centred care (Walton and Hamilton, 1995). That is to say, in their role of being 'with women', the midwives' role is to preserve their client's autonomy in order to facilitate and support woman-centred care. The Royal College of Midwives in a position statement articulates this role in the following way:

> 'Woman-centred care' is the term used for a philosophy of maternity care that gives priority to the wishes and needs of the user, and emphasises the importance of informed choice, continuity of care, user involvement, clinical effectiveness, responsiveness and accessibility. (Royal College of Midwives [RCM], 2001)

There is little dissent in the midwifery literature from the view that the midwife's role is to empower women through providing them with choice. For example Crabtree notes that: 'The midwifery model of care ... is grounded in supporting women's choice' (Crabtree, 2008, p. 106), while Pairman (1998) uses the term 'professional friend' to describe how midwives go about supporting women to give birth in the way they have chosen and believe to be right for them and their babies.

Underlying this discourse of choice is a related discourse of normality, that is by exercising their choice women will choose the most normal or natural birth (Edwards, 2006; Graham and Oakley, 1981; Newburn, 2006; Walsh and Newburn, 2002).

The discourse around choice and safety

The statutory body responsible for the conduct of midwifery in the UK, the Nursing and Midwifery Council (2008) in its professional code for midwifery practice endorses the key role which midwives should play in empowering women albeit the Code does not explicitly use the term choice; the nearest it gets is in the following statements that:

> You [the midwife] must listen to the people in your care and *respond to their concerns and preferences* ... You [the midwife] must uphold people's rights to be *fully involved in decisions* about their care. (Nursing and Midwifery Council, 2008, paras 8 and 14 emphasis added)

However the Code also placed a major emphasis on the effective use of professional expertise in ensuring safety. The Code required midwives to 'maintain the safety of those in your care', to manage risk and use the best available evidence (Nursing and Midwifery Council, 2008, paras 22, 32–34 and 35–37). In its Midwives Rules and Standards (2004 – the last Rules and Standards to provide a definition of midwifery care), the Council defined midwifery care as a means of ensuring safety and preventing harmful outcomes through:

> *preventative measures,* the *detection of abnormal* conditions in mother and child, the *procurement of medical assistance* and the *execution of emergency measures* in the absence of medical help. (emphasis added, Nursing and Midwifery Council, 2004, p. 36)

Thus there is within the Council's discourse a potential tension between actions justified by the scientific expertise of midwives and those based on the choices made by pregnant women. When the two are not aligned and if the midwife anticipates that the woman's choice is risky, that is could result in harm to the women and her baby, the Council makes it clear that the midwife should intervene by counselling the woman about the risks and if the woman persists in her choice, referring her to a superior and documenting the anticipated 'outcome':

> If you judge that the type of care a woman is requesting could cause significant risk to her or her baby, then you should discuss the woman's wishes with her; providing detailed information relating to her requests, options for care, and outlining any potential risks, so that the woman may make a fully informed decision about her care. If a woman rejects your advice, you should seek further guidance from your supervisor of midwives to ensure that all possibilities have been explored and that the outcome is appropriately documented. The woman should be offered the opportunity to read what has been documented about the advice she has been given. She may sign this if she wishes. (Nursing and Midwifery Council, 2004, p. 17)

To summarise, at national level the dominant policy discourse centres on choice, with advocates of choice arguing that providing women choice over birthing will both empower them and lead to better outcomes. However underlying this dominant discourse is a discourse about risk and safety that is most clearly articulated in the professional regulatory body, which seeks to qualify the freedom of choice. In this discourse midwives have a duty to intervene if a pregnant woman proposes to exercise her choice in a way which the midwife judges will expose the woman and her unborn child to excess risk. In the next section we will focus on the discourse which midwives use when they judge women's choices are creating preventable risk.

Midwives' discourses: choice and risk

In everyday midwifery practice there were two potential areas in which midwives' assessment of risk did not align with mothers' choices. Midwives could assess a pregnant woman as low risk and therefore recommended a setting in which medical intervention was unlikely such as home or a midwife-led birthing unit while the mother wanted a more medicalised birth in an obstetric unit even an elective caesarean section. In this case the midwives anticipated the risk of unnecessary and harmful medical intervention. In contrast midwives could assess a mother as high risk with an increased probability of an adverse outcome and recommend that she give birth in a more protected setting such as a consultant-led obstetric unit and the mother wanted to give birth in a birthing unit or even at home. We will start this section by considering the role which choice played in the discourse of practising midwives and then explore the ways in which they manage the tension between rhetoric of risk and choice.

Midwives' choice discourse

Choice played a central role in midwifery discourses. For example when we invited Cindy, an experienced midwife to describe her role as a midwife she centred her description on choice and her role in enabling women to have choice: 'your *whole* role is to support women and be the women's advocate' (emphasis added).

Similarly, when another experienced midwife, Gail, reflected on the role of choice she defined her role as empowering women by providing information enabling them to make informed choices:

> I think informed choice is exactly what it says it is. That women ... have the right to choose what they want to choose and believe. And if you have given them all the facts and all the information and they still choose their way of doing things. Their method of birthing or their decision, then more power to their elbows. You know.

However in practice midwives recognised there were limits to choice. As Hope, a senior midwife, put it:

> Some of the constraints ... I mean there are criteria ... and no matter what the woman chooses she won't be allowed, if it isn't thought to be appropriate. The midwife doesn't have any control over that or any say in that nor does the woman.

In the remainder of this section we will explore these constraints.

Low risk and high safety

Practising midwives were aware of the harm which medical intervention could cause and sought to avoid the cascade of intervention which could result if women gave birth in a medicalised environment. Such intervention could result in iatrogenic or medically created harm. For example Fay an experienced midwife described this in the following way:

> You see where I've been banging on about things, like not putting women on monitors, mmm, just not going down that cascade of intervention – you know, that sort of thing, making it all abnormal – well, now all the evidence is coming out to support all that.

However while they were aware of iatrogenic harm and despite their commitment to inform pregnant women of all risks, they often chose not to highlight the danger of birthing in a medicalised environment. For example Hope described the nature of these 'man-made' risks and then noted that she did not feel it was her professional role to tell women about them:

Hope:	There is a risk to going to unnecessary intervention and the cascade of intervention, erm, of being in an obstetric unit when actually there is no need to be there. Or even if you have a need to be there, there is still risks of unnecessary intervention and the consequences involved in that.
Mandie (Researcher):	And is it the midwife's role to explain those risks to the woman?
Hope:	[long pause] Mmm, it probably should be but, erm [pause], I don't know whether it is. The thing is, there is just so many risks, there is risks to everything so you have to balance it all out and make sense of it all, it is like, oh I don't know, if you think about it too deeply [pause]. I think risk management is about more check-ups, more scans, that sort of thing.

Thus Hope did not see it as her responsibility to inform the woman about the evidence of the iatrogenic risks associated with unnecessary hospitalisation, which may include major abdominal surgery. She did not identify these as risk that she had a responsibility for mitigating. Instead she focused on the intrinsic vulnerability of the women's body and subjecting it to surveillance and control through 'more check-ups, more scans'.

All the midwives involved in our research were well versed in the iatrogenic risks associated with the medicalisation of birth, and it was a topic that commonly came up in group conversations, which took place in staff spaces where only midwives were present. As Gail noted in the following interview these iatrogenic risks were discussed amongst midwives but not with pregnant women:

Gail: People can see that doctors can cause problems by over intervention, lack of communication, etc ... that is, causing, introducing risk and I think everybody would accept that. Or I don't know if everybody would but I think that would be accepted [sigh]. I think, I think yeah. I don't think that idea is too marginalised ... I think that amongst midwives, I think that's perhaps the predominant. No, I don't know, I don't know, mmm. You will find out [laughs]. I think it is probably a widely held view and I think that the majority of midwives think, see that, that iatrogenic risk and they understand that.

Mandie (Researcher): Where would you hear that? Would it be expressed to the women?

Gail: Probably not. They might express it to each other in the coffee room mightn't they? Sort of [pause], you know. I think in the coffee room. They might at labour ward forums. I think that could be I think a lot of it would be unexpressed and taken as a given. Unexpressed or to colleagues really.

While midwives recognised that 'trading-up' using more technology or giving birth in a more medicalised setting exposed pregnant women to iatrogenic risks, they did not actively discourage the mother from taking such risks. They were effectively silent about them. The following extract from our field notes shows the willingness of Miranda, the midwife who was providing the care, to accept a mother's request for monitoring technology that was not clinically indicated and to which the woman was not legally entitled:

Fieldnotes from a high-risk, obstetric unit Pregnant women was admitted in early labour. On admission, the midwife, Miranda, explained the observation procedures she would have to carry out as part of her routine care and assessment. The mother, however, was not satisfied with the list of surveillance procedures and questioned Miranda, saying:

'What about the foetal monitor? I want to have my baby monitored just for peace of mind.'

Miranda responded to this by reassuring the mother that continual foetal monitoring was not necessary as she was low risk.

'If anything happened and there was a clinical indication', Miranda explained, 'then, of course, the baby would be monitored'.

'But I cannot possibly do this without the monitor. I just need to know everything is Okay ... I couldn't relax otherwise', protested the mother.

Miranda acquiesced, leaving the room to discuss the request with the midwife in charge of the shift and returning with the appropriate equipment to carry out a continual electronic foetal monitoring, in line with the mother's request.

When we later discussed this situation with Miranda, she did not refer to the possibility that such monitoring could start a cascade of intervention rather she placed the responsibility for the decision on the patient and within the sphere of patient choice, arguing:

It's not up to me, is it? I mean we live in a world where ... well, women are entitled to choose, aren't they?

High risk low safety

Midwives discourses around 'high-risk' women in setting suitable for 'low-risk' women such as the home or birthing centres, were different. They discussed such choices as a challenge and a problem they had to manage. Every 'high-risk' pregnant woman who chose to birth away from the high-risk, obstetrically run birthing environment, was formally talked-to by a midwife who explained the dangers of her proposed choice and identified the things that might go wrong in either the mother's or the baby's body. The midwives stressed the importance of recording this discussion in the mother's hand-held maternity notes. Interestingly, this formalised discussion and documentation was not a professional responsibility which we could locate anywhere in the Trust's protocols or guidelines, as Mary an experienced midwife noted in her discussion of the 'advice' which was offered to such women:

We usually write on their birth plan words to the effect: 'Aware no doctors, no epidurals, reasons for transfer' ... I don't think there is any formal guidance on this and now you mention it I don't know how I know to write that!

When midwives talked about high-risk mothers who chose to birth in settings the midwives considered suitable for low-risk mothers they tended to see the issues in moral terms. In the following extract Lindi highlights the 'dishonest' aspects of the woman's behaviour and her sense of professional and personal affront at being made to run 'round like an idiot':

> What I don't like is when, we had an incident not so long ago when somebody was, erm, wanting a home birth had had rupture of membranes, all explained to her, she decided she didn't want to go into the high-risk unit, which is fine. I have got no problem with that but then we were trying to send midwives in to check that everything was okay and she was pretending not to be at home. So she wasn't, so she didn't actually call them until she was in labour. Now I feel that woman had every right to make that decision; what makes me cross is that when we were running round like idiots after her.

Lindi clearly saw this as a challenge to her professional identity highlighting the moral implications of the decision, implying that the pregnant woman would be unwilling to take full responsibility for the decision including the blame if something went wrong:

> My line [to the woman] would then be: 'I am more than happy for you to have your home delivery, I am more than happy to leave you alone. If you take that decision and something happens to your baby would you ever forgive yourself?' And I think that makes somebody really think about it so that that would be my way of dealing with it.

Lindi was convinced that if things did go wrong and the outcome was not good for mother or baby, then it would be the midwife who took the blame. She said:

> They [parents who did not accept her advice] are not prepared to actually go to the bottom line and say: 'Okay, I understand that is a risk and *if anything happens I will not blame you*'. (Emphasis added)

Lindi's story illustrates the unsettled ground upon which the client's right to choice is placed within the maternity care setting. Although service user autonomy has been endorsed through health policy for almost 20 years, how this is allowed to be expressed is strictly policed through routine midwifery practice, which revolves around the selective identification of risks. Those women who choose options that are not on the presubscribed menu of choices that have been carefully set out by the midwife, create, through their choices, a site of tension where professional understanding of human rights and risk collide and where professional commitment to the possibility of normality is undermined. It is at these points of collision that the moral loading of risk crystallises into a discourse of deviance and, once loaded in this way, operates to fracture relations between the midwife and her client.

This is evident in Cindy's experience of caring for a woman who had been diagnosed as morbidly obese. Having had two normal vaginal deliveries before in a hospital setting, this woman decided, largely for personal reasons, to opt for a home birth. Following her NHS Trust protocol, which states that women with a 'body mass index at booking of greater than 35 kg/m should be excluded from delivering at either a midwifery-led birth centre or at home', Cindy tried her best to persuade this mother to have her baby at an acute, obstetrically run site. When the mother refused to accept this advice, tensions arose within the relationship. As Cindy described the situation in the following way categorising the mother as irresponsible because she would not follow her advice as other mothers ensuring they did what was 'right for them and the baby':

> She, erm she, understood that but she was very, well [pause] very adamant that she was going to have a home birth and nothing was going to stop her. She was very challenging in that she was defensive, argumentative, rather than sort of going through the risks with me, and us making a plan together that we were both happy with. She was making clear that it was her that she was going to do exactly what she wanted to do ...
> I mean usually women, if you explain to them the reason why they need to do that and the other, they, they are happy to do that because they want to do what is right for them and the baby. But for this case it was really difficult because I knew what I was suggesting according to policies and guidelines was, erm [pause] was the right thing for her, erm [pause] and she was just disagreeing with me at every moment.

Cindy was confident here that she had provided this mother with all the information she needed to make the right choice. In her professional opinion, therefore, this mother was in a position to make a fully-informed decision about where to give birth to her baby. Clearly, Cindy had fulfilled her professional duty of care as it is set out by the Nursing and Midwifery Council and the Trust's protocols in relation to the risks this mother was choosing to take. However as the woman rejected Cindy's advice, Cindy felt she was rejecting her professional expertise:

> The way she reacted made me feel like she didn't care what I thought as a professional. Erm, it almost made me feel like I didn't know why she was coming to see me! It felt like she wasn't listening to any of my advice, she didn't want any of my advice and it made me feel a bit, erm, useless, I suppose.

There was effectively a power struggle. Cindy expected to have authority over what and how risks should be understood, and these, in her professional opinion, should reflect her Trust's policies and protocols. This meant that when her client refused to accept her authority the relationship became virtually pointless in her eyes. The tension created by her deviant client's assertiveness seemed to make Cindy feel uncomfortable, vulnerable even, suggesting that professional identity and her right to authoritative knowledge heavily coincide. When her recommendations were ignored, the basis of her professional confidence fractured. At that point, her role as a midwife was severely compromised, since this role depended upon her maintaining a status gap between them, where she was placed in a position of authority. As Cindy explained:

> When you feel ... that everything you're advising [is rejected], it is very hard then to be that woman's advocate because you don't understand what she, what she wants, and what she is saying. You don't understand where she is coming from and it is really hard to go to support her in her decision.

To summarise, the midwives we spoke to during this research were very keen to explore the risks associated with the physical process of birth with the women in their care and actively used this information to guide women through their decision making. These were the risks which evoked notions of professional responsibility and accountability and, ultimately, fear of blame. By contrast, those iatrogenic risks associated with the hospital environment remained predominantly unvoiced. These risks seemed to have, at most, tenuous links to understanding of professional responsibility and accountability. Indeed, many of the midwives were uncomfortable talking about such matters with their clients. The moral loading of risk involved a systematic bias, with some risks being highlighted, while others were obscured through midwifery activity.

Discussion

Underpinning the discourse we have analysed in this chapter are two types of potential harm, one grounded in the intrinsic uncertainty of childbirth. Even when there appear to be no warning or danger signs, that is pregnancy and childbirth appear to be normal and low risk, things can go wrong, and this likelihood is increased when there are warning signs and woman's pregnancy and birth is not categorised as low risk. However there is a low probability of things going wrong even in the 'riskiest' option, the home birth. A recent national study conducted by the National Perinatal Epidemiology Unit designed to measure birth outcome against choice of place of birth showed a small but significant increase in negative outcomes for first-time mother choosing to have home birth but none for other mothers (Brocklehurst et al., 2011). Importantly, if things do go wrong then the consequences can be catastrophic, serious harm even death to the baby. Therefore midwives' discourses around childbirth are grounded in the need for constant vigilance and care (Scamell and Alaszewski, 2012).

The other type of harm comes from unnecessary medical interventions, especially caesarean section. According to the findings from the Birthplace study, risk of medical intervention for all women is significantly decreased in out of hospital birth settings. Some obstetric interventions do prevent worse harm, particularly for the baby. However what the birthplace study does show is that it is difficult to estimate the level of 'life-saving' interventions. The World Health Organization (1985) has made an informed guess that an optimal caesarean section rate would be between 5 and 15% of all births, however rates in the UK exceed 25% (HSCIC, 2013). Iatrogenic interventions associated with hospitalised births represent a relatively high probability risk compared to the complications of 'normal childbirth' in the non-medicalised birthing setting of home (Brocklehurst et al., 2011).

Thus in their discourses midwives are balancing two types of risk, the low probability/high consequence risk of normal childbirth versus the high probability/lower consequence risk of medical intervention. In the national discourse over the role of midwives the balance is clearly in favour of normal birth over medical intervention. A review for the Cochrane data base clearly articulates this:

> The philosophy of midwife-led care ... is normality, continuity of care and being cared for by a known and trusted midwife during

labour. There is an emphasis on the natural ability of women to experience birth with minimum intervention. (Sandal et al., 2013)

Yet when the midwives in our study were faced with two situations, one in which a birthing woman was requesting more intervention than the midwife judged to be appropriate and the other in which the woman was requesting less, they acquiesced without much resistance to the request for increased intervention but were upset by and resistant to the request for reduced levels of monitoring and intervention.

The differences in discourses reflect the power relations. The midwives work in a medicalised environment in which benefits of medical intervention are embedded in documents such as their employers' policies. Midwives select those risks which coincided with Trust protocol priorities, the first-order risks associated with birth, leaving other more controversial man-made risks unvisited in their conversations with their clients or as Kirkham and Stapleton (2004) put it, just 'going with the flow'.

However the moral guiding of maternal choice through risk selection does not appear to be just a case of midwives passively submitting to protocols over which they have little control. Rather, this is a practice those involved in this study actively pursued out of a consciousness that such careful risk selection was seen as being part of their role as a responsible midwife.

Such selections reflected midwives' narratives around the fragility and untrustworthy nature of normality and their professional duty to be ever vigilant. In contrast to the certainty around medical interventions such as C-sections, which would have harmful side effects but would ensure the safety of the mother and baby, normal childbirth could only be judged to be safe once it was over. As Mary a senior midwife put it:

> But I always have here, in the back of my mind, that things can go wrong so, that's how, that's how I practice as a midwife. That you know, it can be wonderful but it's wonderful when it is finished. You must be alert to things that can happen. Because I watch very carefully and unpick things and I check everything and erm because things happen. I would put her (the mother) in the bracket of 'at risk' of any risk until, until it is over.

However the uncertainty of normality does not fully explain midwives' discourse around mothers requesting less intervention than the midwives judged necessary. Their discourses articulated the mothers' actions as a personal as well as a professional challenge. Not only did they describe how these mothers were rejecting their expert advice they were also rejecting their values and the personal relationship, for example by hiding from the midwife. As Cindy noted there was a breakdown in the relationship as she could not act as an advocate for her client. Midwives described this rejection as hurtful. They could no longer relate to the mother as a moral and competent human being.

In such discourses the moral underpinning of professional practice becomes visible. As Douglas (1990) has argued in contemporary societies, risk may appear as a neutral technical concept but this conceals its moral basis. Indeed it is this apparent neutrality that makes it so attractive, and provides legitimacy to the experts who use it. The unease which underpins midwives' discourses in relationship to mothers requesting less intervention than the midwives judged necessary comes from this exposure of the moral basis of their work.

Conclusion

In this chapter we have analysed a range of discourses around midwifery practice, risk and place of birth. We have shown that the purpose and meaning of texts cannot be considered in isolation but only becomes evident when the relationship between different texts is considered and the creation of these texts is placed within their social contexts. Thus the discourses of mothers' choice and the normality of birth are prominent within both national and practice texts. Yet despite these discourses the proportion of births subject to medical intervention grows and the proportion of births in the least medically controlled environment remains static. By exploring midwives' discourses around mothers who choose to have more or less intervention than the midwives judge necessary we have been able to explore the moral and ideological underpinning of midwives' discourse. Midwives tended not to challenge those mothers who wanted more medical surveillance and intervention as this tended to go with the flow of their medicalised work. In contrast, mothers who wanted less were treated as both a professional and personal threat and as women who were not behaving morally or responsibly.

References

Brocklehurst, P., Hardy, P., Hollowell, J., Linsell, L., Macfarlane, A., McCourt, C., Marlow, N., Miller, A., Newburn, M., Petrou, S., Puddicombe, D., Redshaw, M., Rowe, R., Sandall, J., Silverton, L., and Stewart, M. 2011. Perinatal and maternal outcomes by planned place of birth for healthy women with low risk pregnancies: The Birthplace in England national prospective cohort study. *BMJ* 2011; 343:d7400.

Crabtree, S. 2008. Midwives constructing 'Normal Birth'. In S. Downe ed. *Normal Childbirth: Evidence and Debate*. London: Churchill Liningstone, p. 85.

Department of Health Report of Expert Maternity Group. 1993. *Changing Childbirth*. London: HMSO.

Douglas, M. 1990. Source risk as a forensic resource, *Daedalus*, 119, 4, pp. 1–16

Edwards, N. 2006. Why are we still struggling over home birth? *AIMS Journal*, 18, 1, p. 3.

Fairclough, N. (1992) Discourse and Social Change. London, Polity Press.

Graham, H. and Oakley, A. 1981. Competing ideologies of reproduction: Medical and maternal. In H. Graham ed. *Women, Health and Reproduction*. London: Routledge Kegan & Paul.

Health and Social Care Information Centre (HSCIC). 2013. *NHS maternity statistics – England, 2012–13*. Available at http://www.hscic.gov.uk/catalogue/PUB12744 Accessed 28 November 2014.

House of Commons Health Committee. 2003. *Choice in maternity services*. HC 796-1, London: HMSO.

Kirkham, M. and Stapleton, H. 2004. The culture of the maternity services in Wales and England as a barrier to informed choice. In M. Kirkham ed. *Informed Choice in Maternity Care*. London: Palgrave, p. 117.

Newburn, M. 2006. What women want from care around the time of birth. In L. Page and McCandlish eds *The New Midwifery: Science and Sensitivity in Practice*. 2nd ed. Philadelphia: Churchill Livingstone, p. 3.

Nursing and Midwifery Council. 2004. *Midwives rules and standards. Professional rules and standards*. London: NMC Professional Regulation Body.

Nursing and Midwifery Council. 2008. *The Code: Standards of conduct, performance and ethics for nurses and midwives*. London: NMC, revised version.

Pairman, S. 1998. Women-centred midwifery. Partnership or professional friendship? In M. Kirkham ed. *The Midwifery Mother Relationship*. Basingstoke: Palgrave Macmillan, P207.

Royal College of Midwives. 2001. Position paper No 4a *Women centred care*. London: RCM.

Sandal, J., Soltani, H., Gates, S., Shennan, A. and Devane, D. 2013. Midwife-led continuity models versus other models of care for childbearing women. *Cochrane Database of Systematic Reviews*, Issue 8.

Sarangi, S. and Candlin, C. 2003. Categorization and explanation of risk: A discourse analytical perspective. *Health, Risk & Society*, 5, 2, pp. 115–124.

Sarangi, S. and Roberts, C. 1999. *Talk, work and institutional order*. Berlin: Moutton and Gruyter.

Scamell, M. and Alaszewski, A. 2012. Fateful moments and the categorisation of risk: Midwifery practice and the ever-narrowing window of normality during childbirth. *Health, Risk & Society*, 14, 2, pp. 107–115.

Walsh, D. and Newburn, M. 2002. Towards a social model of childbirth. *British Journal of Midwifery*, 10, 9, pp. 540–544.

Walton, L. and Hamilton, M. 1995. *Midwives and changing childbirth*. London: Books for Midwives.

Weedon, C. 2004. *Identity and culture*. Maidenhead: McGraw-Hill International.

World Health Organization. 1985. Appropriate technology for birth. *Lancet*, 2, pp. 436–437.

Part II
Communicating Risk in Legal Processes

6
Risk, Law, and Security

Pat O'Malley

'Post-risk' security in an uncertain world

Thirty years ago, it was possible to see the dawning of a new age of predictive governance of crime and security based on probabilistic 'risk' modelling. Now especially since 9–11 and the rise of the 'new terrorism', many challenges are argued to be generating 'post-risk' responses. Probabilistic risk techniques rely on two related conditions: the building up of a large body of data from which statistically based predictions may be calculated; and an environment that is sufficiently stable into the future for such predictions to apply. As Ulrich Beck (1992) suggested some time ago, these conditions are no longer to be relied upon. For example, terrorist incidents are said to be too few in number and too dispersed and diverse in nature to permit the required accumulation of mass statistical data; the increased autonomy of terrorist cells and individuals render surveillance and interception much harder; and terrorists take steps to avoid conforming to predictable patterns – for example by selecting operatives who do not fit risk-profiles (Beck 2002). Faith in preventive government appears to have given way to the emergence of an array of non-probabilistic manoeuvres that do not rely on prevention – or that change its meaning considerably. These 'post-risk' forms of governance can be grouped into several broad, and increasingly familiar, categories.

'Preparedness' is perhaps the most straightforward. By this, Collier and Lakeoff (2008:11) mean 'a form of planning for unpredictable but catastrophic events ... the aim of such planning is not to prevent these events from happening, but rather to manage their consequences'. They argue that while this is not a new form of governance, it has increasingly been foregrounded and enhanced as an institutionalised response to the uncertain environment post 9/11. Collier outlines enactment as key technique of preparedness, in which expertise is not so much drawn upon to calculate as to *imagine* catastrophic futures and plan optimal responses.

> Rather than drawing on an archive of past events, enactment uses as its basic data an *inventory* of past elements at risk, information about the *vulnerability* of these elements and a model of the threat itself – the *event* model. And rather than using statistical analysis, enactment 'acts out' uncertain future threats by juxtaposing these various forms of data. (Collier 2008:226)

This owes much to war gaming as practised in military circles for many years. In the simplest model, this may involve imagining scenarios and placing multiple transparent overlays on maps in order to simulate responses and outcomes, with the aim of giving emergency service planners a foundation on which to prepare. Recently, more sophisticated models have been developed in which hazard, inventory, vulnerability, and loss are deployed through the development of computer modelling based on analogous past events, rather than accumulated statistical data. With terrorism risk, for example, estimation of the timing and nature of the threat is 'elicited' from experts (cf. Ericson and Doyle 2004:150–1). While 'elicitation' is no more than expert-informed guesswork, as Collier points out, the use of analogous models of natural disasters, military damage assessments, and nuclear reactor failures are likely provide reasonable approximations to imaginable harms terrorism will produce (Collier 2008:242).

'Precaution' is a second major response to increasing uncertainty, operating through imagining worst case scenarios. As Francois Ewald (2002:288) suggests, precaution is not just inviting one 'to take into account doubtful hypotheses and simple suspicions' but even 'to take the most far-fetched forecasts seriously, predictions by prophets, whether true or false'. Since the turn of the 20th century it has been argued that waiting for sufficient evidence to establish a calculable risk to emerge is a luxury that cannot be afforded. Thus the US '9/11 Commission' noted that before 9/11 there had been considerable speculation about catastrophic scenarios, including that of the destruction of the Twin Towers. Given this, the Commission asked why no preventive measures had been taken. In response it argued that imagination should be bureaucratised in the name of governing terrorist risks. If it can be imagined, it must be governed (Bougen and O'Malley 2009, Carlen 2009). In a highly uncertain environment, imagination thus directs or even replaces the calculation of risk.

This is closely related to 'speculative pre-emption', used for example in justifying the war in Iraq. Of course pre-emptive strikes as such are not new, but these have largely been based on clear evidence of threats. However, as the US National Security Strategy makes quite explicit, recent developments offer a radical departure from conservative risk models:

> The greater the threat, the greater is the risk of inaction – and the more compelling the case for taking anticipatory action to defend ourselves, *even if uncertainty remains as to the time and place of the enemy's attack*. To forestall or prevent such hostile acts by our adversaries, the United States will, if necessary, act pre-emptively. (National Security Council 2002 quoted by Cooper 2006: 125)

Speculative pre-emption relies precisely on high uncertainty for its rationale and this is argued to differentiate it from historical precursors. Previously, pre-emptive strikes would be founded on clear evidence that an attack was imminent. In 'the risk society', however, high uncertainty has become a justification in its own right. Cooper (2006:125–7) has argued that the logic of post-9/11 speculative pre-emption – to intervene in emerging events precisely *because of their uncertainty* – is clearly illustrated by the Iraq war, and by the subsequent furore over the relevant governments' lack of evidence of the existence of weapons of mass destruction.

Most recently, 'resilience' has become another major response to an understanding that risk-based prevention will imminently fail. A plethora of discourses and strategies for increasing resilience have emerged at the organisational, national, and transnational levels. Individuals, infrastructures, cities, and even economies must be made capable not only of surviving but of 'bouncing back' from catastrophic damage arising from unspecified, unpredictable events (Rogers 2012, O'Malley 2010b). In contrast to other responses to security problems, resilience is not organised around a specific or even imagined kind of threat: and in this sense reflects belief that we are in an era of extreme uncertainty. For example, new forms of warfare against terrorism have involved significant restructuring of the military, including the installation of mandatory resilience training in most western militaries. This focuses on such 'skills' as: rethinking problems as opportunities; forming an 'active' management style based on information gathering and consultation; developing 'cognitive flexibility' in problem-solving activities; training in 'emotional intelligence'; developing skills in forming and drawing upon social support networks; and so on (O'Malley 2009).

It cannot be disputed that such 'post-risk' discourses and techniques dominate much of contemporary political debate and awareness, at least in the West. And no doubt a politics of fear (bearing sometimes harshly on members of unpopular minorities such as Muslims) has been effectively ramped up (Furedi 2008). But thus far these new and revived manoeuvres against illegalities directly impact on a relatively small number of people, although admittedly even this is out of proportion to the miniscule numbers of people in the West actually harmed by terrorist actions (Walklate and Mythen 2014). Risk regimes, on the other hand, have played and continue to play a major role in reshaping key institutions of everyday security.

The emergence of preventive justice

In the late 1980s and early 1990s, a number of criminologists began to note the increasing role played by risk and preventive frameworks in sentencing and justice policy (e.g. Cohen 1985, Feeley and Simon 1992, O'Malley 1992). This 'new penology' was marked by a shift away from the reformist foci of penal modernism, with its emphasis on correctional-therapeutic interventions tailored to the expert diagnosis of individual offenders. Instead, the emphasis was seen to be on 'the replacement of a moral or clinical description of the individual with the actuarial language of probabilistic calculations and statistical distributions applied to populations' (Feeley and Simon 1992). In their conceptualisation of this as 'actuarial justice', Feeley and Simon (1994) linked together increased use of statistically based tariffs in sentencing, displacement of correctionalism by incapacitation, and a managerial emphasis on maximising throughputs: the three elements together being regarded as optimising 'effective control' of risky populations.

Risk had been developed previously as a tool in parole and probation decision making in some American states, albeit in a limited way. Now, however, risk had expanded its scope to include sentencing, and with that had come a risk-based refocusing of justice onto protection of the community rather than reform of the offender. In this process social welfare and judicial expertise was marginalised in favour of risk reduction. Thus whereas the problem of sentencing people for their prospective risks had long been recognised as unjust because some of them would not have re-offended, this objection was now swept aside by a concern with statistical probability rather than individual justice. Formally, at least, judicial resistance to risk had been overcome by legislatures mandating long sentences aimed at incapacitating repeat offenders. More broadly, interpreters such as Rose (2000) and Baumann (2000) saw the deployment of risk as creating a bifurcated criminal justice system, in which high-risk offenders are consigned to exclusionary incapacitating gulags while the lower-risk population is channelled into circuits of surveillance and containment (such as electronic tagging or house arrest). As Rose (2000:333) puts it 'in these exclusionary circuits the role of custodial institutions is redefined. They are understood and classified not in terms of their reformatory potential, but in terms of the secure containment of risk.'

It is easy to read-off directly from such practices to a more or less irresistible trend, but this would be to discount the effects of resistance. There is little doubt the development of risk-based sanctions corresponds in time with the massive increase in imprisonment in many countries, especially the United States. It is quite possible, if not officially admitted, that much of this expansion flows from a changed attitude toward the role of prisons as merely containing risks. But as Simon and Feeley (1995) themselves later recognised this cannot be attributed to the impact of actuarial justice, for it has not taken over as the new penology. They suggest that effective resistance has stemmed from an enduring attachment to individualised, just desserts penal philosophies, and to sentencing as a spectacle of moral denunciation – whereas risk management strategies are dispassionate, seemingly amoral, and technical. Ironically much resistance of this sort has stemmed from conservative sources. Research, has certainly suggested that criminal justice officials played a key role in this resistance to risk techniques. Freiberg's (2000) analysis showed that judges did not appreciate having their discretion overridden by risk-based sentencing schedules, or having their commitment to the principle of proportionality of offence and punishment violated. His work indicated that the judiciary retain significant discretion even in the face of mandatory sentencing regimes. This may be achieved, for example, by judicial redefinition of the offence into a category that does not attract risk-based sanctions. As well, Austin and his colleagues (Austin et al. 1999) have found that resistance also stems from criminal justice officials' recognition of the practical limits to implementation – for example in the form of the finite capacity of prison systems. The result of such resistance has been uneven. While some of the initial momentum toward risk-based justice was lost, sentencing has been markedly affected with respect to deeply unpopular categories such as repeat sexual and violent crime. Many jurisdictions specify long sentences – or in Britain, life sentences – for such offenders in the name of risk reduction (Kemshall and Maguire 2001).

In risk's most striking manifestation 'Sarah's Law' in Britain and 'Megan's Laws', operative in the majority of US states, legislation mandates some form of continued surveillance and restrictions on released sex offenders, and often requires public notification of their whereabouts. In Britain and some states of Australia released sex offenders are required to notify the police of their home address. Together with the probation service, police undertake a risk assessment of the offender, and where the risk is considered high enough they may notify other organisations, individuals, or even the local community. According to the courts this does not constitute a further punishment after completion of the sentence, but rather is a practice that allows communities to be aware of the risk in their midst and to take appropriate precautions (Levi 2000). Risk schedules have also been widely introduced into parole and probation decision making. However, in these procedures, Kemshall and Maguire (2001) note that that while there are actuarial risk-assessment schedules available for determining post-release conditions, in Britain these are never the sole basis upon which decisions are made. Rather they play a key role in the initial filtering process, while final decision making is heavily reliant on the case record approach. Rather than actuarial risk techniques simply displacing expert judgment, what is emerging are new combinations of the two, so that actuarial techniques become one element, but still within a broader risk-assessment exercise.

It is also clear that risk has reshaped provision of correctional services within prisons, particularly in the form of 'risk-needs' analysis. One of the standard criticisms of correctional interventions was that they were wasteful of resources, providing all manner of services that may or may not have been effective in correction. Beginning in the late 1980s, the work of psychological criminologists such as Andrews and Bonta (Andrews and Bonta 1989) began developing risk-assessment instruments that identified the 'criminogenic needs' of offenders. These are 'needs' linked to the particular offender's risk factors for recidivism, treatment of which will reduce the risk of re-offending. As part of this thinking, only those risks regarded as 'manageable', that is, on which intervention will have a measurable impact, will be recognised. As Hannah-Moffatt (2005) has shown in her studies of Canadian prisons, risk-needs have reshaped the major approaches to offender management. Rose's 'secure containment of risks' is not simply associated with incapacitation but 'now involves efficient, rational calculations of need' (Hannah-Moffatt 2005:34).

In sum, risk, and its associated discourses, have become much more pervasive in criminal justice even though it has not developed into anything resembling a 'new penology' as was once envisaged. Risk plays a secondary role in sentencing and while incapacitation may be a major rationale of imprisonment in some jurisdictions this is not always attributable to risk.

Risk, policing and crime prevention

Until the 1980s crime prevention was a minor aspect of policing. Most forces had a crime prevention officer, but this role usually was restricted to limited data gathering and public relations functions. The growth of prevention in this decade and thereafter was partly an effect of insurance industry pressures (O'Malley and Hutchinson 2007) and perhaps for this reason, the first area in which risk was registered in the field of prevention was with respect to situational crime prevention. This approach, directed mainly at property crime, stressed redesign of statistically high-risk 'criminogenic' environments and settings to render them less vulnerable. Early advocates of situational crime prevention argued that spiralling crime rates demonstrated that attempts to control crime through rehabilitation had failed and that it was time to move away from this model (e.g. Geason and Wilson 1989). In a fairly aggressive self-description, the American National Crime Prevention Institute developed 'the contemporary perspective in criminology' which took as its guiding principles the idea that

> Prevention (and not rehabilitation) should be the major concern of criminologists; no one is sure how to rehabilitate offenders; punishment and/or imprisonment may be relevant in controlling certain offenders; criminal behaviour can be controlled primarily through the direct alteration of the environment of potential victims; crime control programs must focus on crime before it occurs rather than afterward; and as criminal opportunity is reduced, so too will be the number of criminals (National Crime Prevention Institute, 1986:18).

This was not just a rethinking of crime as a preventable risk, for it was a very specific formulation of risk technique. It has been seen already that parallel developments such as risk-needs analysis put great faith in (risk-driven) rehabilitation as a preventive technology. Situational crime prevention, on the other hand, is much closer to actuarial justice in the sense that it is unconcerned with reform, and pays no attention to the biography, circumstances or character of the offender – who is regarded as a rational choice actor (O'Malley 2010a). Both approaches share something in common with neo-liberal governance, for this also put considerable faith in economic models of behaviour, and had a corresponding antipathy to the social sciences.

While it is not obvious that situational crime prevention represents a form of risk-based policing, I suggest that its function is precisely this, for it replaces a police presence – whether public or private in nature. Like the copper on the beat, or the 'silent policeman' of old – the dome placed in the road to keep traffic from cutting corners at intersections – it enforces the law by making offending more risky and less attractive. There is now a multitude of developments of this kind where statistical data have been gathered to identify 'criminogenic situations', and architectural or engineering solutions created to reduce their riskiness. Street lighting is improved, speed humps built, bushes removed in order to improve lines of sight, closed circuit television (CCTV) is installed to deter offending in elevators or corridors, suburbs designed with limited access to outsiders, and so on. While some of this is surveillance, much of it is simply behavioural channelling, sending offending behaviour elsewhere or making it too difficult to perform at all. The fact that many CCTV monitors are either not connected to any recording apparatus (or are dummies, not connected to anything at all) is a clear example of these forms of deterrence (Smith 2014). On the other hand, CCTV frequently provides prevention through other means. Constantly monitored television scanners can allow security guards to be literally panoptic, seeing around corners and to quite distant places, allowing instant despatch of security personnel where needed. Even where monitors are not manned, the recorded video information allows security and police later to identify both the modus operandi of offenders (triggering further situational crime prevention redesign) and in some cases the identity of the offenders themselves. Much situational crime prevention effectively is passive risk-policing based upon security information.

In many ways, however, police themselves resisted enlistment in crime prevention, and much of their preventive activity may be regarded as peripheral to mainstream policing: vulnerable to the accusation that it is largely about police getting others to carry out risk-reduction measures. Participation in crime prevention panels and campaigns, organising Neighbourhood Watch and Shop Watch programmes, promotion of home security and community awareness practices, property marking schemes, the operation of 'police shops' (often aimed at reducing fear of crime as well as promoting security consciousness), school education programs and so on, are typical of these. However, such an interpretation may be misleading, for all these developments link police to information gathering and dissemination in the name of risk-reduction. It is this informational characteristic of risk that is said to be transforming the work of public police. In their classic work on *Policing the Risk Society*, Ericson and Haggerty (1998) argue that the bulk of state police work is no longer detection and enforcement, but the gathering, creation, and processing of security related information. Police have always been well placed to gather security data, but in the current 'risk society', Ericson and Haggerty argue, this characteristic becomes definitive. Led by the insurance industry, many commercial, financial, educational, health and state welfare agencies demand security related information tailored to their needs. The result is argued to be that the police are governed by the 'risk knowledge formats' of such institutions. For instance, incident report sheets are structured in such ways that no matter what event police confront, it is recorded in pre-coded ways that make risk and security the focus. Another effect is that the risk focus has placed police forces in the centre of a new security-based configuration of governance that traverses all manner of contexts and issues. Thus the Canadian 'Shield of Confidence Home Security Program' involves the police, in conjunction with home-builders, insurance companies, and government building-code regulators, in a programme that certifies the security technologies and design features of new homes. Builders are required to contact the police crime prevention branch on multiple occasions in the process of construction, and police are required to inspect and monitor the construction of the building (Ericson and Haggerty 1998:157). Work of this kind brings a variety of diverse parties together in a risk network governed in new ways by police. In such ways policing is being transformed through the requirements imposed on them by the risk demands and frameworks of others.

While pointing to the impact of risk, Ericson and Haggerty may overstate the case. It is clear for example that pressures for security information are not the only forces impacting on police. Changes inspired by market-focused governments require strengthening of many traditional service and public order activities linked with audits of the public's satisfaction with 'their' police. Such demands by the public may serve to reinforce many elements neutral or even hostile to risk regimes. For example, crime fighting has not necessarily been reined in from its dominance in the 1960s by new concerns with risk and prevention. Many pressures keep crime fighting very prominent. Both police working culture and mass media representations ranging in focus from 'The Bill' to 'CSI' do much to preserve a primary place for the crime fighting model. Likewise, as Johnston (2000) has argued, the increased emphasis on risk and security ironically generates a level of insecurity among police 'customers'. This strengthens demands for visible reactive policing – as is indicated by the high profile police crazes for 'zero tolerance', 'street sweeping', and other hard-line crime fighting strategies. It may yet be that a convergence of recurring media and government demands for 'crackdowns' merge with public preferences for a visible police presence and rank and file police preferences to push policing in a crime fighting direction rather than further subordination to a risk society regime.

From preventive justice to mass preventive justice

While the earliest criminological recognition of risk-based justice dates from the late 1980s (Reichman 1986, Simon 1987) risk was in fact already beginning its transformation of justice in the 1960s. While traffic accidents and injuries had been recognised as a problem for decades, this issue largely had been thought of in terms of individual accidents and tragedies. In large measure the principal motoring offences, concerned with speeding, were regarded as 'technical offences', and often denigrated merely as taxes – since the offence of dangerous driving existed to govern questions of risk. Even then, dangerousness was very much a matter of subjective judgment by police on a case by case basis. However, by the early 1960s the exponential increase in traffic density and volume meant that road deaths and injuries appeared as both a statistical phenomenon ('the road toll') and one of national importance. Over the ensuing decade traffic safety was established primarily in the form of aggregate risk rather than individual danger. Thus, risk assessments and legal measures that embraced and embodied these statistical realities proliferated around the road toll: drink-driving, road engineering, vehicle design and equipment, and of course speeding were all transformed in a relatively short span of time around measures of aggregate risk – and mostly at a time when other aspects of crime prevention more generally were still marginal, low status tasks assigned to superannuated police officers (O'Malley 2010a).

Speeding regulation was transformed into preventive justice in two principal ways. First, speeding was identified as a risk factor. That is, like blood alcohol content some years later, speed was made to appear as objectively linked to rates of traffic accidents, death, and injury. Speed in some measure, like drink-driving, began to lose its standing as a 'merely technical' offence and came instead to be seen as objectively harmful: high speed was closely correlated with high risk of death and injury. The higher the speed, the greater the statistical risk of harm. Driving well above the speed limit now attracted greater punishment not simply because it represented a greater disregard for law but at least equally because it represented a greatly elevated *scientifically* demonstrable harmful potential. This model was exactly paralleled with drink-driving where precise calculations of blood alcohol content (BAC) were correlated with data on driving impairment and casualty rates. 'Speeding' and 'drink-driving' had now been transformed into *offences against risk*. The indicator of the risk became the offence. This was radically illustrated with drink-driving where the offence came to be a BAC level per se instead of behavioural and visible signs such as 'walking the white line'. The same thing had occurred with speeding through the transformed meaning of speed (O'Malley 2010b).

Once speed became a precisely calculable risk factor for death and injury, it ceased to be merely a 'technical' offence, at least in the eyes of the law. At the same time it came to be associated with new enforcing technologies that operated in the name of a criminal law based in the prevention of harm. New instruments in the form of speed and red-light cameras – dubbed 'safety cameras' – now could measure crime objectively, and for that matter remotely. They could detect *and* calibrate offending, courtesy of risk's objectification of the speed-harm nexus. These apparatuses had the potential to detect historically unprecedented volumes of offending. Almost unnoticed by criminology a sea change without comparison in the history of criminal justice had occurred: the creation of 'mass preventive justice' – a justice that was in part produced and justified by risk's measurability (O'Malley 2013). Under a discourse of 'safety' such apparatuses register and record offences and infringements, and transmit their data to central computers that, in turn, calculate and issue penalty notices or more rarely trigger pursuit and apprehension. All are connected with risk-based models that calculate, anticipate, and modulate flows and circulations of traffic. All actively put into effect what are held to be preventive safety measures against risky offences and offenders.

Through such technologies, in the 21st century motorists are subjected to a digitised regime governed by codes and protocols in the name of harm minimisation, risk, and security. Bar-code scanners and registration plate readers passively register vehicle and driver identifiers. RFIDs (radio frequency identification devices), actively transmit identifying codes and perform protocols that provide the identification of risky vehicles. In some instances, such devices are operated by police – or increasingly by cheaper security employees – either from moving or stationary vehicles or from roadside sites. All of this, in turn, is justified by a 'jurisprudence of risk' (O'Malley 2013). Fairly typical is the New South Wales Road Transport Authority (RTA), that headlines its webpage on mobile speed cameras with the caption 'Capturing speedsters. Anywhere. Anytime', which is immediately justified by reference to the road toll in terms of numbers killed and injured. (Mobile Speed cameras' at http://www.rta.nsw.gov.au/roadsafety/speedandspeedcameras/avespeedsafetycameras/index.html.) In this light, 'mobile speed cameras have been introduced because they are recognised internationally as a best practice road safety countermeasure to reduce speeding, leading to a reduction of crashes. The introduction of mobile speed programmes in Queensland and Victoria has reduced casualty rates in those states by at least 25 percent'. With respect to fixed safety cameras – which may be speed cameras, red-light cameras or both – the claim is made by the RTA that 'the use of cameras to enforce speeding has proven road safety benefits', and points out that 'independent research found that where cameras have been installed there has been a 70 per cent reduction in speeding resulting in a 90 per cent decline in fatalities and a 23 per cent reduction in injuries'. In turn, it is claimed that safety cameras are installed at sites determined by the number of crashes and the 'cost to the community' of these crashes. Further, it is argued that evaluations of Point to Point enforcement have shown a 50% reduction in fatal and serious injury crashes. (http://www.rta.nsw.gov.au/roadsafety/speedandspeedcameras/avespeedsafetycameras/index.html). Justice has not only be transformed by calculable risk, but its jurisprudence also rendered scientific and quantified

The relevant enforcing technologies associated with such justice register measurably 'risky' behaviours. In the era of informatics, they attach these registrations to identifying codes: the digital images of registration plates, bar codes, and so on mentioned above. These codes attach the behaviour, calibrated as an offence, to a legal subject. But that subject is not the complex subject of liberal criminal law. The object of regulation is not so much the individual driver – as may be perceived by the fact that anyone may pay the drivers' fines – but the statistical distribution of driving offences understood as a volume of risk. It is 'mass' preventive justice not merely because of the volume of cases, but because of the mass object of its regulation. But this shift also transforms what we understand as 'justice'.

Mass preventive justice more or less requires a shift away from individual justice because the sheer volume of mass disobedience detected made familiar processes of trial and even summary justice unworkable in practice. As Fox (1996) has demonstrated the mounting pressure of court business forced the 'streamlining' of procedure. The 'owner' of a vehicle came to be the subject of 'on-the-spot fines' for which penalties were fixed – mitigation and aggravation practically disappeared. An effective reverse-onus justice was imposed, and strong financial incentives in the form of discounted fines were put in place to discourage opting-in to court procedure. For administrative convenience, 'owners' were assumed to be the offending drivers, unless they could prove otherwise, and were identified by their driver's digital licence number. Individuals only rarely – statistically speaking – appear in the operation of mass preventive justice. For the most part they remain throughout preventive justice procedures as electronically coded entities: what Deleuze (1995) refers to as 'dividuals', corresponding to drivers, owners, operators, and so on. They are detected as codes, 'judged' and sentenced as codes, and through coded informatics they expiate their sentence – overwhelmingly through electronic payment of a fine or cancellation of licence. For the most part, this electronically coded dividualisation is what mass preventive justice relies upon in order to work.

Last but not least, mass preventive justice has undone the age-old assumption that justice must be seen to be done. It is of course possible for offenders to insist on their day in court. But few do, and the system has been designed to discourage court appearances. For example, apart from obviating the cost and nuisance of attending a court hearing, on-the-spot fines require payment of an amount that is heavily discounted from the maximum – whereas attendance in court brings this maximum into play. Payment of on-the-spot fines means that while the offence is recorded, there is no conviction per se. And of course, the system displaces the potentially shaming ceremonial of court appearance. Such inducements to settle have roll-on effects. It is already clear that public detection, charge, and conviction are displaced by machine routines. In a real sense, no one has witnessed my offence nor my detection. But there is another key matter. Because the principal sanction is payment of a fine, the penalty can be paid remotely and invisibly, by cheque or credit card through the mail, or by electronic transfer of funds online. In the latter case, increasingly the norm, justice has now become virtual: detection, charge, sentence and expiation all occur electronically. Thus is justice unseen. This is even the case with licence suspension or cancellation, for the allocation of demerit points and the cancellation of licences are all effected virtually (O'Malley 2010b).

It is possible to imaginatively project this diagram of apparatuses and routines into a totally governed future, a self-funded infinitely expandable justice, all in the name of risk and safety. How could resistance form, especially when subjects are anonymous and justice invisible, and where for the most part only money fines are at stake? As suggested already, the danger with all such accounts is that they are projected forward into epochal visions. Processes that are developing are made to appear as if they will inexorably unfold according to a technologically determined logic. In fact, political critiques and new political discourses and solidarities have emerged around such developments, facilitated by the very e-media and technologies upon which the new control technologies themselves depend. In a few cases, there have been challenges to law in the courts. In others, political parties have picked up these 'sectional' politics and aligned them with their programmes. (Introna and Gibbons 2009, Wells 2011) But perhaps more interesting are the innumerable online discussion forums, blogs, online media publications, websites and so on that focus in part or entirely on the issue of new traffic-tracking technologies.

As I have documented in detail elsewhere (O'Malley 2013) the internet has proven an enormously effective medium for facilitating resistance that has linked up with a critical politics, the effect of which has often been to limit mass preventive justice in significant ways. To begin with, the internet has operated as a site for gathering and broadcasting information that challenges the safety effects of camera technologies. Many speed cameras have been shown not to reduce accidents, while a few seem to have increased accidents on site. Such evidence has seen the withdrawal of cameras in some areas of the UK. Red-light cameras may reduce broadside crashes, but have been shown to increase nose-to-tail accidents, and in the United States this has influenced their removal in some states. It has also been argued that safety cameras are a cheap option for governments seeking to show they are 'doing something', whereas more expensive engineering options (such as dual carriageways, speed humps, etc.) are demonstrably more effective. In turn, this has led governments to direct local administrators to scrap or reduce reliance on camera technology. While money seems to be a key resource for such justice, it is also its Achilles Heel. Campaigns have been raised and effected against governments on the grounds that safety cameras are cash cows, and are placed where accident reduction cannot be attributed to them. In NSW, for example, the state government has withdrawn more than 40 such cameras after public pressure. Finally, the cameras themselves have been challenged for being systematically inaccurate especially (but not only) in adverse weather conditions and at night. Again, major reductions or suspension of such programmes have occurred as a result, for example with respect to the Melbourne Ring Road. This is not to suggest that such opposition will lead to the unravelling of mass preventive justice. Far from it: electronic 'virtual justice' likely to extend its reach simply because of the mass phenomena it regulates. However, neither is it true that a kind of technological determinism will have its way unchecked.

Conclusions

Even in the domains of criminal justice, security, and safety, risk appears as a contested technology and discourse of governance. Indeed, some opponents, such as Ulrich Beck (1992) have called for a democratisation of risk. Confronted by a world in which, it is argued, ungoverned technological development has created risks to the entire planet, Beck calls for democratic controls to be applied to science: for in many respects risk has become a technology dominated by scientific expertise. This democratisation does not, involve the subordination of scientific findings to populist politics. Rather, the democratic concern is with defining which problems should be defined as risks, which risks to govern, and by what means? What would a government of crime and security risks look like if governed democratically along such lines?

The issue can only be gestured toward in this conclusion. But one answer has been the subject of experimentation by Shearing (2001) and his colleagues (Johnston and Shearing 2003). People in the township environment in South Africa were audited on what they regarded as the main risks to their security. Perhaps surprisingly, these did not overwhelmingly relate to crime, and in some degree police were regarded as a security threat rather than a solution. The researchers persuaded the ruling ANC to redirect some of the funding for police to 'peace committees'. This involved a participatory and informal approach along the lines of restorative justice in which communities addressed the resolution of conflicts that they felt threatened security. In this democratised risk-oriented environment, Shearing notes that 'justice' took on a rather different meaning from that familiar to courts. While the procedures strongly resembled restorative justice, justice was not so much about the backward-looking moral condemnation of offenders, the attribution of blame, and the delivery of punishment. Rather people focused on forward-looking questions of risk – on the attempt to ensure that unwanted events and actions did not happen again. In this process expertise was subordinated to the task of developing and proposing solutions to problems of risk defined by the communities.

The proposals they made were in turn subject to debate and approval. Crime prevention was being reshaped as a democratic process of risk reduction.

The point is not whether one likes or dislikes the approach taken by Shearing or of any brand of restorative or re-integrative justice. Nor is it meant to understate the difficulties of such a course. It is rather that this illustrates a kind of 'risky' innovation that opened up new directions and took 'risk' away from a pre-set course in which expert definitions went unchallenged. Such work pursued new directions that could offer an alternative to the present. In their nature such innovations in risk are dangerous: any change we advocate may be co-opted, may be blocked or may open up developments that have unanticipated negative consequences .One possibility is that unpopular minorities may get still more marginalised. But to argue that this is inevitable should make us question the point of a 'critical' discipline. Nothing is more disempowering than theoretically driven pessimism. A risk-taking criminology may just be a better alternative.

References

Austin, J., Clark, J., Hardyman, P. and D. Henry (1999) 'The impact of three strikes and you're out' *Punishment and Society*, 1:131–162.
Baumann, Z. (2000) 'Social issues of law and order' *British Journal of Criminology*, 40: 205–221.
Beck, U. (1992) *Risk Society: Toward a New Modernity*. New York: Sage.
Beck, U. (2002) *World Risk Society*. London: Polity Press.
Bougen, P. and P. O'Malley (2009) 'Bureaucracy, imagination and US domestic security policy' *Security Journal*, 22: 101–118.
Carlen P. (ed.) (2009) *Imaginary Penalities*. Devon: Willan.
Cohen, S. (1985) *Visions of Social Control*. London: Polity Press.
Collier, S. (2008) 'Enacting catastrophe. Preparedness, insurance, budgetary rationalization' *Economy and Society*, 37: 224–250.
Collier, S. and A. Lakoff (2008) 'Distributed preparedness: Space, security and citizenship in the United States' *Environment and Planning D: Society and Space*, 26:7–28.
Cooper, M. (2006) 'Pre-empting emergence: The biological turn in the war on terror' *Theory, Culture and Society*, 23:113–35.
Deleuze, G. (1995) 'Postscript on control societies' in G. Deleuze (ed.) *Negotiations 1972–1990*, New York: Columbia University Press, pp.177–182.
Ericson, R. and A. Doyle (2004) 'Catastrophe risk, insurance and terrorism' *Economy and Society*, 33:135–73.
Ericson, R. and K. Haggerty (1998) *Policing the Risk Society*. Toronto: University of Toronto Press.
Ewald, F. (2002) 'The return of Descarte's malicious demon. An outline of a philosophy of precaution' in T. Baker and J. Simon (eds.) *Embracing Risk*. Chicago: University of Chicago Press, pp.273–301.

Feeley, M. and J. Simon (1992) 'The new penology: Notes on the emerging strategy of corrections and its implications' *Criminology* 30:449–74.
Feeley, M. and J. Simon (1994) 'Actuarial justice: The emerging new criminal law' in D. Nelken (ed.) *The Futures of Criminology*. New York: Sage, pp.173–201.
Fox, R. (1996) *Criminal Justice on the Spot*. Infringement Penalties in Victoria Canberra: Australian Institute of Criminology.
Freiberg, A. (2000) 'Guerrillas in our midst? Judicial responses to governing the dangerous' in M. Brown and J. Pratt (eds.) *Dangerous Offenders*. Punishment and Social Order, London: Routledge, pp.51–70.
Furedi, F. (2008) *Invitation to Terror. The Expanding Empire of the Unknown*. London: Continuum.
Garland, D. (2001) *The Culture of Control*. Oxford: Oxford University Press.
Geason, S and P. Wilson (1989) *Designing Out Crime: Crime Prevention Through Environmental Design*. Canberra: Australian Institute of Criminology.
Hannah-Moffatt, K. (2005) 'Criminogenic needs and the transformative risk subject' *Punishment and Society*, 7:29–51.
Introna, L. and A. Gibbons (2009) 'Networks and resistance. Investigating online advocacy networks as a modality for resisting state surveillance' *Surveillance and Society*, 6:233–58.
Johnston, L. (2000) *Policing Britain Risk, Security and Governance*. London: Routledge.
Johnston, L. and C. Shearing (2003) *Governing Security. Explorations in Policing and Justice*. London: Routledge.
Kemshall, H. and M. Maguire (2001) 'Public Protection, "partnership" and risk penalty' *Punishment and Society*, 3:237–254.
Levi, R. (2000) 'The mutuality of risk and community: The adjudication of community notification statutes' *Economy and Society*, 29:578–601.
National Crime Prevention Institute (1986) *Understanding Crime Prevention* Louisville: National Crime Prevention Institute.
O'Malley, P. (1992) 'Risk, power and crime prevention' *Economy and Society*, 21:252–7.
O'Malley, P. and S. Hutchinson (2007) 'Reinventing prevention. Why did "crime prevention" develop so late?' *British Journal of Criminology*, 47:439–454.
O'Malley, P. (2009) *The Currency of Justice. Fines and Damages in Consumer Societies*. London: Glasshouse Press.
O'Malley, P. (2010a) *Crime and Risk*. New York: Sage.
O'Malley, P. (2010b) 'Simulated justice: Risk, money and telemetric policing' *British Journal of Criminology* 50:795–807.
O'Malley, P. (2013) 'The politics of mass preventive justice' in A. Ashworth and L. Zedner (eds.) *Prevention and the Limits of the Criminal Law*. Oxford: Oxford University Press pp.273–296.
Reichman, N. (1986) 'Managing crime risks: Toward an insurance based model of social control' *Research in Law and Social Control*, 8:151–172.
Rogers, P. (2012) *Resilience and the City*. London: Routledge.
Rose, N. (2000) 'Government and control'. *British Journal of Criminology* 40: 321–339.
Shearing, C. (2001) 'Transforming security: a South African Experiment' In: Strang, H. and Braithwaite, J. (eds.). *Restorative Justice and Civil Society*. Cambridge: Cambridge University Press, 14–34.

Simon, J. (1987) 'The emergence of risk society: Insurance, law, and the state' *Socialist Review*, 95:61–89.

Simon, J and M. Feeley (1995), 'True crime. The new penology and public discourse on crime' in T. Blomberg and S. Cohen (eds.) *Law, Punishment and Social Control: Essays in Honor of Sheldon Messinger*. New York: Aldine de Gruyter, pp.147–180.

Smith, G.J. (2014) *Opening the Black Box. The Work of Watching*. London: Routledge.

Walklate, S. and G. Mythen (2014) *Contradictions of Terrorism. Security, Risk and Resilience*. London: Routledge.

Wells, H. (2011) 'Risk and expertise in the speed limit enforcement debate: Challenges, adaptations and responses' *Criminology and Criminal Justice*, 11:225–241.

7
'Making a Raise' and 'Dusting the Feds': Contextualising Constructions of Risk and Youth Crime

Joe Yates

Introduction

This chapter challenges some of the assumptions which inform the contemporary policy trajectory and the risk research paradigm which underpins it. It draws on a piece of ethnographic research which was conducted over a 20-month period and involved observations, focus groups, and interviews with a range of professionals, including Housing Officers, Community Workers, Youth Justice Professionals and Community Safety practitioners, young people and members of the local community living, or working, in a community on the edge of a large northern city in England – the Estate. The research adopted an appreciative approach (Yates, 2006) which sought to give voice to the perspectives of young people in order to generate insights regarding their perspectives around risk and crime from their position in social structural hierarchies. In critically appraising risk, the chapter focuses on the 'intersection' between what Wright Mills referred to as the 'personal troubles of the milieu' and the 'public issues of social structure' (1959:7–8). As such the chapter highlights the importance of context not only in shaping the risks which young people face, in a manner which extends beyond the constructions of risk in the risk factor paradigm, but also how they negotiate them in situ in the context of the sub-cultural milieu they inhabit.

The hegemony of 'risk'

Within Youth Justice, across the western world, the positivistic risk factor research paradigm has become hegemonic (Case and Haines, 2009). It has gained pre-eminence in policy making circles – informing and shaping a range of policy and practice frameworks. This body of work stressed the capacity of a range of risk factors which create the propensity for individuals to become involved in criminality. This research sought to classify these and in doing so constructed them into a number of risk artefacts such as hyperactivity and high impulsivity, low intelligence and school failure along with poor parenting. These were considered alongside other factors such as peer influence, school and community influences, and socio-economic deprivation (narrowly understood as low family income and poor housing). Within this model the presence of these factors were purported to play a significant role in propelling individuals into a criminal career. In doing so, through the model of classification adopted, risk became constructed as uni-directional with little consideration of the contextual influences or structural constraints which shaped young peoples' lived experience. The often crude reductionism associated within this paradigm also ensured that the meanings associated with various forms of criminal action were negated – with young people silenced in their methodological construction as passive recipients of decontextualised risk factors.

The reasons for the ascendancy of the risk factor paradigm at this particular juncture related to its perceived utility in understanding youthful transgression in simple and easily understood terms (see Armstrong, 2004; Case and Haines, 2009; Goldson and Yates, 2008; Jamieson and Yates, 2009; Kemshall, 2008 for further discussion). Indeed, its provision of standardised, quantifiable, individualised, and static metrics of risk, coupled with powerful claims around their predictive capacity chimed with the political moment (Armstrong, 2004; France, 2008; Yates, 2004). However, these attempts to standardise and quantify risk into a series of measurable variables, which were perceived to propel children and young people into criminal pathways, served to re-orientate 'thinking about youth crime in favour of a focus upon those psychogenic antecedents of criminal behaviour which (were) believed to lie in the immediate social environment of the child (rather than in the structural characteristics of society itself)' (Armstrong, 2004:103). Thus risk and involvement in criminality became refracted through the prism of the individual – responsibilising young people (Kemshall, 2008). This factorisation of risk, and in particular its tendency to focus on psycho-genic factors, in the realm of individuals and their families, has been criticised as being implicitly reductionist (Armstrong, 2004), for obfuscating the complex aetiology of youth crime (Goldson and Yates, 2008) and for viewing young people as malfunctioning 'passive' recipients' of risk (Case and Haines, 2009). All of which have served to marginalise attempts to understand the complex subjective meanings attached to various forms of criminal activity by individual social actors in the context of the broader structural constraints which shape their lived experiences.

However, whilst influence of the risk factor paradigm, in terms of both policy and the research agenda itself is indisputable (Armstrong, 2004; Kemshall, 2008) there are, as Case and Haines (2009) identify, a range of questions regarding the research base which underpins it in terms of; its static and uni-directional conceptualisation of risk; its methodological rigour; the efficacy of utilising meta-analytical frameworks and the problems associated with this method; the limitations of its approach to factorisation; its tendency towards determinism; its reductionist tendencies; its tendency to over generalise and the extent to which it has obfuscated the complex aetiology of youth crime. Indeed as Case and Haines argue 'critical analysis has exposed the limitations, weaknesses, fallibilities and fallacies of risk factor research' and raised 'serious doubt about the legitimacy and validity of the methods, analysis, findings and conclusions of much risk factor research and its claims to be atheoretical, value free and scientific' (2009:310).

Re-engaging structure, meaning and the contours of risk: A view from the Estate

'There is only one way to do this, and that is through the insights afforded by the people themselves. It seems strange in this post-postmodernist time to have to point this out, and yet the richest source of information is often neglected. It is as if we have not learnt to listen, or even recognised the importance of listening' (Cullingford, 1999:25)

Due to its focus on factorisation – young people, and particularly marginalised young people, have been objectified within the narrow confines of artefact risk research (Armstrong, 2004), their voices 'pathologised, muted or silenced altogether' (Griffin, 1993). Here little attention is paid to their own perspectives regarding 'risk' or how they interpret, construct, negotiate adolescent transitions. Ethnographic studies have the capacity to generate appreciative in-depth data which can provide deep insights into crime and the realities of life in marginalised communities, which would be difficult if not impossible to obtain through any other method (see Hobbs, 2001 for an overview). This reflects what Bottoms has argued is 'a particular strength of the ethnographic tradition, rarely found in other types of criminology ... its ability to uncover some of the deep cultural meanings and normative bonds which are so important in everyday life' (Bottoms, 2000:30).

The Estate, where the research was conducted, was identified as a 'deprived neighbourhood' and an area where the problems associated with disadvantage and marginalisation were concentrated. Indeed the city, on the edge of which the Estate was located, was one of the most disadvantaged local authorities in the country, with over 50% of its wards included in the 10% most deprived nationally.[1] The city itself was also in the top ten most deprived local authorities, within which the Estate had the highest levels of income, employment, and educational deprivation and the highest levels of multiple deprivation.[2] As such the Estate, with its spatial concentration of interlocking and intersectional poverty and disadvantage, could be seen as representing a site where 'risk' in terms of structural disadvantage and poverty was concentrated (see Fyson and Yates, 2011; Yates, 2010) impacting on people's lives in a myriad of ways.

However, in terms of hegemonic professional risk discourses on the Estate there was little meaningful consideration of the broader structural issues which shaped young peoples' lived realities or the complex manner in which they interacted and negotiated these risks. Furthermore, where dissenting views were expressed, by professionals – primarily Youth Workers, there was evidence that these were silenced or side-lined as providing 'excuses' for young people's offending. Here there was evidence that the discourses of risk emanating from the portals of the risk factor paradigm shaped the power interplay between different professionals on the Estate.

When poverty and disadvantage, did appear in professional discourses on the Estate, these issues were communicated in terms of the moral otherness of the Estate or the individual families and young people residing there. Resonating with Right Realist perspectives (Muncie, 2000) inflected with the language and categorisation of 'risk', young people were identified as a problematic residuum disconnected, morally and economically, from the rest of society and in need of control. Here 'risk' factors were articulated in relation to poor parenting and welfare dependency – factors constructed as key to understanding young people's criminal proclivities. The Estate was portrayed as being populated by 'inbred' 'villains' who were out of work and 'wanted something for nowt' (Hosing Officer). Here the Estates residents were presented as 'undeserving' poor people' pathologically separate from 'us' 'spatially, socially and morally' (Young, 2007:6). Other professionals communicated 'risk' as being a direct result of the lack of effective controls describing the young people as 'out of control'. Here the failings of informal and formal control mechanisms, including 'parents, school, police' (Senior Police Officer), were highlighted and the need for more coercive control stressed. This particular respondent went further stressing the role of inadequate parenting and, again, welfare dependence stating that young people 'just do what they want – to be honest they don't give a fuck – mum and dad, if there is a dad, are signing on' (Community Centre Worker). It was clearly evident that within this hegemonic discourse attention was directed at people who are poor rather than poverty itself – with little attention paid to the processes which cause poverty (Davidson and Erskine, 1992:12).

Other professionals were more specific about the impact of intergenerational patterns of offending which they identified as being passed down through individual families – identifying the problems with crime as lying with 'toe rag families'. Here the focus was on psycho- and socio-genic factors and the need to target these. As one professional noted 'getting into that cycle and checking it and dealing with them' (Housing Officer). Here the risk assessment tools, employed to identify 'at risk' subjects and to target interventions, were derived from the risk factor paradigm. Whilst this provided a perceived sense of legitimacy, due to its portrayal as non-ideological, value neutral and evidence based (see Case and Haines, 2009 for discussion), it was clear that the factorisation of risk limited any attempts to secure a more holistic understanding of young people's behaviour. Indeed, the sense of young people as passive recipients of risk, whose behaviour could be understood through the application of pre-determined factorisation failed to take into sufficient account 'the role of accelerated social and economic change in engendering and concentrating risk factors in destabilised neighbourhoods among their inhabitants' (Webster et al., 2006:18) or the meanings young people associated with crime.

The perception of the Estate as 'problematic' (Youth Worker), as a 'breeding ground for anti-social behaviour' (Housing officer) and as a 'difficult place to manage' (Housing Officer) overshadowed any attempt to meaningfully engage with complexity of life there. This represented a failure to 'fully grasp and interpret social action, interaction and reaction' which Scraton argues requires 'the inter-weaving of the "personal", the "social", and the "structural"' (Scraton, 2004c:179). Thus the problems experienced on the Estate were located as problems with the Estate and the pathology of its residents, who were presented as 'other' 'disconnected from us ... not part of our economic circuit: ... an object to be pitied, helped, avoided, studied ... perceived as a residuum, a superfluity'

(Young, 2007:5–6). Here risk factorology and the 'deployment of risk discourse', as Coleman and Sim argue, marked 'a resurgent positivism' (2005:104) enmeshed with 'amoral assessments of the other associated with risk indicators, and 'dangerisation' (Lianos and Douglas, 2000:104 Cited in Coleman and Sim, 2005:101–102). Here the Estate became symbolic 'a signifier of character, a metaphor for the state of mind' (Higson cited in Munck, 2003:14).

Growing up and getting by

As Mooney observes communities such as the Estate are 'all too frequently portrayed as intense, uniform locales of unrest, housing homogeneous groups of 'socially excluded' people. Little thought is given to their internal social and spatial differentiation and heterogeneity, to the strategies which residents adopt to 'cope', and to residents' day-to-day struggles with public agencies (Mooney, 1999:73). It was clear that there were complex networks of support on the Estate and that it was, in many respects, felt to be a safe place to be by young people (see Yates, 2006).

Young people communicated that risk discourse, articulated through the language of moral otherness, had an exclusionary potential – adding another layer to the complex landscape of 'risk' they encountered. As one young person noted; 'They judge the Estate as rough. Well we're part of it so they're classing us as rough as well. They judge by the place where you live, like sometimes they're like oh you rob cars cos you live on the Estate or you're a trampy bitch' (YR:10). Here risk and crises became shifted onto the life history of the individual. As one young person noted – 'they think the people here cause the problems themselves' (YR:08). This stigmatisation of the Estate was reported by a large number of young people. As one young person noted; 'If I go to someone I've never met before like someone like you, they ask me where I'm from and I say the Estate, they immediately give me the stereotype' (YR:13). Another added 'Because we live around here and some of our mates are criminals they think that we're criminals and it feels crap'. (YR:09). This negative portrayal, and these stigmatising blaming discourses, not only impacted on young people's perceptions of self-worth but also, for some, resulted in cementing a place based identity and a form of defensive territoriality which was reactive and inward looking. As one young man noted 'they can say what they want, I'm from the Estate and we do things different, if they don't like it then fuck em' (YR:09). However, perhaps more worrying there appeared to be an element of self-fulfilling prophecy evident (Merton, 1957). In the words of one young man; 'kids are told that's what they're going to be like so they think "fuck it I'll just do it then"' (YR:03). There was evidence that, as Haines and Drakeford observe, 'individuals come to durable dispositions and particular ways of behaving through the cumulative ways in which messages are communicated about their worth and prospects' (1998:30). As such, for some young people, identity of resistance became forged as a defensive strategy (Collins et al., 2000:208). Adopting a 'delinquent solution' (Downes, 1966) which could serve to further entrench their marginalisation simply further excluded them.

In structural terms as Kemshall argues, 'Life for young people in contemporary society is both challenging and uncertain' and that the life course of individual young people is no longer 'mapped out and predictable' (Kemshall, 2008:21). On the Estate it was clear that life and transition for young people was particularly challenging and uncertain and that they faced 'new risks' (Furlong and Cartmel, 1997:8) which were only glimpsed by previous generations (Kemshall, 2008). These risks were associated with the specific socio-economic and cultural context of neo-liberal Britain and how these forces played out on the Estate (see White and Cuneen, 2006; Yates, 2010). For most, traditional transitions to adulthood, such as securing a job, proved difficult to negotiate as they encountered limited opportunities to secure meaningful employment. Poverty or more specifically lack of opportunity to 'get on' was an important issue. For example as YR:04 noted

> like when you've got no job and you're bored and you have no money, it's easy to think oh fuck it. You see them with money to burn or do what they want with and you've got fuck all. The big thing for me is getting a job but that's not that easy unless you want a shit paid job with an [employment] agency

Young people residing in these 'pockets of acute marginalisation' (Hall et al., 2008:22) experience the most protracted and fractured transitions into adulthood and this was clearly evident on the Estate.

However, in addition, as White and Cuneen argue 'Many young people in "modern" and "advanced" industrialised societies are not simply marginal to the labour markets, they are literally excluded from it – by virtue of family history, structural restrictions on education and job choices, geographical location, racial and ethnic segregation, stigmatised individual and community reputation' (2006:19–20). This is evidenced by the following young man. For him, securing employment and negotiating transition was more challenging because of the stigmatisation of the Estate.

> I wouldn't say growing up on the Estate has been hard but it could have been easier [What do you mean?] Well it's like down to the reputation again, it's like hard, the reputation makes it hard, hard to get through it, hard to prove yourself and hard to get by. My brother he couldn't get an (job) interview when he put the Estate and the post code down as an address. (YR:05)

Here the risks associated with unemployment and economic inactivity become compounded by the perception of the Estate and its residents as being amoral, pathological, other and risky.

Understanding crime in the margins: 'Making a Raise' and 'Dusting the Feds'

> Reputation that's what they do it (car crime) for ... and to make money ... they want to make themselves feel better and get more friends and they do. (YR:09)

For young people residing on the Estate crime impacted on their lived experiences in a variety of ways (see Yates, 2006). It was also clear that various forms of youth crime served very different functions and had different meanings associated with them. In this section I will focus on two particular functions – the instrumental function ('to make money') and the expressive function ('to make themselves feel better'). The following sections, drawing on the perspectives of young people themselves, illustrates that involvement in criminal activity, rather than being the outcome of decontextualised risk factors, should be understood as a 'resource' in the negotiation of what Young refers to as the 'two stigmas which the poor confront': first, 'relative deprivation described as poverty and exclusion from the major labour markets and secondly 'misregonition' described as lower status and lack of respect (Young, 2007:51).

'Making a raise'

It was clear that exclusion from the labour market, discussed earlier, limited young people's opportunities to achieve the material goals set out for them in a legitimate manner (see Merton, 1957). Within the Estate it was, in Young's words, 'not so much the process of ... being simply excluded but rather one which was all too strongly included in the culture but, then, systematically excluded from its realisation' (Young, 2007:25). In this context, vehicle crime and in particular theft from a vehicle or sale of a stolen vehicle, either as a whole or in part, served an economic function. It enabled young people, growing up in a marginalised context with limited opportunities to legitimately generate finance to 'make a raise'. Engaging in a range of entrepreneurial, albeit illicit, activities facilitated participation in consumer culture. As one young person described, it was about 'just taking the chance, showing initiative – taking that chance, making a raise' (YR:03). Another noted 'I love flogging things that's why I know loads of people ... People ask me if I can get things and I can 'cos I know people so I sort it. I make money on it' (YR:09). Contrary to the crude individualisation, and denial of the import of structural factors, inherent in the risk factor paradigm (Armstrong, 2004) it was clear that broader economic changes manifested in unemployment and social exclusion played a significant role in young people's decisions to become involved in various forms of instrumental offending. Rapid de-industrialisation, unemployment, casualisation and the multiple deprivations they experienced meant that the ability to 'make a raise' was key for the young respondents.

The level of sophistication, and financial gain which could be secured, i.e. whether a 'proper raise' could be made, was predicated on connections young people had with the alternative economy and the ability to secure outlets for the merchandise – again stressing the importance of local context in terms of the role of the alternative economy on the Estate (see Yates, 2006). As one young person noted; 'I knew people who would buy the gear. I had some connections, which meant ... I could get rid of the stuff. Like the others knew the same people I knew it's just that they wouldn't deal with them' (YR:04). However, they were also aware that the real financial benefits would be secured by adults who would make the 'proper money'. As one young person noted:

> Like I know where to sell the gear, like if we rob a Subaru most of the Twockers wouldn't know what the fuck to do with it, they'd have the radio and that off, maybe the alloys. But I know where I can take that car and get it sorted [ringing]. We won't make proper money off it just a few quid but the bloke I sell it to will make some proper cash mate ... thinking about it we're just small time really. (YR:05)

The motivations for involvement here were clearly financial. In short it allowed young people to generate money to fuel their participation in consumer culture through an easily accessible avenue. Resonating with Merton's strain theory (1957) this provided an opportunity to generate finance in a context where there were limited opportunities to do this legitimately. However, rather than a form of oppositional identity or as an indication of their moral otherness this is best considered as participation in the dominant consumerist values. As one young person noted he had used the proceeds to 'buy a hundred and something pound pair of trainers' (YR:04) which would have been out of his reach through legitimate means.

In socially excluded working-class communities, such as the Estate, the traditional adolescent route from school to work, economic maturity and the traditionally ascribed male role of worker and breadwinner had become unobtainable for most and drastically elongated for the majority. In this context, many of the young male respondents, such as those quoted above, sought individual solutions to these broader societal problems in the form of criminal activity. This served as a mechanism to negotiate their way through the obstructed paths to adult statuses, identities and activities and also provided finance for them to participate in consumer culture. Thus, rather than the result of decontextualised biological and psychological factors, immorality, or limited self-control, as Scraton and Haydon argue 'young people's "offensive" or "offending" acts may be ways of coping with, or reacting to, their experiences of social injustice' (2002:325). Here crime acted as a mechanism to manage the risks associated with social exclusion and poverty and illustrates how risks associated with structural factors find form in young people's offending.

Reputation matters: 'Dusting the Feds'

On the Estate reputation mattered and it was within the context of peer groups that these reputations were achieved and maintained (Campbell, 1993). For some young people, and in particular young men, involvement in vehicle crime, and particular riding stolen motorbikes and stolen cars, was identified as a key resource for engendering status and respect among peers. The motivations here were expressive rather than instrumental. Putting on 'shows', through conspicuously driving the vehicles in full view of others on the Estate, was seen as a key mechanism to engender status and respect. Indeed being seen in this space and being seen to exert control and power over both the vehicle and the space was infused with meaning. Here the Estate was a paradoxical space symbolising both 'the systematic powerlessness so often felt by the individuals who live in such environments' and as site of 'risk consumption' (Hayward, 2002:86). It was clear that the majority of young people lived their 'collective life' on the street (Collins et al., 2000:243) and that this space which was becoming increasingly regulated (Jamieson, 2006). Vehicle crime in this context, and within this space, provided a form of resistance to this regulation whilst simultaneously engendering status in the sub-cultural milieu of the Estate. As one young man noted 'The police come up here thinking it's their control ... we (are) like yeah fuck off pricks ... like we were like let's go an get a car or a bike or sommit and fucking have a laugh' (YR:31). For young people, who were relatively powerless, vehicle crime, and in particular TWOCking, served as a mechanism to momentarily assert control over the space and to exert control over their relationship with the police. Here the motivation was the opportunity to manage, or exert control over, interactions with an increasingly regulatory state. Being able to 'dust' or escape from the feds (police), leaving them metaphorically with only dust in their faces, was also seen as a significant achievement – not only for the young people involved but also young people who were spectators. As one young man noted 'I like the shows, when they put the shows on – they fucking rag them about, they rally them around and then they make knobs out of the feds – dustin them ... like they can dust the feds and make them look like Faggots' (YR:32).

Here crime as a form of structured social action was infused with meaning. Far removed from its conceptualisation in the risk factor paradigm it related to status but also to power, resistance, protest, and controlling social space. It was also heavily laden with gendered meanings associated with the situational accomplishment of masculinity. However, it also provided a delinquent solution to status in another respect. As one young person noted 'from a lot of people you get quite a bit of respect ... it's important for the lads to be in a click, a crew' (YR:02). However, young people were also aware that their involvement could serve to further marginalise them. As one young man noted 'Like I reckon I've fucked my life ... I come out of school with no grades, cos you didn't get qualifications for black boxing cars and breaking into buildings, if you did I would have got an A plus ... so pretty much Joe I've got fuck all' (YR:04).

For many of the young people TWOCking also provided excitement or what they referred to as the 'buzz'. This was clearly seductive (Katz, 1988) and police pursuit along with the risks of being caught were key. As one young person described '[It's] *Just a buzz, your getting chased ... it's a buzz you don't know what's going to happen next*' (YR:03). Central here was the sense of risk and danger. Indeed, the young people involved in TWOCking were clearly aware of the risks which they took including the risks to their own safety, but weighed this against the 'buzz' which the act offered them. Being willing to take risks was seen engendering respect – this was not linked with impulsivity but rather needed to be understood in relation to the situational construction of respect in the context of the Estate.

However, there were a variety of other meanings associated with varieties of vehicle crime by individual young people: meanings which would be obfuscated through crude factorisation which fails to appreciate the complexity of social actors in their structural, cultural and spatial contexts (MacDonald, 2007).This is illustrated by the following accounts from a young 14-year-old male. It should be noted here that this young man was a higher educational achiever and had high aspirations to attend university to pursue a career in computing. In short he would not be identified as an 'at risk' subject in terms of the risk factor paradigm. He came from a family who were not involved in criminal activity and he also had a range of fluid friendship groups with a range of young people from outside the Estate – who were not involved in criminal activity. For him involvement in vehicle crime was not only about taking risk it was also about managing risk. For him the risk he encountered was the 'risk of being seen as one of those wanky lads who are never doing anything – like not going out and taking a risk and stuff like that'. He decided to manage this by becoming involved in riding stolen motorbikes – to gain acceptance. He noted 'as I got on it (the stolen motorbike) people stopped taking the piss out of me' noting that involvement in this type of activity, on the Estate, meant that 'People do treat you differently' (YR:13). For him involvement was not a form or resistance or an expression of an oppositional identify. Rather it was a form of managing the risk of not being a part of the youth networks on the Estate. Here as France (2000) argues being seen as 'different' presented risks – risks, for him, which needed to be managed. His focus was not on the thrill or excitement of transgression, although this undoubtedly was seen by him as a secondary benefit, but rather the opportunity engagement in the act offered to counter perceptions that he was 'geeky' as he was studious. He wanted to be seen as 'one of the lads' a feeling which he identified was a 'liberating experience'.

He identified the risks that these two parts of his life posed to one another 'if I get caught on a motorbike I would get a record, if I get caught I'm fucked for like my career getting a job and that'. In negotiating this he made informed choices he noted that he steered clear of some of the young people involved in more serious offending as these where 'people you don't want to know' – he saw their behaviour as too risky 'you don't want to go there' (YR:13). Thus there were people and activities which he would not get involved with noting 'further up ... is proper crime like breaking into houses and that is something I would definitely not do' (YR:13), clearly making decisions regarding his behaviour in the context of the sub-cultural milieu he inhabited.

Conclusion

This area of the field, in terms of understanding the subjective meanings associated with crime and risk by young people, along with how this is shaped by cultural and structural contexts, is currently neglected within a mainstream administrative criminology dominated by adult and academic concerns (France, 2000:328). In this context, risk has become 'individualised' and social problems associated with poverty presented as 'individual shortcomings rather than as a result of social processes' (France, 2000:317). This process is described by Case and Haines (2009) as a 'simplistic over simplification'. Here, in the words of C. Wright Mills, the analysis tends to 'slip past structure to focus on isolated situations' (Mills, 1959:534). As a result, the discussion does not consider poverty and disadvantage as reflecting broader structural class based inequalities in society but rather as individualised issues or de-contextualised risk factors – the fault of the individual, the result of poor choices. As Kemshall argues 'In the "risk society" individuals are framed as shapers of their own worlds "making decisions according to calculations of risk and opportunity" (Petersen, 1996:47), but facing blame and punishment if they get their choices wrong' (2008:22).

It was clear from this research that young people's involvement in crime was infused with meaning and the result of various and divergent motivations. Crime, in this respect, could not be understood as the result of a number of decontextualised risk factors which propelled them into criminal pathways but rather was the result of a number of relational, interpersonal, and intersubjective processes which could only be understood in situ and in the context the macro structural factors which shaped, rather than determined, their lived experiences. For young people, different forms of criminal action had different functions and meaning and that the same criminal acts could engender very different meanings for different young people. Thus the reasons for involvement could be diverse and their engagement with such activities could be multi-functional. Listening to the voices of young people and the challenges they experienced 'growing up' and 'getting by' provides insight into a complexity of risk in their worlds – a complexity which is far too easily obfuscated by the crude factorisation of risk which characterises the risk factor paradigm.

Notes

1. References removed to maintain anonymity.
2. Anonymous reference, taken from a report on indices of deprivation produced by the local authority.

References

Armstrong, D. (2004). 'A Risky Business? Research, Policy, Governmentality and Youth Offending'. *Youth Justice.* 4 (2), pp.100–116

Bottoms, A. (2000). 'The Relationship Between Theory and Research in Criminology'. *In:* R. King and E. Wincup (eds), *Doing Research on Crime and Justice.* 1st ed. Oxford: Oxford University Press. pp.75–117.

Campbell, B. (1993). *Goliath: Britain's Dangerous Places.* 1st ed. London: Methuen.

Case, S. and Haines, K. (2009). *Understanding Youth Offending: Risk Factor Research, Policy and Practice.* 1st ed. Cullompton: Willan.

Coleman, R. and Sim, J. (2005). 'Contemporary Statecraft and the Punitive Obsession: A Critique of the New Penology Thesis'. *In:* J. Pratt, D. Brown, M. Brown, S. Hallsworth, and W. Morrison (eds), *The New Punitiveness: Trends, Theories and Perspectives.* 1st ed. Cullompton: Willan. pp.110–115.

Collins, J., Noble, G., Poynting, S. and Tabar, P. (2000). *Kebabs, Kids, Cops and Crime: Youth, Ethnicity and Crime.* 1st ed. Annandale: Pluto Press.

Cullingford, C. (1999). *The Causes of Exclusion: Home, School and the Development of Young Criminals.* 1st ed. London: Kegan Page.

Davidson, R. and Erskine, A. (1992). 'Introduction'. *In:* R. Davidson and A. Erskine (eds), *Poverty Deprivation and Social Work: Research Highlights in Social Work 22.* 1st ed. London: Jessica Kingsley Publishers. pp.11–21.

Downes, D. (1966). *The Delinquent Solution: A Study in Subcultural Theory.* 1st ed. London: Routledge.

France, A. (2000). 'Towards a Sociological Understanding of Youth and their Risk Taking'. *Journal of Youth Studies.* 3 (3), pp.317–331

France, A. (2008). 'Risk Factor Analysis and the Youth Question'. *Journal of Youth Studies.* 11 (1), pp.1–15

Furlong, A. and Cartmel, F. (1997). *Young People, and Social Change: Individualisation and Risk in Late Modernity.* London: Open University Press.

Goldson, B. and Yates, J. (2008). 'Youth Justice Policy and Practice: Reclaiming Applied Criminology as Critical Intervention'. *In:* J. Yates, with B. Stout, and B. Williams (eds), *Applied Criminology.* 1st ed. London: Sage. pp.103–117.

Griffin, C, (1993). *Representations of Youth: The Study of Youth and Adolescence in Britain and America.* 1st ed. Cambridge: Polity.

Haines, K. and Drakeford, M. (1998). *Young People and Youth Justice.* 1st ed. London: Palgrave.

Hall, S., Winlow, S. and Ancrum, C. (2008). *Criminal Identities and Consumer Culture: Crime, Exclusion and the New Culture of Narcissim.* 1st ed. Cullompton: Willan.

Hayward, K. (2002). 'The Vilification and Pleasures of Youthful Transgressions'. *In:* J. Muncie, G. Hughes, and E. McLaughglin (eds), *Youth Justice: Critical Readings*. 1st ed. London: Sage. pp.80–95.

Hobbs, D. (2001) 'Ethnography and the Study of Deviance', in P. Atkinson, A. Coffey, S. Delamont, J. Loftland, and L. Loftland, (eds) *Handbook of Ethnography*, London: Sage.

Jamieson, J. (2006). New Labour, Youth Justice and the Question of Respect. *Youth Justice*. 5 (3), pp.180–193

Jamieson, J. and Yates, J. (2009). 'Young People, Youth Justice and the State'. *In:* R. Coleman, J. Sim, S. Tombs, and D. Whyte (eds), *State, Power, Crime*. 1st ed. London: Sage. pp.76–90.

Katz, J. (1988). *Seductions of Crime: Moral and Sensual Attractions in Doing Evil*. New York: Basic Books.

Kemshall, H. (2008). 'Risks, Rights and Justice: Understanding and Responding to Youth Risk'. *Youth Justice*. 8 (1), pp.21–37

MacDonald, R. (2007). 'Social Exclusion, Youth Transitions and Criminal Careers: Five critical reflections on "Risk"'. *In:* A. France and R. Homel (eds), *Pathways and Crime Prevention: Theory, Policy and Practice*. 1st ed. Cullompton: Willan. pp.0–0.

Merton, R. (1957). *Social Theory and Social Structure*. 1st ed. New York: Free Press.

Mills, C. (1959). *The Sociological Imagination*. New York, NY: Oxford University Press.

Mooney, G. (1999). 'Urban "Disorders" '. *In:* S. Pile, C. Brook and G. Mooney (eds), *Unruly Cities?*. 1st ed. London: Routledge. pp.0–0.

Muncie, J. (2000). 'Pragmatic Realism? Searching for Criminology in the New Youth Justice'. *In:* B. Goldson (ed.), *The New Youth Justice*. 1st ed. London: Russell House Publishing. pp.14–35.

Munck, R. (2003). *Reinventing the City? Liverpool in Comparative Perspective*. 1st ed. Liverpool: Liverpool University Press.

Scraton, P. and Haydon, D. (2002). 'Challenging the Criminalization of Children and Young People: Securing a Rights Based Agenda'. *In:* J. Muncie, G. Hughes and E. McLaughlin (eds), *Youth Justice: Critical Readings*. 1st ed. London: Sage. pp.311–329.

Scraton, P. (2004c) 'Speaking Truth to Power: Experiencing Critical Research', in M. Smythe, and E. Williamson, (eds) *Researchers and their 'Subjects': Ethics, Power, Knowledge and Consent*, London: Policy Press.

Webster, C. MacDonald, R. and Simpson, M. (2006). 'Predicting Criminality: Risk/protective Factors, Neighbourhood Influence and Desistance'. *Youth Justice*. 6 (1), pp.7–22

White, R. and Cuneen, C. (2006). 'Social Class, Youth and Crime'. *In:* B. Goldson and J. Muncie (eds), *Youth Crime and Justice*. 1st ed. London: Sage. pp.17–30.

Wright Mills, C. (1959). *The Sociological Imagination*. 1st ed. New York: Oxford University Press.

Young, J. (2007). *The Vertigo of Late Modernity*. 1st ed. London: Sage.

Yates, J. (2004). 'Evidence, Group Work and the New Youth Justice'. *Groupwork*. 14 (3), pp.112–132

Yates, J. (2006). 'You Just Don't Grass': Youth, Crime and 'Grassing' in a Working Class Community. *Youth Justice*. 6 (3), pp.0–0

Yates, J. (2010). 'Structural Disadvantage: Youth, Class, Crime and Poverty'. In: W. Taylor, R. Hester, and R. Earle (eds), *Youth Justice Handbook: Theory, Policy and Practice*. 1st ed. Cullompton: Willan. pp.5–23.

Yates, J. and Fyson, R. (2011). 'Anti Social Behaviour Orders and Young People with Learning Disability'. *Critical Social Policy*. 31 (1), pp.102–25

Part III
Communicating Risk in Social Care

8
Communicating Risk in Youth Justice: A Numbers Game

Stephen Case

We live in a 'risk society' characterised by rapid social changes and globalisation (see Beck 1992), which presents the concept of 'risk' as universally negative – as a threat or harm to be managed or mitigated, rather than, for example, a positive sensation to be pursued (see Katz 1988; Kemshall 2008). The mass media, politicians, academics, and the general public have conspired to present risk as a social harm, particularly the risk that *young people* allegedly pose to themselves and others (Haines and Case, in press; Case and Haines 2009; Goldson, in Hendrick 2003). The anxieties of governments in the industrialised Western world over burgeoning rates of crime and antisocial behaviour by young people, the perceived failures of systemic responses to youth crime (e.g. deterrence, community sentences, custody) and the spiralling costs of youth justice (Kemshall 2007) have proved to be a catalyst for *Risk Factor Research* – a positivist research movement intent on identifying factors in childhood and adolescence that purportedly 'predict' or increase the 'risk' of offending in later life. Westernised youth/juvenile justice systems have tended towards understandings of offending behaviour by young people that have been framed by a discourse of *neo-conservative correctionalism* – seeking to 'correct' and reduce the 'problem' of offending by (adults) targeting perceived deficiencies in the individual (child), including the choice ('agency') to offend and/or to not attempt to resist risk, for which the child is held responsible (*neo-liberal responsibilisation*). Neo-conservative correctionalist youth justice practice (a new form of conservative ideology that emphasises rational choice and the responsibilisation of children who commit crime) measures, assesses, and responds to risk as a collection of statistical, quantifiable, aggregated (group-level) 'factors' that exert a deterministic influence on offending that is simultaneously irresistible to young people yet amenable to targeted intervention by adults (see Case 2006, 2007).

The discourse of risk promulgated by neo-conservative correctionalism is retributive, punitive, coercive, interventionist and can have deleterious consequences for children, such as criminalisation and deviancy amplification (cf. Bateman 2011; McAra and McVie 2007; O'Mahony 2009; Paylor 2010). Furthermore, tailoring prevention projects around the risks presented and experienced by children 'can serve to consolidate negative representations of the risk posed by young people (or "define deviance up") and give credence to the notions of choice and intractability that underpin punitive policies. Meanwhile, alternative justifications for youth provision are silenced' (Kelly 2012: 101). By linking risk inextricably to negative behaviours and outcomes, neo-correctionalist and neo-liberal discourses engender a reductionist, regressive youth justice that individualises explanations for behaviours and outcomes, responsibilises children for these behaviours and outcomes, whilst eschewing more progressive discourses and practice models that facilitate children's strengths, potentialities, and capacity for prosocial and positive behaviours and outcomes through their consultation, participation, engagement, and enablement (by adults) to access universal entitlements and rights (see later – *AssetPlus*).

This chapter critically evaluates the hegemonic discourse of 'risk' fostered by youth justice practice, typically manifested in risk assessment and risk-focused intervention, and argues that risk-based practice constitutes a reductionist exercise that has over-simplified, distorted, and invalidated extremely negative understandings of young people's lives and the systemic responses to them. Invalidity has been an artefact of the crude factorisation and aggregation of the complexity of risk and the individualisation of responsibility for offending and desistance. Furthermore, these processes can result in the labelling, stigmatisation, and marginalisation of young people due to their negative, deficit-focus, and interventionist tendencies, whilst simultaneously deprofessionalising and disempowering practitioners through their biased, prescriptive nature. The uncritical methods used to establish and represent risk are illustrated empirically by discussion of *Risk Factor Research* and its corollary, the *Risk Factor Prevention Paradigm* and illustrated practically by *Asset* risk assessment and the *Scaled Approach* assessment and intervention framework in the Youth Justice System of England and Wales. The chapter concludes by outlining the progressive alternative risk discourse promised by the *AssetPlus* assessment and intervention framework, which emphasis young people's needs, strengths, perspectives, and positive behaviour, alongside practitioner discretion – a welcome subversion of neo-correctionalist risk discourses and practices.

From *Risk Factor Research* to the *Risk Factor Prevention Paradigm*

Risk Factor Research has been dominated by positivist methodologies, employing highly structured surveys (questionnaires, interviews), psychometric tools, secondary data, and statistical (regression) analyses to measure risk as a statistical, quantifiable, objective, and value-free scientific 'fact' with a predictable and reliable (consistent) relationship to offending. Identified 'risk factors' for offending relate to young people's characteristics, behaviours, attitudes, and experiences – typically psychological/individual (attitudinal, emotional, cognitive) or in the immediate social domains of their lives (family, education, neighbourhood, lifestyle). These 'psychosocial' risk factors and their associated discourses have populated an expanding and widely replicated evidence base for the development of youth justice policy and practice (see Case 2010; Farrington 2000). Most notably, the identification of risk factors and their subsequent targeting through risk-focused intervention has shaped and driven youth justice systems internationally (see Görgen, Kraus, Taefi, Bernuz Beneitez, Christiaens, Meško, Perista, and Tóth 2013; Dunkel, Grzywa, Horsfield, and Pruin 2010) – an approach known as the *Risk Factor Prevention Paradigm* (see Hawkins and Catalano 1992). The *Risk Factor Prevention Paradigm* offers a straightforward and logical approach to the prevention of offending by young people:

> Identify the key risk factors for offending and implement prevention methods designed to counteract them. There is often a related attempt to identify key protective factors against offending and to implement prevention methods designed to enhance them.
> (Farrington 2007: 606)

The approach has captivated politicians and policymakers, offering an ideal practical solution to their concerns regarding transparent, defensible, cost-effective and 'evidence-based' practice to communicate and address the youth crime 'problem' in the modern risk society. The *Risk Factor Prevention Paradigm* constitutes multi-purpose 'risk management' – allowing policy (makers) to manage the risks that young people present to themselves and to society, concurrent to managing the risks that youth justice practitioners may exercise excessive discretion and understand risk in ways not prescribed by government. Accordingly, the *Risk Factor Prevention Paradigm* is considered to be a common sense approach that is 'accepted by policy makers, practitioners and the general public' (Farrington 2007: 606), based on 'a consistent risk management methodology ... a cautious and defensive response to the challenges of modern society' (Stephenson, Giller and Brown 2007: 3–4).

> **Case in focus: Animating the *Risk Factor Prevention Paradigm* through the *Scaled Approach***
>
> Youth justice practice with all young people (aged 10–17 years old) who enter the Youth Justice System of England and Wales is underpinned by the Scaled Approach, an assessment and intervention framework that animates the Risk Factor Prevention Paradigm. The assessment portion of the framework is driven by Asset, a structured, standardised risk assessment instrument completed by Youth Offending Team[1] practitioners, based on their interviews with young people. Using Asset, Youth Offending Team practitioners are required to rate a young person's risk of reoffending (more accurately, their risk of reconviction) by measuring their (current or recent) exposure to risk factors in 12 psychosocial domains: living arrangements, family, and personal relationships, education, training, and employment, neighbourhood, lifestyle, substance use, physical health, emotional and mental health, perception of self and others, thinking and behaviour, attitudes to offending and motivation to change, alongside additional sections measuring positive (protective) factors, indicators of vulnerability, indicators of risk of serious harm to others and a self-assessment 'What do you think?' section (Youth Justice Board 2000). The risk domains and additional sections within Asset each contain a series of risk-based statements that are rated 'yes' or 'no' by Youth Offending Team practitioners to indicate the presence or absence of that risk factor in the young person's life. Practitioners then quantify the extent to which the risks in each risk domain are associated with 'the likelihood of further offending': 0 = no association, 1 = slight or limited indirect association, 2 = moderate direct or indirect association, 3 = quite strong association, normally direct, 4 = very strong, clear and direct association. Although the risk rating for each domain can be accompanied by a qualitative, narrative explanation within a small, summative 'evidence box', the Asset risk

> assessment process is predominantly reductionist – reducing risks to statistical, quantifiable factors. To compound this reductionism, the risk ratings for each domain are then totalled (giving a score from 0–64, based on 16 sections) and this total risk score dictates the frequency, nature, and intensity of the intervention giving to the young person: standard (risk score of 0–14), enhanced (risk score of 15–32), or intensive (risk score of 33–64). This is the Scaled Approach, which necessitates 'tailoring the intensity of intervention to the assessment'. (Youth Justice Board 2007: 4)

The reductionism of risk assessment

The *Risk Factor Prevention Paradigm* that underpins risk assessment in youth justice systems transforms the concept of 'risk' into a measurable, quantitative 'factor' amenable to probabilistic calculation (France 2008), thus promoting a reductionist, restricted, and restrictive, deficit-based discourse (see Candlin and Crichton 2011). The 'transformation' of risk is more akin to a process of dumbing down, depriving risk of its essential quality, meaning, depth and detail. In short, a potentially dynamic, complex, and subjective interplay of experiences, perceptions, and responses is reduced to a static, single, isolated quantity. Each step along a youth justice assessment process that utilises the *Risk Factor Prevention Paradigm*, therefore, takes the measurement of risk further away from how it is understood and experienced by the young person. The process of converting risk to a factor/quantity, otherwise known as 'factorisation', serves to carve risks into blunt, crude, over-generalised variables that may not represent either the phenomena being measured or the lives of the individuals who experience risks and to whom intervention will be applied as a result of this invalid 'assessment'. Such invalidating reductionism has been facilitated by the vague, ambiguous, generic and indefinite phrasing of risk statements across *Risk Factor Research* and youth justice risk assessment tools (see, for example, Haines and Case 2008). For example, the *Asset* risk assessment instrument used in the Youth Justice System in England and Wales asks practitioners to rate (yes/no) whether 'significant adults fail to communicate with or show care/interest in the young person'. However, this statement is insensitive to who the significant adult(s) may be (e.g. parent? both parents? carer? sibling? teacher? police officer? Youth Offending Team worker?), whether the issue is failure to communicate and/or to show care/

interest, what the nature, duration, and intensity of the issue is and how it was experienced by the young person, whether it is still an issue, how it may interact with other family-related issues or issues in other risk domains, and so on. Factorisation produces statistical artefacts whose relationship with offending is understood (more accurately, 'imputed') in generic, abstract terms and solely in relation to the generic, abstract (psychosocial) factors programmed into statistical analyses. The nature of the risk factor-offending relationship is poorly theorised, ambiguous, and contentious (e.g. variously 'understood' as causal, predictive, correlational, indicative, reciprocal, proxy, indeterminate or even illusory – Case and Haines 2009; see also Case 2010) – underpinned by imputation and misguided confidence in a widely replicated body of methodologically limited artefactual *Risk Factor Research*, leading O'Mahony (2008) to disdainfully characterise risk factors as 'vague, inadequate proxies for putative causal processes'.

Risk assessment in the field of youth justice has been undermined by a 'staged' reductionism – with practitioners (typically through interviews with young people and significant others such as parents and teachers) compelled to reduce risk to a quantifiable factor, then to total (aggregate) risk assessment outcomes in order to give a risk score, which can be used to allocate young people to a 'scaled' risk category for intervention (e.g. the *Scaled Approach* in England and Wales). This staged reductionism has been exacerbated still further by a reliance on *aggregation*. The *Risk Factor Research* that has informed the application of the *Risk Factor Prevention Paradigm* across youth justice systems has been grounded in the identification of differences in levels of risk exposure between aggregated groups of young people (e.g. offenders compared to non-offenders), rather than being grounded in the measurement of the risk exposure of individual young people. The 'de-generalisation' and 'rounding down' of risk from group profiles to individual young people in this way uncritically assumes that group-level (aggregated) risk factors are equally applicable to all members of the group. Therefore, a young person who is a member of a 'high risk' or 'at risk' group (e.g. on the basis of demographic and/or geographic characteristics) by default is allocated the group's risk profile/status and their perceived need for responsive, risk-focused interventions, regardless of personal exposure to group-level risk factors. Consequently, aggregation eschews investigations of within individual change that would offer a more valid exploration of causality (cf. Farrington 2007). Of course, aggregation has been born of practical and financial necessities to analyse large numbers of young people in order to offer generalisable findings to inform policy and practice that is underpinned by neo-conservative explanations for youth crime. However, such aggregation begets the *homogenisation* of large (often disparate) groups of individuals and is thus insensitive to socio-demographic individual differences such as age, gender, ethnicity, nationality, culture, social economic status and locality, not to mention personal experiences and perceptions. Aggregation is a heuristic device that washes away individual differences – a further reductionist step towards producing invalid measures and thus, negative, partial, and potentially invalid discourses of risk.

A further example of the reductionism of risk assessment lies in the *psychosocial bias* of risk assessment tools and the negative influence on the interventions that result. As stated, the *Risk Factor Research* that guides risk assessment has privileged individualised risks in the psychological and immediate social domains (i.e. psychosocial risk factors) to the exclusion of socio-structural, political, and historical understandings of risk (see, for example, Armstrong 2004; France 2008). Such psychosocial reductionism and bias precipitates *partial* (limited and biased) understandings and discourses of risk and its relationship to offending. Psychosocial reductionism has been driven by governmental demands for amenable, practical policy targets that are simultaneously politically attractive in that they do not implicate governmental practices that may have contributed to the generation of risk factors and offending behaviour or the failure to rectify specific (social) problems such as social inequalities. In other words, psychosocial bias and reductionism have fostered the *individualisation* of risk understandings, pointing the finger of responsibility towards young people's apparent failure to resist and negotiate their exposure to risk. Psychosocial risk factors locate risk and offending within individual pathologies and deficits that cohere risk discourses around a 'morality of blame' (Armstrong 2004) and that suggest short-term, cheap, quick fix responses that can label and stigmatise young people as simultaneously and paradoxically passive/helpless and responsible (agentic) for their own exposure to risk (Case 2007). Once again, O'Mahony (2009: 100) summarises the situation incisively, castigating youth justice risk assessment as:

> Reductionist bias [that] gives virtually no consideration to key issues like justice, equality, human agency, moral development, human rights and restrictions on the state's power over the individual.

The reductionist (over) simplification of complex and dynamic personal and social information, processes and interactions into static, quantitative factors washes away the subjectivity, perceptions, meanings, and understandings of the young people who experience exposure to risks. The over-riding factorisation within risk assessment ignores, or at least underplays, the potential for young people to shape, resist, and negotiate their exposure and responses to risk as active constructors of their own lives (Case 2007; Kemshall 2011), preferring to view these young people as passive 'crash test dummies' (Case and Haines 2009) careering helplessly towards the deleterious consequences of risk exposure in the absence of adult intervention. Ironically, the adults who are charged to intervene (i.e. Youth Offending Team practitioners) are restricted by the prescribed processes and content of risk assessment practice to the extent that their potential contribution to assessment and intervention (like that of young people) is ignored, or at least downplayed. According to Pitts (2001, 2003), youth justice practice has become a technical, automated, 'routinised' undertaking that deprofessionalises practitioners, due in large part to the hegemony of restricted and restrictive quantified risk assessment processes conducted under the umbrella of the *Risk Factor Prevention Paradigm*. Therefore, understandings and explanations of risk are not the only areas artificially reduced by risk assessment – so are the contributions, perceptions, and expertise of the young people and practitioners who actually animate the risk assessment process! The marginalisation of qualitative perspectives and personalised constructions of risk within assessment processes, largely due to factorisation and aggregation of risk and the technical, prescribed completion of quantitative assessment tools by adult practitioners, has pursued risk discourses that provide 'a prescription without a consultation' (Case 2006: 174). This is the case for both young people and for practitioners whose expertise, experience, and discretion is marginalised by reductionist, heavily technical assessment procedures (see also Bateman 2011; Case 2010; Pitts 2003).

> **Case in focus: Psychosocial reductionism in *Asset* risk assessment**
>
> The domains of risk within the *Asset* risk assessment instrument used with young people in the Youth Justice System in England and Wales are entirely psychosocial in nature, disregarding sociodemographic, socio-structural, socio-economic, political and historical risk influences. *Asset* is intended as a tool to support practitioner expertise and judgment in planning interventions, but it engenders

a restricted and restrictive risk discourse by addressing only a narrow range of possible risk factors that could influence young people's behaviour. Despite cautioning practitioners against explanatory bias by 'relying on a favourite or fashionable theory' (Youth Justice Board 2003: 103–104), *Asset* guidance for practitioners pre-determines that assessments should explain the risk factor-offending relationship based on three restricted, developmental, psychosocial theories: the *criminal careers* model (psychosocial variables predicting offending at different developmental stages – Farrington 1997), the theory of *age-graded informal social control* (risk factors, transitions, turning points and life events influence offending at different development stages – Sampson and Laub 1993) and the *interactional theory* (risk factors and offending have a relationship of reciprocal influence over the life course – Thornberry 1987). The developmental, psychosocial biases of *Asset* marginalise alternative aetiological explanations and discourses, even from within *Risk Factor Research* itself, such as constructivist theories, integrated approaches and ecological models (see Case and Haines 2009) and alternative sources of non-individualised risk (factors and processes), contrary to the assertion that *Asset* risk assessment offers 'comprehensive coverage of relevant risk factors'. (Youth Justice Board 2003: 8)

The discourses of risk created and sustained by *Risk Factor Research*, the *Risk Factor Prevention Paradigm*, and risk assessment in youth justice systems have exhibited reductionist tendencies that have been manifested in factorisation, aggregation, psychosocial biases, the marginalisation of young people's personalised understandings and perceptions of risk, and the neglect of practitioners' experiences and judgments relating to risk and appropriate interventions to ameliorate it. Based on 'shaky scientific credentials' (O'Mahony 2008) and on 'naïve scientific-realist assumptions' about the relationship between risk factors and offending (Armstrong 2004: 111), *Risk Factor Research* and the *Risk Factor Prevention Paradigm* have promoted narrow, reductionist understandings of risk premised on equally narrow, reductionist methodologies. Political appeal and simplicity of execution have been preferred to valid understandings and discourses of 'risk' and valid responses to address it. It is to these responsive interventions that this chapter now turns.

Interventionism: The logical corollary of risk obsession

The quantified, factorised, psychosocial assessment of risk has provided the rationale for a wealth of youth justice interventions that seek to prevent and reduce offending by young people across the Westernised world (cf. Sherman, Gottfredson, Doris, MacKenzie, Eck, Reuter and Bushway 1998). The early intervention focus of the *Risk Factor Prevention Paradigm* is premised on preventive goals to identify the risk factors that (allegedly) predict offending in 'high risk' and 'at risk' groups of young people, in order to tackle these risk factors as early as possible, thus nipping offending in the bud. Ostensibly, risk-focused early intervention affords a practical, common-sense approach to the prevention of youth offending. However, proponents of the *Risk Factor Prevention Paradigm* conveniently overlook both the invalidating reductionism of risk assessment and the startling lack of cogent evidence base for risk-focused intervention and for the risk assessment tools underpinned by its logic. Indeed, some *Risk Factor Prevention Paradigm* zealots have even been moved to concede that 'few tools are available to distinguish those youths who will continue with behaviours that may lead them to become child delinquents' (Loeber, Farrington and Petechuk 2003: 8). *Risk Factor Research* itself makes few claims for evidence that the early targeting of risk factors prevents future offending (see Case and Haines 2009), preferring instead to occupy the safer ground that early exposure to risk factors 'predicts' or is 'associated with' later offending. In reality, the key developmental, artefactual *Risk Factor Research* studies have based their conclusions on samples never subjected to intervention (e.g. Sampson and Laub 1993; West and Farrington 1973) or have produced partial (limited, biased) and inconsistent evidence that risk-focused intervention is effective (e.g. Bottoms and McClintock 1973; Thornberry and Krohn 2003). Despite an alarming lack of evidence base for the purportedly 'evidence-based' practice of risk factor intervention, the *Risk Factor Prevention Paradigm* has gained popularity exponentially across youth justice systems (Dunkel et al. 2010; Görgen et al. 2013), such that:

> The risk factor and prediction paradigms have taken hold of criminology, especially for those interested in crime prevention and crime control policies. (Laub and Sampson 2003: 289)

However, it is possible to take issue with the use of risk-focused (targeted) early intervention as prevention practice in the field of youth justice (i.e. the application of the *Risk Factor Prevention Paradigm*) on the number of fronts (cf. Goldson, in Blyth, Solomon, and Baker 2009):

- **Theoretical** – it promotes deterministic, developmental, and psychosocial biases and discourses to the exclusion of alternative explanations of risk and offending behaviour generally and the risk factor-offending relationship specifically (see also France and Homel 2007; Kemshall 2011, 2008);
- **Conceptual** – it labels young people as 'at risk' and individualises the blame/responsibility for offending, thus justifying their stigmatisation and exclusion, and the need for adult-led intervention and interventionism (excessive use of intervention – Kelly and Armitage 2014; see Muncie 2004);
- **Evidential** – it is devoid of a cogent and replicated/reliable evidence-based (whereas the *Risk Factor Research* that guides risk assessment can at least lay claim to a replicated evidence-based, albeit one that replicates the methodological weaknesses of artefactual *Risk Factor Research*), nor is it clear as to the developmental or behavioural stages at which early intervention should be conducted (see also Case and Haines, in Bateman 2011; Goldson and Muncie 2014);
- **Ethical** – it breaches children's rights (not to mention contradicting the rationale for evidence-based practice) by intervening in their lives in a restrictive (even punitive) and pre-emptive way on the basis of an 'at risk' label that predicts what they 'might do', rather than what they actually 'have done' (see also Goldson 2005; O'Mahony 2009).

Notwithstanding these theoretical, conceptual, evidential, and ethical concerns, the use of (early) intervention targeting 'at risk' groups of young people has continued to characterise much youth justice practice internationally. For example, the *Scaled Approach* to assessment and intervention in England and Wales has the potential to apply disproportionate levels of intervention on young people in the Youth Justice System due to the requirement for intervention levels to match demonstrable risk, rather than, for example, the nature of the offending behaviour or the young person's level of expressed need. Through the *Scaled Approach*, it is possible that excessive (unnecessary, potentially criminalising) levels of intervention may be visited on a young person who has committed a minor and/or one-off offence, yet has measured high for exposure to risk, whilst serious and/or persistent offenders who are assessed as low risk by *Asset* may receive low levels of intervention that are disproportionate to the nature of their offending and their need for ameliorative support.

The widespread adoption of risk-focused intervention by youth justice systems internationally indicates a wilful misreading and misunderstanding of the main findings and conclusions from developmental, artefactual RFR. The rationale for risk-focused interventions with young people actually contradicts the conclusions by the pioneers of developmental *Risk Factor Research* (Glueck and Glueck 1930) and developmental crime prevention (Cabot 1940; Bottoms and McClintock 1973) that young people *grow out of crime* without the need for formal intervention. Indeed, a separate body of research has discovered that formal contact with youth justice systems and the use of risk-focused interventions 'risks damaging the individuals they are designed to assist' (McCord 1978: 289; see also McAra and McVie 2007). Inadvertent and deliberate misunderstandings and misrepresentations of *Risk Factor Research* have been (to a significant extent) politically driven, exacerbated by non-experts within governmental departments and the civil service who have demonstrated an over-inflated and unjustified certainty about risk, based on 'a diet of simplified and partial, yet definitive, conclusions from flawed *Risk Factor Research* which have then been turned into "hard scientific fact"' (Case and Haines 2009: 302). Crude factorisation has been used to turn risks into amenable targets for intervention in order to meet political demands for pragmatic, measurable, transparent, and evidenced results (see Muncie 2004), despite a lack of evidence base for this approach. Lack of evidence begets uncertainty compelling policymakers and practitioners to extrapolate and hypothesise explanations for the risk factor-offending relationship and to impute explanations for how risk-focused interventions may function to prevent and reduce offending. These reductionist and imputed processes have produced 'evidence-based' policies and practices that have been, in evidential terms, 'built on sand' (Case 2006; see also Case 2010), yet pursued in a voracious and interventionist manner.

> **Case in focus: The holistic promise of *AssetPlus***
>
> Growing criticisms of reductionist risk assessment and intervention practice in youth justice systems, particularly that of England and Wales, have prompted the Youth Justice Board to conduct a reflective review of the *Scaled Approach* focused on: developments in assessment practice, theoretical debates around 'risk' and the perceptions and experiences of practitioners and 'offenders' (Baker/ Youth Justice Board 2012). In June 2015, a revised assessment and intervention

framework entitled *AssetPlus* will come into force (Youth Justice Board 2013) – allegedly providing an holistic, complex, contextualised, and dynamic process that prioritises young people's needs (over risks), the perspectives of young people and the discretion of practitioners (over prescribed assessment procedures), strengths (over deficits), and promoting positive behaviours (over preventing negative behaviours). *AssetPlus* has been proposed as a challenge to extant risk management and early intervention mechanisms by providing the conceptual and practical space for positive assessments and future orientated interventions. Early working models of *AssetPlus* portray an ongoing assessment cycle (prevention to custody) driven by practitioner completion of a three-stage, iterative Core Record consisting of 'Information Gathering and Description' to inform 'Explanations and Conclusions' to inform 'Pathways and Planning'. The *AssetPlus* sections will eschew numerical, quantitative ratings and measures, so moving away from the inherent reductionism (through factorisation) of *Asset*.

The revised *AssetPlus* assessment and intervention framework has the potential to affect a culture shift away from measuring and responding to psychosocial risk factors using risk-focused intervention and towards a more explicit emphasis on young people's needs, strengths, and positive behaviours and outcomes. However, the proposed explanatory reliance on assessing 'risk and protective factors' (expressed in the 'Self-assessment' portion of the 'Information Gathering and Description' section) and the rating of 'risk/likelihood of reoffending' within the 'Explanations and Conclusions' section as a means of informing an ostensibly 'scaled' (to risk level) response to offending appear to contradict or at least undermine this culture shift. The proposed changes, therefore, do not do enough to reorientate assessment and intervention and seem intent on amending and augmenting existing risk-focused procedures. Whilst *AssetPlus* could offer a promising advance from the entrenched reductionism of the *Scaled Approach*, it does not attempt a sufficient overhaul of assessment principles, policies, and practices to benefit young people in the Youth Justice System. Just as the *Scaled Approach* before it, *AssetPlus* presents as a technique without a guiding philosophy or purpose (see Haines and 1998 for a more detailed discussion of youth justice philosophies) which poses a significant threat to its potential to refocus youth justice assessment and intervention in a more positive, risk-free direction.

Conclusion: The discourse of risk in youth justice – Reductionist invalidity

It has been asserted in this chapter that the discourse of 'risk' in international youth justice systems has been restricted and reduced by an adherence to overly simplistic (risk assessment and intervention) methodologies underpinned by a body of *Risk Factor Research* that has self-replicated using similarly overly simplistic methodologies. In particular, the conversion of 'risk' into quantifiable, psychosocial 'factors' and aggregated scores has produced assessments, understandings, and explanations of young people's lives that are bereft of necessary nuance, sensitivity, quality, detail, relevance, and meaning to the young people who actually experience and interact with risk and to the practitioners whose expertise and experience has been overlooked in its diagnosis and treatment. Reductionist risk assessment exercises have encouraged similarly invalid (risk-focused) interventions as a result. Staged and prescribed processes of reductionism have not only invalidated and biased perceptions and understandings of risk, but have also fostered a rampant and misguided interventionism that constitutes an invalid and unjustified extrapolation of the existing evidence base for *Risk Factor Research* and the *Risk Factor Prevention Paradigm*. The use of the *Risk Factor Prevention Paradigm* in youth justice systems has offered an 'oversimplified technical fix' for a complex social reality (Stephenson et al. 2007), that 'strips away the dynamism, life-force, nuance and immeasurable complexities of risk' (Case and Haines 2009: 303). Furthermore, the predilection for risk-based approaches in youth justice systems represents and responds to young people in negative, damaging ways as risky, harmful, dangerous, feckless and helpless – collections of risks and potential negative behaviours that need to be managed and prevented through adult-enforced intervention, restriction, and punishment.

Note

1. Youth Offending Teams are multiagency youth justice teams in all local authority areas in England and Wales, consisting of representatives from the police, local authority (e.g. social workers), health service, and probation service. Youth Offending Teams were made a statutory requirement by the Crime and Disorder Act 1998 and are monitored by the Youth Justice Board of England and Wales, a quasi-autonomous non-governmental organisation also created by the Act.

Bibliography

Armstrong, D. (2004) A risky business? Research, policy, governmentality and youth offending. *Youth Justice*. 4(2): 100–116.

Baker, K. (2012) *AssetPlus Rationale*. London: YJB.
Beck, U. (1992) *Risk Society: Towards a New Modernity*. London: Sage.
Bateman, T. (2011) Punishing poverty: The scaled approach and youth justice practice. *The Howard Journal of Criminal Justice*, 50(2): 171–183.
Blyth, M., Solomon, E. and Baker, K. (2007) *Young People and Risk*. Bristol: Policy Press.
Blyth, M., Solomon, E. and Baker, K. (2009) Young people and 'risk'. Bristol: Policy Press.
Case, S.P. and Haines, K.R. (2009) *Understanding Youth Offending: Risk Factor Research, Policy and Practice*. Cullompton: Willan.
Bottoms, A.E. and McClintock, F.H. (1973) *Criminals Coming of Age. A Study of Institutional Adaptation in the Treatment of Adolescent Offenders*. London: Heinemann.
Cabot, R. (1940) A long-term study of children: The cambridge-somerville youth study. *Child Development*, 11(2): 143–151.
Candlin, C. N. and Crichton, J. (Eds) (2011). *Discourses of Deficit*. Basingstoke: Palgrave Macmillan.
Case, S.P. (2006) Young people 'at risk' of what? Challenging risk-focused early intervention as crime prevention. *Youth Justice*, 6(3): 171–179.
Case, S.P. (2007) Questioning the 'evidence' of risk that underpins evidence-led youth justice interventions. *Youth Justice*, 7(2): 91–106.
Case, S.P. (2010) Preventing and Reducing Risk. In: W. Taylor, R. Earle and R. Hester (Eds) *Youth Justice Handbook*. Cullompton: Willan.
Case, S.P. and Haines, K.R. (2014) Risk Management and Early Intervention. In: B. Goldson and J. Muncie (Eds) *Youth, Crime and Justice*. London: Sage.
Dunkel, F., Gryzywa, J., Horsfield, P. and Pruin, I. (2010) *Juvenile Justice Systems in Europe, Vols. 1–4*. Monchengladbach: Forum Verlag Godesberg.
Farrington, D.P. (1997) Human Development and Criminal Careers. In: M. Maguire, R. Morgan and R. Reiner (Eds) *Oxford Handbook of Criminology*, 2nd Ed., Oxford: Clarendon Press, pp. 361–408.
Farrington, D. (2000) 'Developmental Criminology and Risk-focussed Prevention. In M. Maguire, R. Morgan and R. Reiner (Eds) *The Oxford Handbook of Criminology*, 3rd Ed. Oxford: Oxford University Press.
Farrington, D. P. (2007) Childhood Risk Factors and Risk-Focused Prevention. In M. Maguire, R. Morgan and R. Reiner (Eds) *The Oxford Handbook of Criminology*, 4th Ed. Oxford: Oxford University Press.
France, A. (2008) Risk factor analysis and the youth question. *Journal of Youth Studies*, 11(1): 1–15.
France, A. and Homel, R. (2007) *Pathways and Crime Prevention. Theory, Policy and Practice*. Cullompton: Willan.
Glueck, S. and Glueck, E. (1930) *500 Criminal Careers*. New York: Alfred Knopf.
Goldson, B. (2005) Taking Liberties: Policy and the Punitive Turn. In H. Hendrick (Ed.) *Child Welfare and Social Policy*. Bristol: Policy Press.
Goldson, B. (2007) Early Intervention in the Youth Justice Sphere: A Knowledge-Based Critique. In: M. Blyth and E. Solomon (Eds) *Prevention and Youth Crime*. Bristol: Policy Press.
Görgen, T., Kraus, B., Taefi, A., Bernuz Beneitez, M., Christiaens, J., Meško, G. Perista, H. and Tóth, O. (2013) *Youth deviance and youth violence: Findings from a European study*. http://www.youprev.eu/pdf/YouPrev_InternationalReport.pdf (Accessed January 2014).

Haines, K.R. and Case, S.P. (in press) *Positive Youth Justice: Children First, Offenders Second*: Bristol: Policy Press.

Haines, K.R. and Drakeford, M. (1998) *Young People and Youth Crime*. Basingstoke: Macmillan.

Haines, K.R. and Case, S.P. (2008) The rhetoric and reality of the Risk Factor Prevention Paradigm approach to preventing and reducing youth offending. Youth Justice, 8 (1), 5–20.

Hawkins, J.D. and Catalano, R.F. (1992) *Communities that Care*. San Francisco: Jossey-Bass.

Hendrik, H. (2003) Child welfare and social policy. Bristol: Policy Press.

Katz, J. (1988) *Seductions of Crime*. New York: Basic Book.

Kelly, L. (2012). Representing and preventing youth crime and disorder: intended and unintended consequences of targeted youth programmes in England. *Youth Justice: An International Journal*, 12(2): 101–117.

Kemshall, H. (2011) Crime and risk: Contested territory for risk theorising. *International Journal of Law, Crime and Justice*, 39(4): 218–229.

Kemshall H (2008) Risks, rights and justice: Understanding and responding to youth risk. *Youth Justice*, 8(1): 21–37.

Kelly, L. and Armitage, V. (2014) *Diverse Diversion: Youth Justice Reform, Localised Practices and a 'New Interventionist Diversion'?* Paper presented to the British Society of Criminology annual conference, Liverpool, UK.

Kemshall, K. (2007) Risk Assessment and Risk Management: The Right Approach? In M. Blyth, E. Solomon and K. Baker (Eds) *Young People and Risk*. Bristol: Policy Press.

Laub, J. and Sampson, R. (2003) *Shared Beginnings, Delinquent Lives. Delinquent Boys to Age 70*. London: Harvard University Press.

Loeber, R., Farrington, D. and Petechuk, D. (2003) Child delinquency: Early intervention and prevention. *Child Delinquency Bulletin Series*, May 2003.

McAra, L., McVie, S. (2007) The impact of system contact on patterns of desistance from offending. *European Journal of Criminology*, 4(3): 315–45.

McCord, J. (1978) A thirty year follow-up of treatment effects. *American Psychologist*, March, 284–289.

Muncie, J. (2004) *Youth and Crime*. London: Sage.

O'Mahony, P. (2008) The risk factors prevention paradigm and the causes of crime: A deceptively useful blueprint. *Youth Justice 2008: Measuring Compliance with International Standards Conference*, Cork University, Republic of Ireland, April 2008.

O'Mahony, P. (2009) The risk factors paradigm and the causes of youth crime: A deceptively useful analysis? *Youth Justice*, 9(2): 99–115.

Paylor, I. (2010) The scaled approach to youth justice: A risky business. *Criminal Justice Matters*, 81(1): 30–31.

Pitts, J. (2001) *The New Politics of Youth Crime*. Basingstoke: Palgrave.

Pitts, J. (2003) *The New Politics of Youth Crime: Discipline or Solidarity?* Lyme Regis: Russell House.

Sampson, R.J. and Laub, J.H. (1993) *Crime in the Making: Pathways and Turning Points Through Life*. Harvard: Harvard University Press.

Sherman, L., Gottfredson D., MacKenzie, D., Eck, J., Reuter, P. and Bushway, S. (1998) *Preventing Crime: What Works, What Doesn't, What's Promising*. Department of Criminology and Criminal Justice, University of Maryland, Baltimore.

Stephenson, M., Giller, H. and Brown, S. (2007) *Effective Practice in Youth Justice*. Cullompton: Willan.
Thornberry, T.P. (1987) Toward an interactional theory of delinquency. *Criminology*, 25(4): 863–92.
Thornberry, T.P. and Krohn, M.D. (2003) *Taking Stock of Delinquency: An Overview of Findings from Contemporary Longitudinal Studies*. New York: Kluwer.
West, D.J. and Farrington, D.P. (1973) *Who Becomes Delinquent?* London: Heinemann.
Youth Justice Board (2000) *ASSET: Explanatory Notes*. London: YJB.
Youth Justice Board (2003) *Assessment, Planning Interventions and Supervision*. London: YJB.
Youth Justice Board (2007) *Youth Justice: The Scaled Approach*. London: YJB.
Youth Justice Board (2013) *Assessment and Planning Interventions Framework – AssetPlus. Model Document*. London: YJB.

9
Working with Risk in Child Welfare Settings

Tony Stanley

Introduction

In carrying out everyday duties, statutory social workers make sense of risk. Understandably then, the language of risk is drawn on discursively to help define, classify, and decide a course of action to protect children from danger and harm (Coppock & McGovern, 2014; Featherstone, 2013; Kemshall, 2010; Stanley, 2013; Webb, 2006). Social workers do this on behalf of the state where the rights of children to be free from harm and abuse are enshrined in law. This is a legitimate activity for statutory social workers – as their work is primarily about making sure that children are protected, as best they can be, from abuse and neglect. The work is tough, with significant decisions needing to be made about children and their families. Risk communication is central to these practice decisions – and, the subject of this chapter.

Social workers' decisions can have a significant impact for families, and some children do need to be placed out of the family home because the situation is too risky, or too dangerous. But how is this determined? Is it always the case that a child must leave a risky situation? And how much risk is too much? We need to understand how practice is influenced discursively by risk if we are to find ways to bring about a more compassionate child welfare system. This is important because families are often left out of decision making (Stanley, 2013).

Young Muslim teenagers heading to Syria from England and Australia highlight this point. The anxiety around radicalisation is fuelling a contemporary moral panic that serves to position social workers as the guardians of risk on behalf of the state (Dodd, 2014; Mohammed & Siddiqui, 2013). Social services agencies are being called on to investigate families where the risk of childhood radicalisation is reported or feared. So the espoused value base of acting humanely by helping families is easily relegated by a dominant and anxious presentation of radicalised childhood risk (Coppock & McGovern, 2014) and a 'child rescue' ideology is promoted. Where risk is viewed as a calculable and predictable object, this constructs it as a controllable phenomenon, easily attached to individuals, rather than connected to the social structures or contextual influences (Culpitt, 1998; Webb, 2006).

Our 'risk thinking' is influenced by messages from politicians, media headlines, and our experiential knowledge (Tullock & Lupton, 2003). But what do we mean by 'a child at risk of radicalisation?' The risk of some children being radicalised into terrorists, while others are reported to be at continued risk of sexual exploitation now fuels an English media hungry for sensationalised childhood risk narratives. Social workers need to understand how their practice is shaped and influenced accordingly. We need to think about how practice can remain a moral and ethical enterprise and not something turned into a soft arm of policing or an extension of the surveillance state (Dodd, 2014). Social constructionist theories are helpful, because they assist us to explain how risk, need, and responsibility discourses circulate, mobilise, and organise practitioners, families, and importantly, the decisions made about children. An argument for state intervention is rendered easier through the rhetorical use of 'high risk'. The problem here is that some families will be pre-empted or wrongly classified (Lupton, 1999).

This chapter shows how working with risk and uncertainty must involve family members in a way that helps us to co-construct understandings about risk, even in the most extreme or worrying of situations. This will improve communication about risk and means we are not overly reliant on reports and papers, data, and information flowing between and within agencies. Rather, a better understanding about how risk comes to be discursively located and positioned should lead to better decision making because in doing so a humane partnership approach to practice is encouraged. Following an introduction to the English child protection working world, and an argument that this is a muscular system characterised by rigidity and authoritarianism, I show how social constructionist ideas are helpful to interrogate risk as it organises, influences, and shapes our practice. Brief practice illustrations are used to help me do this.

Child welfare social work in England

I am the principal social worker for a large-inner city social service department in London. I hold senior responsibility to raise professional practice standards across the social work teams (Stanley & Russell, 2014). I hold a caseload and work as part of the senior management team. I see excellent practice on a daily basis, and work alongside committed social workers who strive to do the best job possible. I also see some workers struggling to reconcile the authority inherent in the role with working in partnership with families. What this means for some families is a more heavy handed experience of social work. Workers draw on their risk thinking to have conversations with families about child abuse and neglect. If the risk thinking is narrow, conversations will be narrow – with solutions even narrower (Stanley, 2013).

In Anglophone countries, social workers are encouraged to set out and locate risk, and then manage it (Featherstone, White & Morris, 2014). A set of practice dilemmas for social workers and families has emerged because of this. The problem for social workers is that communication about risk tends to encourage objectified notions of risk, resulting in rather narrow risk thinking about the available options for children reported to be 'at risk'. The problem for children and families in the English system is that they are often left out of the professional meetings and discussions where important decisions are made. Risk communication tends to be dominated by professionals. This needs overcoming if we are to deliver social work services that are morally informed and delivered in more humane ways.

The English system has developed through a mix of policy and practice reactions to high-profile child deaths, responsive legislative imperatives, and further public condemnation each time a child abuse tragedy played out (Jones, 2014); it is characterised by professional decision making. Following a social work determination that a child is at risk of significant harm, a child protection conference is convened. This is the main forum where a child protection plan is developed through multi-agency input with an aim of ensuring safety for the child. The child protection plan sets out how the child will be protected, and who will do what in order that this is sustained. These are professionally driven and professionally led meetings (Connolly, 2004), with families invited in to participate. In the English system risk definitions tend to be the work of the professional within the child welfare system – the social worker and their manager, along with a child protection coordinator. The coordinator of the child protection conference has an influential voice in this risk work, regardless of not meeting the family or children. Their work is informed by the risk communications of other professionals. The family members have the least power and afforded the least influential voice. Social justice principles are easily lost within a system characterised by professional dominance and muscular authoritarianism (Featherstone, White and Morris, 2014). A risk-averse culture operates because the child welfare system seeks to control (the reported) non-normalising child and family. This example illustrates how requests for help can turn into exercises of monitoring.

> A parent noticed a change in her youngest child's behaviour and sought advice from her doctor. The doctor referred the child to the local children's services department, as a precaution, in case the behaviour was caused by abuse or neglect. The child was cared for by relatives during the day. A social worker visited, and recommended that the child attend a form of organised day care. The worker argued that it is better if the child is seen regularly by professionals.

The notion that children need to be 'monitored' is unfortunately a common response, influenced by notions of distrust, and one in which social workers are encouraged to hold 'a suspicious eye'. In the example, a referral for help gets translated into an investigation to be carried out by suspicious social workers, resulting in a recommendation that a child is monitored. This practice typifies a rigid English child welfare system where families tend to receive the same service irrespective of their presenting need and risk (Featherstone, White & Morris, 2014). Families become subjects of the service, rather than the service being flexible enough to offer the help that they need. Ideas of helping and working *with* families are therefore secondary to a focus on the individual child's right to be safe. Active trust is replaced by active mistrust (Beck, 2002) and risk is calculated in order to do this (Zinn, 2008).

Discourse, power and risk practice

The construction of risk as a calculable object enables and legitimates decisions about statutory interventions. It is understandable then that social workers construct those who become associated with risk as needing to be governed and controlled (Horlick-Jones, 2005). This thinking influences decisions, because decision making rests with the professional team (social worker, manager, child protection coordinator) – they hold the power. Discourses of 'high risk' are applied simplistically and risk gets located as an inherent parental feature, a fixed trait in the family system and this can contribute to risk aversive and oppressive practice. Children, it is argued, need protection from *their* family. Ideas of working with families to manage risk are easily relegated by child rescue ideologies. Thus, professional risk discourses are powerful.

> One of the important ways in which the meanings of children's lives may be constructed and become powerful is through the assumptions underpinning professional discourses, but such frameworks may reshape and arguable empty out the moral and political aspects of experience and suffering. (Ribbens McCarthy, 2013:331)

Professional discourses and assumptions that aim to uphold societal norms may actually reshape and pressure family life through normalisation discourses (Rose, 1999) with precautionary approaches adopted to govern social problems (Aradau, & van Munster, 2007). Social workers find themselves governing normalisation at a distance from the state but acting on behalf of the same state. A family where a convicted terrorist resides is now considered a risky situation for children. Reports have been made to some children's services departments about children living in homes with fathers who have convictions under United Kingdom anti-terrorist legislation. A significant number of men have been convicted of offences under this legislation, many being fathers. Typically on release a request is made to reunite with family. Reports made to children's services suggest that the children of these men are themselves 'at risk' of becoming radicalised toward holding extremist views or, more worryingly, being groomed toward terrorist activities. But what happens when this family chooses to home school their children? Should we be more worried about this? How is risk communicated between professionals when we do not see children? A further distancing of these families occurs when they are constructed as being at even higher risk because children are outside of professional systems like schools. Families that experience a sense of pressure because of this surveillance may choose to go underground or, in some cases leave the country. Risks are weighed up and choices are made. Paradoxically, a family may place their children in greater harm with a decision to leave England and head to a conflict zone like Syria. So, how can we avoid drawing simple conclusions about these choices?

Making sense of risk

Two sociological sources help us to take a more critical approach to how risk discourses operate: social constructionism and the sociological imagination. A social constructionist approach to understanding social knowledge examines the ideologies and assumptions central to how risk is understood in any given context. This position states that as knowledge arises from interactions between people and their social environment, what comes to be understood as 'risky' is not a reference to innate qualities, rather, to sets of social meanings (Berger & Luckman, 1967). Social constructionism can help explain these influences as they shape risk, safety, need, and responsibility discourses, and help us to understand how they can mobilise and organise practitioners and families. Social constructionist ideas can also be used to explain a more supportive and helpful agenda for working with risk, one that, by emphasising the socially generated nature of knowledge and its influence on behaviour, sets out to construct safety and solutions that are helpful – working together in collaboration with families. Strengths and safety-oriented approaches attempt to manage risk in ways that contribute to collaborative working relationships with families, offering rigorous and humane approaches to work that help to guide the protection of children.

Another tool that can help to understand risk thinking is the sociological imagination. Social workers often work to connect the personal situation with the social context when they engage in practice thinking. This is the sociological imagination (Mills, 1959). The drawing together and making sense of where and when we are, who the influential people around us might be, and what messages we become subject to, often without us noticing, are important ideas for us to understand how our practice comes about. Likewise, examining the political influences or dominant political messages of the day is important to enable the social worker to connect the political context they are working in with their individual value and belief base, thus enabling an empowerment approach to working with risk (Featherstone, Broadhurst, & Holt, 2012).

Implying that assessing risk takes place in a disembodied, objective fashion denies the interconnected nature of the social construction of knowledge in any given micro-context, such as the organisation we work in, and within a given socio-political context. The sociological imagination and social constructionist theorising are practice tools that can help us take a more critical look at the sources of the discourses used to understand risk in social work. Muslim families known to children's services that leave for Turkey are suspected as heading to Syria, and so their children are classified as high risk and subject to a child protection enquiry. This is happening irrespective of any proof about where they might be. Thus the child is 'at high risk' beyond borders, while in the care of parents and family. This classification occurs before we even meet them.

If numerous conflicting discourses exist that present us with a choice of interpretations, then how do we pick the right one? And what can happen if we don't question that multiple discourses operate? The most obvious outcome to not questioning this is that families are more likely to be monitored, and not helped. State surveillance of radicalising families now extends beyond national borders for those children of families who have decided to leave.

The power of risk factor science

Well intentioned goals of intervening in reported child abuse and neglect actually start prior to the meeting of any child or their family. A series of interrelated practice steps begins the constructing of a social work case, and as I have argued elsewhere (Stanley, 2013), social workers can reach probability judgments that legitimise subsequent statutory intervention. The counting up of risk factors from a range of sources happened by the social workers – these included initial reports of at risk children, case histories held in case files, in supervisory conversations and health colleagues' reports – thus contributing to risk being understood and named, and following this, determining professional thresholds for action. This method of counting up risk factors contributes to risk being regarded as something professionally determined, and so risk becomes a virtual object, something workers see as needing to be resolved in order that children are made safe (Stanley, 2013). Individuals could then be held accountable for resultant harm to a child. And, this includes workers being blamed for failing to protect a child. A precautionary approach to practice is therefore understandable. For children and social workers, the stakes are high if something goes wrong.

There are a number of consequences when risk is constructed through risk factor counting. This construction encourages a narrow risk paradigm, and a 'logic of risk' that acts as a framework where intersecting knowledge and attitudes narrow the aim of practice to those aspects of family and individual behaviour that can be assessed as 'risky' (Broadhurst et al., 2010). Problems are thus constructed as resulting from individual choices and responsibilities. This view weakens the practice imperative of collaborating with parents, as they can be easily be seen as untrustworthy, viewed only as a source of risk, and hence unequal partners in the work. This compromises the social worker-client relationship by encouraging a focus on deficits in a working context that will blame a social worker if things go wrong (Firkins & Candlin, 2011; Hall & Slembrouck, 2011; Jones, 2014). So we guard against this, and practice is influenced accordingly. Parents are expected to turn up to meetings and appointments and if they don't they are seen as failing on their part to be active, and insightful good citizens. The normalising project continues (Rose, 1999). Risk measurement regimes based on actuarial methods influence a 'get things right' approach, with a focus on certainty and 'knowing for sure', leaving socially determined or socially constructed notions of risk at the practice margins. While these tools can be useful, they cannot replace actual understandings of the family's current relationships, their functioning, their hopes, dreams, and wishes for their children. Further, a neglect of the social and political context is easily reinforced and so political action by some Muslim families and radicalisation become easily conflated in terms of childhood risk.

Mostly, risk discourses still tend to be understood by social workers in line with Beck's (1992) influential ideas of risk as a negative and 'all pervasive' and, as described earlier, this can play out in practice through risk-averse decision making. Two problems arise when this narrow focus operates. First, as discussed, the work of risk assessment and risk analysis tends to be regarded as a task for professionals, with families less likely to play a part in determining what comes to be known as risky, dangerous or safe enough. Second, some professionals outside of the child welfare system place pressure on child protection workers to use the care system as a method of risk resolution. Some child protection workers will accept this and argue that a child being removed is an acceptable way to manage risk. Beck argues in his 'risk society' thesis that as people become more aware of risks, and internalise the idea that they must manage personal risks, people perceive risks as both all-encompassing and yet able to be controlled by the individual (Beck, 1992). Social work practice is influenced by this thinking, as the implicit construction operating in this model is of risk of being pervasive and needing to be avoided, and that professionals should aim for control and order. This practice can leave families out; while at the same time encourage professional anxiety.

> After six years in prison for terrorist related activities, a father reunites with his family. Married with four children, the father quickly resumes his public protesting against the UK for the war in Syria. His older children are seen with him at a number of protest rallies that the police deem illegal rallies. All of the children are home schooled. Police refer the family to children's services. A social worker was asked to assess the risk of the children being radicalised. After meeting the

> social worker once, the parents withdraw parental consent to continue the assessment under the Childrens Act (s17, 1989). The social worker has two options: close the case, or elevate the risk to a child protection enquiry (s47, 1989) that is not dependent on parental consent.

By predetermining 'at risk' classifications social workers can undermine family involvement and participation in the social work process (Stanley, 2013). This is partly because of the preparatory work that is done in the office prior to meeting families, or as shown in the case study parents withdrawing consent to an assessment can leave social workers thinking they have no choice but to assess without consent. Risk is then easily reified into an objective matter that needs resolution. Social workers can draw on probability judgments about risk and present these to their supervisors for ratification rather than using this as an invitation for critical or an engaged challenge to how the risk case was actually established (Stanley, 2013). Further, risk evidence is always actively sought from external professionals with health colleagues and police highly regarded. These new forms of evidence get countered along with recorded case histories and this contributes to a codification of 'higher risk' that legitimises statutory intervention. Workers are encouraged not to work with uncertainty, but rather to seek out the evidence to help them bring clarity to situations (Stanley, 2013). So, 'risk thinking' can be understood as a product of a particular way of constructing risk, one reliant within a social constructionist perspective, on the dominant discourses of the time (Munro, 2010). In the case study above, police referring to children's services can position the social worker as an arm of the state aligned with the surveillance project. The rights of families become relegated by the rights of the state to protect society from the as yet unknown terrorist events, yet to become.

Opening up our risk thinking

So, how can we resist the tendency to view and present risk in such narrow ways and clearly at odds with the humanistic and social justice aims of social work? How can we ensure that we are not using risk discourses in ways that leave families out from decision making? This leads us to ask how social workers can make better use of their risk thinking, and be encouraged to find new ways of approaching risk, more humanely, in order that they avoid narrow options. They can do this while still responding to the very real harms experienced by children who are abused by their parents.

Traditional models of assessment have tended to focus on deficit and problem identification approaches. Offering balance to this, strengths-based practice emerged in the 1980s formulised as a set of practice principles in response to the pathology-laden treatments available for people living with mental illness (Weick et al., 1999). According to Saleebey (1997), operating from the strengths-based perspective means that our work is about helping the discovery and exploration of client's strengths and resources and turning these into tools and assets to aid the helping process. Not prescriptive in the skills used, this perspective orientates the social worker to open up their thinking to accommodate a wider lens for thinking critically about the client and their situation. This is a framework through which workers are invited to see their client's situation as more than a set of problems, and it asks for a profound belief by social workers that people can and want to have different lives. Applications of this perspective are found in the sex offender treatment literature in the 'Good Lives Model' (see Barnett & Mann, 2011; Ward & Mann, 2004), and recently in the interface between English statutory social work and health visiting practice through the signs of safety approach (Stanley & Mills, 2014). Some argue that social work has always considered strengths and positive aspects for client lives (Connolly, 2004). However, as discussed earlier, it is too easy for the dominance of professional risk discourses to lead narrow approaches to problem resolution.

The 'signs of safety' approach is premised on a strengths-based philosophy that emphasises the importance of establishing clients' views about their lives, and respecting clients as 'people worth doing business with' (Turnell & Edwards, 1999, p.42). It avoids pathological or psychodynamic analyses of personal problems. The signs of safety approach rest on the general premise of social constructionist therapeutic approaches that contend that if language constructs reality, rather than reality being an objectively existing entity, then the ways people interpret and describe social experience are powerful: they create our lived experience, and can result in future behavioural change (Berg & Kelly, 2000; Milner & O'Byrne, 2002). The signs of safety approach utlilises social constructionist concepts by showing how particular discourses are drawn on to construct harm and safety, and this invites a deeper thinking about risk. In this way our risk thinking can be opened up.

Putting risk thinking into practice

Removing perpetrators or having men with convictions for terrorist offences monitored might leave the agency management feeling better about the reported risks. However, as Featherstone et al. (2014) argue, we need to work proactively with risk because we all live relational lives. The argument here is that terrorist offenders are also fathers, husbands, and partners, and social workers need to think about the sets of relationships we may be asking them to exit. Thinking relationally will help social workers theorise the sets of meanings people bring to relationships, like fathering, and hold multiple discourses in mind. Fathers are easily constructed in rather narrow terms, dominated by concerns of being 'violent' or 'dangerous' (Featherstone, 2013). Violent partners often stay in relationships despite the advice and warnings from social workers. Moreover, moving on to another relationship is not uncommon (Featherstone et al., 2014). The risk problem is not resolved, but now involves a new family. Managing risk in resolution terms always fails.

Working actively with risk in these cases, that is talking about risk in its many variances and possibilities opens up a more collaborative working relationship where talking about what people might want to do is welcomed. I do this in my practice, in case supervision meetings, and with my peers – it helps me to get confident in having tough conversations about what needs to happen. It needs to be encouraged as a way of thinking about risk differently – that is risk as a co-producer of options followed by clear decision making. Families will understand the decisions we reach when we work this way. Risk is a useful discourse for us to reflect on, because it shows us how power operates in and through our work. We can invite a human rights-based approach in to our work, at the same time as we use our authority skilfully and respectfully. This will not feel easy or comfortable for everyone.

Some social workers will still prefer a more traditional (professionalised) approach to risk work – that is through the counting up of risk factors and defining how much risk is too much. It may feel safer for them. Some commentators argue that scandal politics about child abuse tragedies get played out in the media and thus fuel arguments for risk averse and risk factor science approaches (Ferguson, 2004). The point here is that we need to understand the way we approach risk work. The practice methodologies we use need to be understood as discursively operating and influential (Forrester, 2010). Acting unethically or misusing statutory power easily follows if we can't or won't stop to reflect on 'what informs the way I think about risk and need, safety and strengths'. A rights-based practice continuum is one helpful tool to invite people to see where they are along the child rescue orientation toward a family focussed one (Connolly & Ward, 2008).

Human rights-based approaches to practice mean resisting the construction of a child as an individual in need of protection or rescue.

The child's right to be free of harm and abuse and remain part of family life is a tension for social workers to negotiate daily (Connolly & Ward, 2008). But the way risk is communicated through public and political rhetoric continues to fuel a media hungry for scandal and blame if things go wrong. A child abuse tragedy with an identified social worker and manager to blame makes compelling headlines. In England, the blame culture does not wait for criminal trials of parents or caregivers – the people who actually injure and kill children. Rather, the social worker is a ready-made folk devil every time a practice tragedy unfolds (Jones, 2014). Risk discourses here are simple and totalising: powerful. Social workers are finding themselves caught up in a new moral panic concerning childhood radicalisation risk; a panic that serves to silence their moral and value based position of helping families.

Through different contexts of time, social work tends to construct risk simply in terms of a moral panic definition of 'the enemy' – it is the same approach, just different names and faces. Understanding risk as a discursive construction that can be classified through the practises of work is therefore important – thus the professional practises of social work need to be critically engaged with. Beck (1992) made the case for this thinking some time ago:

> [Risks] only exist in terms of the (scientific or anti-scientific) knowledge about them. They can be changed, magnified, dramatized or minimized within knowledge, and to that extent they are particularly open to social definition and construction. (Beck, 1992, p. 23)

I am talking about deepening our understanding about risk – to reflect on how best to work with risk, how to communicate about it in ways that help us to open up thinking and then in turn opening up our practice to involve and include families. To date, social work has lacked a critical framework where our risk work can be prised open. The strengths-based approaches to practice discussed earlier do offer a way forward. This requires a considerable level of skill by the practitioner and manager. They must also believe that this is *the right thing to do*.

Conclusion

The radicalisation risk for children as a particularly acute issue has been raised to illustrate risk communication as a discursive process. In order that social workers work with risk *as* risk, we need to understand this, so that we make best use of the available risk discourses – that is, the sets of meanings that operate beneath and around the language of risk. To halt risk anxiety influencing decision making we need to marshal our risk arguments about rights, power and discourse. A high degree of practitioner skill is needed to do this. The sociological imagination helps me to connect family life with war, decision making to powerlessness and invites me to understand how power operates in and through my practice.

Risk discourses are affected by our 'risk thinking' and dominant policy discourses, however, the use of safety oriented approaches, such as the signs of safety approach, can offer a significant way forward for practitioners to help empower families and bring about change that means a child's life is improved. Social workers need to think critically, and be encouraged to open up their risk thinking. Practice settings need to promote families and children as partners in the work, in the true sense of the word. As such, the moral endeavour of social work is too important to leave to discursive chance.

References

Aradau, Claudia and van Munster, Rens (2007). Governing terrorism through risk: Taking precautions, (un)knowing the future. *European Journal of International Relations*, 13(1), 89–115.

Barnett, G. D. and Mann, R. E. (2011). Good lives and risk assessment: Collaborative approaches to risk assessment with sexual offenders. In: H. Kemshall and B. Wilkinson (eds) *Good Practice in Assessing Risk: Current Knowledge, Issues and Approaches*. London: JKP.

Beck, U. (1992). *Risikogesellschaft: Risk Society: Towards a new Modernity*. London: Sage.

Beck, U. (2002). The terrorist threat world risk society revisited. *Theory, Culture & Society*, August 2002 19(4), 39–55.

Berg, I. K. and Kelly, S. (2000). *Building Solutions in Child Protective Services*. London: W.W. Norton and Company.

Berger, P. L., and Luckmann, T. (1967). *The Social Construction of Reality: A Treatise in the Sociology of Knowledge*. London: Penguin.

Broadhurst, K., Hall, C., Wastell, D., White, S. and Pithpouse, A. (2010). Risk, instrumentalism and the humane project in social work: Identifying the *informal* logics of risk management in children's statutory services. *British Journal of Social Work*, 40(4), 1046–1064.

Connolly, M. (2004). *Child and Family Welfare: Statutory Responses to Children at Risk*. Christchurch: Te Awatea Press.

Connolly, M. and Ward, T. (2008). *Morals, Rights and Practice in the Human Services*. Jessica Kingslea: London.

Coppock, V. and McGovern, M. (2014), 'Dangerous Minds'? deconstructing counter-terrorism discourse, radicalisation and the 'Psychological Vulnerability' of Muslim children and young people in Britain. *Children & Society*, 28, 242–256

Culpitt, I. (1998). *Social Policy and Risk*. London: Sage.
Dodd, V. (2014). Beware drfit to police state, chief constable warns. *Guardian Newspaper*, 1.
Featherstone, B., Broadhurst, K., and Holt, K. (2012). Thinking systemically – Thinking politically: Building strong partnerships with children and families in the context of rising inequality. *British Journal of Social Work*, 42(4), 618–633. doi: 10.1093/bjsw/bcr080
Featherstone, B (2013). Working with fathers: Risk or resource? In Jane Ribbens McCarthy, Carol-Ann Hooper, and Val Gillies (eds) *Family Troubles?:Exploring Changes and Challenges in the Family lives of Children and Young People*. Policy Press: Bristol.
Featherstone, B., White S. and Morris. K. (2014). *Re-Imagining Child Protection: Towards Humane Social work with Families*. Policy Press: Bristol.
Ferguson, H. (2004). *Protecting Children in time: Child Abuse, Child Protection and the Consequences of Modernity*. Houndmills, England: Palgrave Macmillan.
Firkins, A. and Candlin, C. N. (2011). 'She is not coping': Risk assessment and claims of deficit in social work. In C. N. Candlin and J. Crichton (eds) *Discourses of Deficit* (pp. 81–98). Basingstoke: Palgrave Macmillan.
Forrester, D., (2010). Playing with fire or rediscovering fire? The perils and potential for evidence based practice in child and family social work. In P. Ayre and M. Preston-Shoot (eds) *Children Services at the Cross Road: A Critical Evaluation of Contemporary Policy for Practice*. Plymouth: Russell House.
Hall, C. J. and Slembrouck, S. (2011). Categorisations of child 'in need' and child 'in need of protection' and implications for the formulation of deficit parenting. In C. N. Candlin and J. Crichton (eds) *Discourses of Deficit* (pp. 63–80). Basingstoke: Palgrave Macmillan.
Horlick-Jones, T., (2005). On risk work: Professional discourse and everyday action. *Health Risk and Society*, 7(3), 293–307.
Jones, R. (2014). *The Story of Baby P: Setting the Record Straight*. Policy Press: Bristol.
Kemshall, H. (2010). Risk rationalities in contemporary social work practice. *British Journal of Social Work*, 40, 1247–1262.
Lupton, D. (1999). *Risk*. London: Routledge.
Mills, C. Wright. (1959). *The Sociological Imagination*. New York: Oxford University Press.
Milner, J. and O'Byrne, P. (2002). *Brief counselling: Narratives and Solutions*. Hampshire: Palgrave Macmillan.
Mohammed, J. and Siddiqui, A. (2013). *The Prevent Strategy: A Cradle to Grave Police-State*. Cage: London.
Munro, E. (2010). Conflating risks: Implications for accurate risk prediction in child welfare services. *Health, Risk and Society*, 12(2), 119–130.
Ribbens McCarthy, J. (2013). What is at stake in family troubles? Existential issue and value frameworks. In Jane Ribbens McCarthy, Carol-Ann Hooper, and Val Gillies (eds) *Family Troubles?:Exploring Changes and Challenges in the Family lives of Children and Young People*. Policy Press: Bristol.
Rose, Nikolas. (1999). *Powers of Freedom*. Cambridge: University Press.
Saleebey, D. (1997). *The Strengths Perspective in Social work Practice* (2nd ed.). New York: Longman.
Stanley, T. (2013). 'Our tariff will rise': Risk probabilities and child protection. *Health Risk and Society*, 15(1), 67–83.

Stanley, T. and R. Mills (2014). "Signs of Safety" practice at the health and children's social care interface. *Practice*, 26(1), 23–36.

Stanley, T. and M. Russell (2014). The principal child and family social worker: A Munro recommendation in action. *Practice*, 26(2), 81–96.

Tullock, J. and D. Lupton (2003). *Risk and Everyday Life*. Sage: London.

Turnell, A. and S. Edwards (1999). *Signs of Safety: A Solution and Safety Oriented Approach to Child Protection Casework*. New York: Norton.

Ward, T. and R. Mann (2004). Good lives and rehabilitation of sex offenders: A positive approach to treatment. In: A. Linley and S. Joseph (eds) *Positive Psychology in Practice*. New York: John Wiley.

Webb, S. (2006). *Social Work in a Risk Society: Social and Political Perspectives*. Basingstoke: Palgrave Macmillan.

Weick, Anne. (1999). Guilty knowledge. *Families in Society*, 80(4),327–332. doi: 10.1606/1044-3894.1212.

Zinn, J. (2008). Heading into the unknown: Everyday strategies for managing risk and uncertainty. *Health, Risk & Society*, 10(5), 439–450.

Part IV
Communicating Risk in Environmental Management and Biosecurity

10
Interpretive Environmental Risk Research: Affect, Discourses and Change

Karen L. Henwood and Nick Pidgeon

Introduction

Environmental risk perception and communication research seeks to answer questions about the acceptability and governance of the social and material impacts of environmental changes within late-modernity, as well as the future sustainability of established and newly emerging global socio-technical risks and proposed solutions. It subscribes to the importance of understanding public risk perceptions, the dynamics of everyday experiences of risk including people's psychological investments and meaning-making, communication and dialogue about risk issues and the questions of public value that participation and engagement highlights, and the diverse interpretations that people place on aspects of both risk and uncertainty. Research into these topics helps us to explain the implications of the pace of environmental, socio-technological, and socio-cultural change for people as they live out their lives. It is also a means of elucidating how intractable local environmental problems are often bound up with ambiguous global risk issues as manifest in topics such as chemical pollution, nuclear power, climate change, geoengineering, or low carbon/energy transitions, all topics with high contemporary relevance to science policy, society, and individuals. In this chapter we explain, and exemplify with case studies from our work, the rationale and purpose of interpretive risk research, which we present as part of developments within the socio-cultural and environmental risk field as it articulates with studies of science in society and global change. Our main argument is that conducting in-depth analyses of discourses, with a focus on affect and the dynamics of change, can aid understanding of epistemological and ontological complexities that are endemic when investigating environmental risk issues.

Discourse, interpretation and environmental risk: Theory and practice

While the concepts of risk can be traced back to the 16th Century its formal treatment is more recent, and it was not until the 1960s that the professional practice of risk analysis became institutionalised. Research into the social sciences of environmental risk perception arose not long after, founded principally upon elements of decision theory, cognitive and social psychology, all disciplines with a methodological orientation towards quantitative and experimental methods (see e.g. Pidgeon et al., 1992). A core problematic of this quantitative work, which has come to dominate the academic field in both the United States and large sections of Europe, is to understand the social and psychological reasons why some risks deemed 'small' by expert analysis can become a significant focus in people's perceptions and community controversy, while others deemed highly significant through expert risk assessment could be seen as unimportant by ordinary people. With its broad cross disciplinary base, and as a field of empirical inquiry, risk research has always encompassed multiple conceptual approaches and, increasingly, studies drawing upon the theory and methods of interpretive and discursive social science (Pidgeon, Simmons, and Henwood, 2006). Hence, the 1980s saw the development of social science approaches to environmental risk perceptions problematising *both* expert and public conceptions of 'risk'. The cultural theory of risk (Douglas and Wildavsky, 1982) and the social amplification of risk (Pidgeon, Kasperson, and Slovic, 2003) both highlight the ways in which our favoured societal and institutional arrangements, and the values attached to these, shape the dynamics and perceptions of what comes to be considered risky or not, and caution us that the social and institutional context matter (greatly) in matters of risk assessment and perceptions. This has implications not only for how we conceptualise perceptions in both sociological and psychological terms, but also how expert assessments of risk are inevitably contingent upon the cultural assumptions and social judgments that risk analysts adopt: environmental risk, in both its perceived and assessed guises, has to be viewed as a social construction (Pidgeon et al., 1992).

Coming somewhat later to the debate about how to articulate an understanding of risk subjectivities as embedded with wider societal issues, Beck (1992) and Giddens (1991) have argued that our attempts to conceive and deal with environmental and technological catastrophes, and fears thereof, are bound up with a wider set of structural changes in late-modern society. Growing individualisation of responsibility for risk, alongside the decline of traditionally taken for granted social identities, has generated a new politics with risk at its core and a new need for individuals to become 'reflexive risk subjects'. Related theorists, working within the governmentality tradition of risk, have likewise sought to critique the widespread uptake of techniques of environmental risk assessment and management as technologies of governance embedded within naive narratives of ecological modernisation – the idea that the existing system of global capital can simply be reordered in ways which will afford economic growth simultaneously with protection of the environment (e.g. Oels, 2005; Pidgeon and Butler, 2009). All of these sociologically oriented approaches emphasise how societal and institutional discourses shape our knowledge about, and behaviour towards, risks.

Alongside attempts to conceive of risk in terms of sociological metanarratives, a more anthropologically inspired, locally based and discursively grounded empirical approach to studying environmental risk has developed. Here the emphasis is upon the importance of studying people's understandings of environmental and technological risks in their locality – a chemical plant, waste landfill site, or a nuclear facility. Irwin argues that 'environmental problems do not sit apart from everyday life (as if they were discrete from other issues and concerns) but instead are accommodated within (and help shape) the social construction of local reality' (2001: 175). Most notably socio-cultural, geographical, and political characteristics shape perceptions of locally situated risks, raising questions about what matters to people, including their immediate and long-term concerns, and the values and localised identities that they take up (Henwood and Pidgeon, 2013). A range of studies has illustrated how people construct, perceive, and reflect on their experiences of living under threat of socio-technical and environmental hazards (e.g. Edelstein, 1988; Irwin, Simmons and Walker, 1999), reflecting the emergence of a distinctive *interpretive perspective* within socio-cultural risk studies more generally. This perspective stresses the need to pay attention to the framing and construction of the 'risk object' (Hilgartner, 1992) in terms of people's everyday lives and social worlds (Henwood, Pidgeon, Sarre et al., 2008) rather than in expert knowledge(s) or categories derived from formalised environmental risk assessments. Such a perspective is typically accompanied by particular methodological commitments to more qualitative and *in situ* research designs, paying close attention to the ways risk is constructed in people's own biographical narratives or their discourses elicited in interaction with others (e.g. in focus group conversations).

Interpretive risk studies place emphasis on the need for detailed theoretical and methodological work, most especially for its affordances in examining the diverse forms, modalities, and meanings of data in exploring ambiguities, uncertainties, possibilities and contestations in truth making, and as a means of understanding epistemological tensions already at the heart of the inquiries that many risk researchers undertake. An interpretive approach to risk illuminates a whole range of social and contextual considerations through which people come to comprehend and respond to what they believe is hazardous, or alternatively not hazardous, in the world. On the one hand, the potential dangers, threats, and harms that are characterised as 'risks' are often real in their material consequences (deaths, injury, ecological damage) but, on the other hand, our knowledge of them and the future is never directly given. Risk is socially situated, culturally constructed, and complexly mediated – either through expert scientific representations, other institutions (e.g. the media), and/or lay interpretive/meaning-making practices. In this latter sense, and because discourse is the primary medium of meaning-making practice, our *knowledge(s) of risk* inevitably arrive within discourses and, for this reason, the study of risk knowledge becomes, in effect, the study of risk discourse (Henwood et al., 2010). Explicitly taking up a discursive perspective on risk can, moreover, provide a means of reflexively conceptualising and operationalising risk and ways of following it through as an analytic practice. These methodological commitments inevitably bring to the surface the need to analyse and account for a range of discourses in both data analysis and interpretation. In interpretive and discursive research, then, risk arises within (and, critically, has to be understood in terms of) a series of socially derived discourses, where the material fact of causing or doing harm is one issue upon which others invariably become layered – questions of values, morality, responsibility, blame, governance and trust. These latter considerations also connect questions of environmental risk to the cultural and psycho-social dynamics of risk awareness, identity, and subjectivity (see Henwood and Pidgeon, 2013).

Affect, discourses and nuclear risk

The issue of nuclear technology has been, since the mid-1970s, a paradigm case within environmental risk perceptions research as well as accounts of the Risk Society. Research on perceptions of nuclear power highlights long-standing public concerns in many nations. Major accidents including the 1957 Windscale fire in England, and those at Three Mile Island in 1979 and Chernobyl in 1986, coupled with concerns about waste disposal, served to reinforce such perceptions. Rosa and Freudenburg (1993) report that opposition to building more nuclear power plants in the United States increased from 20% in the mid-1970s to more than 60% in the early 1980s. Psychometric surveys with national samples highlight a range of associations with the term 'nuclear power' and radioactive waste – anxiety and dread, perceived scientific uncertainty, lack of control, a potential for catastrophic outcomes, deep distrust of risk managers and a historic association with atomic weaponry – that underlay perceptions of risk and its acceptability (Slovic, 2000).

Such national studies are, however, a poor proxy for understanding the views of community members at existing nuclear sites – not only for power generation but military facilities, fuel reprocessing, and waste disposal and storage. Some sites such as at Sellafield in the North West of the England have come to host a complex combination of activities. One might expect, based upon the findings of national surveys, that local residents would display extensive and intense symptoms of anxiety. Likewise, Risk Society theory would point to responsibility for risk as both externalised to socio-political institutions and internalised through individualisation, raising new existential anxieties for individuals (Wilkinson, 2001). However, survey work shows how close proximity is often associated with higher levels of support than in national samples (Greenberg, 2009; Venables et al., 2012). One explanation is that acceptance, or at least refusal to overtly criticise, by those living close to an existing facility stems from perceived economic benefits, particularly where a host community is otherwise economically marginalised (Blowers and Leroy, 1994; Wynne et al., 2007 [1993]). By contrast, a discursive approach to nuclear risks highlights how the question of perception needs to be thought of in more nuanced and psycho-social terms. Interrogating the unsayable through close attention to language and discourse can uncover presumed and implied meanings and affectively charged states such as anxiety, frustration and anger.

Studies in close proximity to the nuclear reprocessing plants at Sellafield and at Cap la Hague in France indicate that, even where support and acceptance is overtly expressed this is often highly qualified, with a degree of underlying unease always present in the discourses of local people and nuclear workers (Macgill, 1987; Zonabend, 1993). In her analysis of discourses at Cap la Hague, Zonabend (1993: 124) suggests that anxiety is 'furtive', 'muted' and repressed but always under the surface of people's discourses and 'is not difficult to detect when you are talking to the people of la Hague'. She asserts that 'real anxieties', which she also refers to as latent or furtive, are easy to find once one looks for them. Although denied or dodged by residents they are always under the surface of local discourse and evident in the whispered asides, jokes, and mutterings of the workers and residents (1993: 6, 124).

Drawing in part upon Zonabend's work, our own research has involved conducting narrative interviews with residents living in proximity to nuclear power station sites in England at Oldbury-on Severn in Gloucestershire and Bradwell-on-Sea in Essex. To ensure that the significance nuclear risk has for our interviewees' lives was neither over or understated, we incorporated lessons from literature on 'risk biographies' (Tulloch and Lupton, 2003), environmental value elicitation (Satterfield, 2001, 2002), and narrative methods (Elliott, 2005). We anticipated that, using a narrative approach, interviewees' judgments, decision-making processes, values, and subjective preferences would be rendered more visible by being embedded in meaningful, contextually and morally rich, value laden, and affectively charged stories about risk (see Satterfield, 2002; Henwood, 2008).

What is clear from our analysis is that nuclear power is not seen by our participants as an overwhelmingly 'dreaded' technology. Nor does contemplation of nuclear energy generate an ongoing 'risk awareness' accompanied by individualised uncertainty management (cf. Beck, 1992). Rather, the *apparent* tolerance found in our interviews, a sense of acceptance of the nuclear stations as ordinary and unremarkable parts of the everyday and their locales, co-exist with a more complex set of contradictions (Parkhill et al., 2010), and this account of living with nuclear risk highlights a number of informative points. First, biographical experiences, dynamically unfolding through space and time, can be interrupted by risk events (mediated and direct, real, and symbolic) to disrupt the usually taken for granted normality surrounding a power station's presence in a particular locality; in effect, an ebb and flow of attenuation and amplification of risk perception (Pidgeon, Kasperson, and Slovic, 2003; also Bickerstaff and Simmons, 2009). Interviewees voiced a number of responses to uncertainty and anxiety. Some 'bracketed' it off by refusing to think about it (Wynne et al., 2007 [1993]). For others, a threat was eventually deemed as distant due to the passing of time, or through other risk issues taking precedence in their biography. Third, whilst threat and anxiety might remain, this was an accepted state; that is, some of our interviewees became reconciled to its existence and simply *moved on*. Yet another possibility, subject to in-depth analysis in our own work, is that negative emotions become revealed, resisted, and at times relieved through humour (Irwin, Simmons, and Walker, 1999; Parkhill et al., 2011). All of this suggests that living with nuclear power involves a degree of fragility and contingency.

The intermittent reconstruction by our participants of the power station as an extraordinary, threatening presence is congruent with Masco's discussion of the 'nuclear uncanny' (2006). According to Masco, this refers to 'spaces of perception'; the moments of 'dislocation and anxiety' brought about by the intrusion of thoughts into daily life about 'invisible, life-threatening forces' (2006: 28). Not only is nuclear technology capable of causing un-calculable harm and death, it also symbolises world destroying capability, and with this significant political power. Masco argues that these qualities set the category of nuclear aside from all other developments in science and technology. In this respect, we too found powerful moments of subjectivity where interviewees regarded the nuclear power station as a unique type of development, or where an incident of some sort led them to distinguish it from other forms of hazardous socio-technical development within or near their area (Parkhill et al., 2010). Alongside the dominant narrative of ordinariness, they can at times be preoccupied with the notion that nuclear power is unique, extraordinary, or uncanny, because it *is* nuclear, through both mediated and direct experiences of nuclear or non-nuclear events. Whilst the nuclear uncanny may have powerful psycho-social effects, influencing perceptions of the localised risk that the power station may represent, we have also shown that its sources are far more wide-ranging than the collective memory of the dawn of the atomic age. What constitutes a possible threat is open to renegotiation dependent on cultural, political, geographical, and biographical influences, with our analysis therefore broadening that of Masco, in that the moments of dislocation prompted by nuclear technologies of the taken for granted and familiar in relation to locale and home are not restricted to associations with nuclear weapons and weapons sites alone.

Gender, risk and discourse

In our second exemplar we draw on theoretical and empirical insights from work on interactional risk perceptions and gendered identity dynamics. Here we consider why it is important to conduct inquiries in ways that avoid making oversimplified assumptions about identity and difference. At the outset of our gender and risk study, we simply wished to explore a conundrum that remained largely unanswered in the risk field for over 40 years. Although socio-demographic variables such as age and class do not consistently predict either environmental risk perceptions or their acceptability, a clear exception is gender. A long-standing statistical finding is that male respondents in surveys tend to show somewhat lower levels of concern than women when asked about their perceptions of a range of environmental and sociotechnical hazards, particularly hazards with consequences involving local environmental contamination (Davidson and Freudenburg, 1996). Subsequently research suggested that statistical gender 'differences' may be due in part to a sub-group of men holding much lower risk perceptions than other demographic groups, the 'white male effect' (Flynn, Slovic, and Mertz, 1994). This observation suggests that such gender effects can be related to more general discussion of identities and societal vulnerability (Satterfield, Mertz and Slovic, 2004) together with questions about how to research diverse relationships between gender, risk, and marginality (Henwood, 2008). Insights from within contemporary gender studies help us to understand why gender still exerts such a powerful influence within society, set alongside contemporary thinking in gender studies which tends to eschew essentialist or fixed accounts of gender 'differences'. Rather, gender researchers ask what empirical findings about gender difference might mean, how they relate to power, and competing discourses about men and women's positions in society, and how they relate to people's ongoing life-projects? Such research seeks to arrive at more contextual and in-depth psychological views of masculinity and femininity as cultural binaries, while recognising that such binaries can operate in ways that hide as much as they highlight (Henwood, Gill, and Mclean, 2002).

Reflecting some of these concerns Gustafson (1998) points out that the traditional 'gender differences' explanations of the gender-risk effect are limited by almost exclusive use of evidence drawn from quantitative surveys. She recommends a more interpretive, qualitative approach to analysing how men and women construct their understandings of risks. Our own project exploring these issues started with Gustafson's insight, as part of the UK Economic and Social Research Council's 'Science in Society' programme of research. Returning to a set of reconvened focus group transcripts that had been analysed for other purposes (see Bickerstaff, Simmons, and Pidgeon, 2008) we analysed aspects of cultural and identity work present in them around risk and gender. These data comprised discussions in mostly mixed gender groups about environmental and technical risk issues: including climate change, radioactive waste and nuclear power, genetically modified crops and food, and radiation from mobile telephones and masts. Although these topics would be expected to produce the gender-risk effect in surveys, the group protocols were not planned with this issue in mind.

Initial analytical work involved development of a conceptual framework for studying 'effects made by gender' (Henwood, Parkhill, and Pidgeon, 2008), drawing upon thinking in both the risk analysis and science and technology studies (STS) fields. We benefited from thinking within qualitative social psychology, especially where there is focus on everyday meaning-making as a site for studying gendering processes and the making of identity (Willott and Griffin, 1999). In many fields of social science, putting these broad sensibilities into research practice has involved considering the role of gender discourses and subject/identity positioning in the ways men and women understand and interpret social and scientific issues in everyday life. Within discursive social psychology a key concern is to avoid essentialising (i.e. fixing) accounts of gender 'differences' (Henwood, Griffin, and Phoenix, 1998). In the course of studying discourse dynamics, people are construed as historically situated subjects whose ways of invoking and reworking meanings complicate the social, cultural, and intergenerational transmission of gender and other social identities. Discourse research also focuses upon how people constantly create and claim subject/identity positions for themselves and others during social interaction; not only do these positions provide sites and means of self-perception and reflection, they can be accepted or disputed by others in interaction. Thus how identity positions are sustained or discarded becomes an emergent feature of social interchange. Such qualitative social psychological work is concerned with the 'most worldly of all interpretive practices' (Denzin, 1997) posing questions about men and women's positions in society such as 'How is it possible to account for the pace (often slow) of cultural and psychological change?' For Lohan (2000), working in the field of STS such insights have raised questions about how – in breaking down binary oppositions – spaces can be opened up to create 'tolerance' of gender difference as a form of anti-essentialist practice. Lohan's suggestion that researchers attend to the workings of socio-political understandings as a means of analysing how culture and power operate to hold differences in place influences our own approach to interpreting patterns in empirical data.

Alongside theory from discursive social psychology, we drew upon Faulkner (2000) who analyses technology-gender relationships as social and cultural constructions as shaped by changing historical circumstances and socio-political processes, and which function as regulatory norms of discourse and conduct. Faulkner notes the durability of modernist cultural associations between masculinity and technology – and especially as part of the conviction that social progress is attributable to technological development and its economic role within industrial capitalism. Her approach sees certain forms of technology as culturally gendered in perception, and, we argue here, this also pertains to the gender-risk effect as originally established in relation to large-scale technological or global environmental hazards. We also draw on the philosophical notion of *epistemic subjectivity* (Scheman, 1993). This refers to taking on an identity as an authority on a subject in discourse, or conversely hesitating over expressing personal authority or pleasure in the activity of knowing. Scheman proposes that, while legitimately contested within philosophy, recognisably masculine epistemic frameworks remain culturally placed as the 'best positions to know' (1993: 4). A relevant example in the risk sphere is the tendency for groups exerting powerful influence within society to place epistemic authority in technical-rational approaches to understanding and managing risk problems, and to attach lesser value to worldviews stressing broader social and ecological considerations (e.g. the responsibility/care principle). Following Scheman's argument, different forms of knowing may lead people to construct risk problems in ways that are recognisably gendered.

The account we have developed of the gendered nature of risk perceptions has enabled us to throw light on the epistemic and ethical dimensions of contemporary debates across different risk cases (Henwood, Pidgeon and Parkhill, 2015) as well as in more detail about nuclear energy, a matter of particular policy concern post-Fukushima (Henwood and Pidgeon, 2015). As a substantive finding we have reported the strikingly gender differentiated pattern where epistemic positions taken up reproduced gender stereotypes, especially early in the focus groups. That is to say, in the focus groups we have observed men taking up positions as authorities and knowers – displaying a particular 'technocentric' epistemic subjectivity, in Scheman's terms. Women by contrast expressed more hesitancy and uncertainty and (self-) assumed, initially at least, a lack of knowledge. At the same time, because of the elements of discourse theory foregrounded by our theoretical synthesis, our empirical work has focused on the dynamic role played by technocentric and care discourses in inter-personal interactions where people debate environmental risks.

One distinctive use of discourse theory is as a means of rejuvenating Gilligan's (1982) conceptualisation of gendered moral voices (of justice and care) by de-essentialising it. Examination of such gendered moral voices can amount to studying the cultural encoding and communication of conventional dichotomies between women and men in perceptions of risk and its acceptability. By drawing on available care and technocratic discourses, men and women can display a highly gendered understanding of nuclear power in their interactions either by claiming the moral inferiority or superiority of a technology in relation to caring for the environment (e.g. though producing long-term contamination from radioactive waste within one framing, or alternatively yielding electricity with relatively low carbon emissions within another), or by arguing for or against the economic necessity or efficacy of technical solutions to risk problems. Alternatively, they can engage with a pro-technological or care viewpoint in a non-essentialising way. On some occasions, we found that both women and men were engaged in affective self-expression and modulation, or self-other regulation, for example by deriding, dismissing, or reacting with incredulity to another's remark (Henwood and Pidgeon, 2015). This was particularly the case when remarks were interpreted as examples of male hubris or of women asserting an unwarranted form of moral superiority over the men in the group over how to act with appropriate care. On these occasions, far from merely individual emoting, these highly affecting forms of speech act could be an indirect but no less potent means of putting one's gendered knowledge to work, and could be an effective form of communication by serving to dispute or destabilise what might otherwise be taken for granted as its self-evident claim to a gendered truth.

Implications for practice and suggestions for further research

Deployment of an interpretive and discursive social science approach in this chapter involved a thorough, productive engagement with the core tradition of environmental and technical risk perceptions research, especially its articulation with socio-cultural risk studies more generally. Knowledge of risk perceptions is enhanced when placed in an expanded conceptual field. There are benefits from methodological inventiveness and knowledge practices associated with different modes of empirical inquiry. The result is a strengthening of social science voices across key areas where research and practice interlink.

It is possible to use fundamental knowledge from risk perception studies in formal risk modelling and as an aid to decision making, but our approach in this chapter contrasts with this. We have pursued our research interests not in the sphere of formal risk assessment but as a means of developing research and practice in risk and society studies more broadly, and by showing how such an approach can inform the conduct of data analysis in our own empirical projects. Our own data analysis has pointed to the significance of the discursive realisation of usually hidden anxieties (in the nuclear risk case study) and shown how to remove barriers to understanding the gendering of environmental and socio-technical risk discourses and identity dynamics (in the gender and risk case study). In this concluding section, we invoke two specific reference points – the ESRC Science in Society programme (2001–2009) and the Global Environment Design project (2011–2012) – to contextualise our pointers to new directions.

Science in Society's interest was in the development of rapidly changing relations between-science and wider society. It had a specific aim of understanding how society could develop knowledge and practice for dealing with technological risks and a specific remit to consider the role of science and technology in daily life, as experienced by affected, diverse publics. In effect, the programme considered risks from technical development as an *intervention* in lived relations within society and an issue in people's lives in late-modernity, an intervention that required inventiveness on the part of science and society to provide an appropriately contextualised and dynamic response and one that is linked (as in the risk perceptions field) to a realisation of risk's uncertainties, ambiguities, contingencies, and actual and possible harms. The gender and risk case study that we conducted as part of the Science in Society programme contributed to understanding how people experience, and make sense of, technology in their lives: its particular focus on how risk manifests itself in discourses (along with its psycho-social affects) points to some specific practice implications and suggestions for further research.

Our findings affirmed the need to acknowledge that gender, as a culturally meaningful code, category, or assumption, has currency among the public, and also within the research and broader stakeholder/expert community where it shapes the production of scientific and environmental knowledge(s) and their communication (see ESRC, 2009: 8). Research into the practices of risk communication already features culturally embedded risk perceptions and lay and expert knowledges since 'there are always interpretations involved' (Boholm and Corvellec, 2014), but the evidence we have cited in this chapter points to complex and dynamic influences potentially exerted by 'effects made by gender'. Developments in the field of psycho-social research (Taylor and McEvoy, 2015) could be mined to develop interest in questions about risk, moralities, and situated affect as they connect with the wider cultural dynamics and gendering of subjectivity. Indeed, it may be necessary to do so, if gender is to enhance scientific, technological, and risk discourse, and not simply reject those discourses for promoting gender insensitivity (Henwood and Pidgeon, 2015; also Buck, Gammon, and Preston, 2013). These issues need to be properly accommodated within the risk and communication field. In a similar vein, and for the study of local experiences of nuclear risk, while taking a discursive approach to data analysis and interpretation has engaged with psychosocial aspects of the perception of such risks, there is more to be said about knowing how to interpret under the surface of risk discourse and its affective dynamics.

We wish to finally draw attention to important efforts that are being made to develop a research agenda for a social sciences of global environmental change (see *Transformative Cornerstones of Social Sciences Research for Global Change*, Hackmann and St. Clair, 2012). Our chapter's focus on developing interpretive research into discourse and affect could be important as part of its efforts to 'find new visions for change', or 'practising research for change', and to address 'fundamental questions about the ways and consequences of reframing global change – particularly climate change – as a deep systemic problem in the global environmental arena'. According to *Transformative Cornerstones*, a deep systems framing can bring about new and different questions and narratives about socially desirable change, associated lifestyles, and alternative socio-economic, technological, and political systems. At the same time it is very circumspect about whether alternative visions of the future can be arrived at through processes of social engineering or public-science engagement exercises. From our own inquiries, we would add that questions are likely to arise about the role of co-existing and conflicting discourses (e.g. technocratic and care), and tensions and relations between affect in knowledge, in explaining the dynamics of change. If 'Interpretation and Subjective Sense Making' is to be consolidated as one of the six proposed transformative cornerstones to understand processes generating sustainable responses to the problem of global change (p13), the approach to be taken to analysis is ripe for further examination and development.

Acknowledgement

This chapter was facilitated by grants from the UK Economic and Social Research Council (RES-160-25-0046 and RES-33625-001) and the US National Science Foundation to the Centre for Nanotechnology in

Society at University of California Santa Barbara (cooperative agreement SES 0938099). We wish to thank Terre Satterfield, Barbara Harthorn, Karen Parkhill and Dan Venables.

References

Beck, U. (1992). *Risk society: Towards a new Modernity*. London: Sage.
Bickerstaff, K. & Simmons, P. (2009). Absencing/presencing risk: Rethinking proximity and the experience of living with major technological hazards. *Geoforum*, 40(5), 864–872.
Bickerstaff, K., Simmons, P. & Pidgeon, N.F. (2008). Constructing responsibility for risk(s): Negotiating citizen-state relationships. *Environment and Planning A*, 40, 1312–1330.
Blowers, A. & Leroy, P. (1994). Power, politics and environmental inequality: A theoretical and empirical analysis of the process of peripheralisation. *Environmental Politics*, 3, 197–228.
Boholm, Å. & Corvellec, H. (2014). A relational theory of risk: Lessons for risk communication. In J. Arvai & L. Rivers III. *Effective Risk Communication*. (pp. 1–22) London: Earthscan.
Buck, H.J., Gammon, A. & Preston, C.J. (2014). Gender and geoengineering. *Hypatia*, DOI: 10.1111/hypa.1208
Davidson, D. & Freudenberg, W.R. (1996). Gender and environmental concerns: A review and analysis of available research. *Environment and Behaviour*, 28(3), 302–339.
Denzin, N. (1997). *Interpretive Ethnography: Ethnographic Practices for the C21*. London: Sage.
Douglas, M. & Wildavsky, A. (1982). *Risk and Culture: An Essay in the Selection of Technological and Environmental Dangers*. Berkley: University of California Press.
Edelstein, M.R. (1988). *Contaminated Communities: The Social and Psychological Impacts of Residential Toxic Waste Exposure*. Boulder: Westview Press.
Elliott, J. (2005). *Using Narrative in Social research: Qualitative and Quantitative Approaches*. London: Sage.
ESRC (2009). *Science and Gender, Ethnicity and Lifecycle: ESRC Science in Society Programme Report*. www.esrc.ac.uk/_images/science_and_gender_tcm8-13538.pdf (Accessed 29/4/2015)
Faulkner, W. (2000). Dualisms, hierarchies and gender in engineering. *Social Studies of Science*, 30, 759–792.
Flynn, J., Slovic, P. & Mertz, C.K. (1994). Gender, race and perception of environmental health risks. *Risk Analysis*, 14, 1101–1108.
Giddens, A. (1991). *The Consequences of Modernity*. Cambridge: The Policy Press.
Gilligan, C. (1982). *In a Different voice*. Cambridge, MA: Harvard University Press.
Greenberg, M. (2009). Energy sources, public policy, and public preferences: Analysis of US national and site-specific data. *Energy Policy*, 37, 3242–3249.
Gustafson, P.E. (1998). Gender differences in risk perception: Theoretical and methodological perspectives. *Risk Analysis*, 18, 805–811.
Hackmann, H. & St. Clair, A.L. (2012). *Transformative Cornerstones of Social Sciences Research for Global Change*. Paris: International Social Science Council.

Henwood, K.L., Gill, R. & Mclean, C. (2002) The changing man. *Psychologist*, 15(4), 182–186.

Henwood, K.L. (2008). Qualitative research, reflexivity and living with risk: Valuing and practicing epistemic reflexivity and centring marginality. *Qualitative Research in Psychology*, 5(1), 45–55.

Henwood, K.L., Griffin, C. & Phoenix, A. (eds) (1998). *Standpoints and Differences: Essays in the Practice of Feminist Psychology*. London: Sage.

Henwood, K.L., Parkhill, K.A. & Pidgeon, N.F. (2008). Science, technology and risk perception: From gender differences to the effects made by gender. *Equal Opportunities International*, 27(8), 662–676.

Henwood, K.A. & Pidgeon, N.F. (2015). Gender, ethical voices and UK nuclear energy policy in the post-Fukushima era. In: B. Taebi & S. Roeser (eds) *The Ethics of Nuclear Energy: Risk, Justice and Democracy in the post-Fukushima Era*. Cambridge: Cambridge University Press.

Henwood, K.L. & Pidgeon, N.F. (2013). Risk and Identity Futures. UK Foresight *Future of Identities Project* Report DR18, Department of Business, Innovation and Skills. http://www.bis.gov.uk/assets/foresight/docs/identity/13-519-identity-and-change-through-a%20risk-lens.pdf

Henwood, K., Pidgeon, N.F. and Parkhill, K. (2014) Explaining the 'gender-risk effect' in risk perception research: a qualitative secondary analysis study / Explicando el 'efecto género-riesgo' en la investigación de la percepción del riesgo: un estudio cualitativo deanálisis secundario, Psyecology: Revista Bilingüe de Psicología Ambiental / Bilingual Journal of Environmental Psychology, 5: 2–3, 167–213. DOI: 10.1080/21711976.2014.977532.

Henwood, K.L., Pidgeon, N.F., Parkhill, K.A. & Simmons, P. (2010). Researching risk: Narrative, biography, subjectivity [43 paragraphs]. *Forum Qualitative Sozialforschung/Forum:Qualitative Social Research*, 11(1), Art. 20.

Henwood, K.L. Gill, R., and Mclean, C. (2002) "The changing man". Psychologist, 15 (4), 182–186.

Henwood, K.L., Pidgeon, N.F., Sarre, S., Simmons, P. & Smith, N. (2008). Risk, framing and everyday life: Methodological and ethical reflections from three sociocultural projects. *Health, Risk and Society*, 10, 421–438.

Hilgartner, S. (1992). The social construction of risk objects'. In: J. Short & L. Clarke (eds) *Organizations, Uncertainty and risk* (pp. 39–53). Boulder: Westview.

Irwin, A. (2001). *Sociology and the Environment: A Critical Introduction to Society, Nature and Knowledge*. Cambridge: The Polity Press.

Irwin, A., Simmons, P. & Walker, G. (1999). Faulty environments and risk reasoning: The local understanding of industrial hazards. *Environment and Planning A*, 31, 1311–1326.

Lohan, M. (2000). Constructive tensions in feminist studies. *Social Studies of Science*, 30(6), 895–916.

Macgill, S. (1987). *The Politics of Anxiety: Sellafield's Cancer-link Controversy*. London: Pion Press.

Maranta, A., Guggenheim, M., Gisler, P. & Pohl, C. (2003). The reality of experts and the imagined lay person. *Acta Sociologica*, 46, 150–165.

Masco, J. (2006). *The Nuclear Borderlands: The Manhattan Project in Post-Cold War New Mexico*. Princeton: Princeton University Press.

Oels, A. (2005). Rendering climate change governable: From biopower to advanced liberal government? *Journal of Environment, Policy and Planning*, 7(3), 185–207.

Parkhill, K.A., Henwood, K.L., Pidgeon, N.F. & Simmons, P. (2011). Laughing it off: Humour, affect and emotion work in communities living with nuclear risk. *British Journal of Sociology*, 62(2), 324–346.

Parkhill, K.A., Pidgeon, N.F., Henwood, K.L., Simmons, P. & Venables, D. (2010). From the familiar to the extraordinary: Local residents' perceptions of risk when living with nuclear power in the UK. *Transactions of the Institute of British Geographers*, 35(1), 39–58.

Pidgeon, N.F. & Butler, C. (2009). Risk analysis and climate change. *Environmental Politics*, 18(5), 670–688.

Pidgeon, N.F., Hood, C., Jones, D., Turner, B. & Gibson, R. (1992). Risk perception. *Risk: Analysis, Perception and Management*. London: The Royal Society, pp. 89–134.

Pidgeon, N.F., Kasperson, R.K. & Slovic, P. (2003). *The Social Amplification of risk*. Cambridge: Cambridge University Press.

Pidgeon, N.F., Simmons, P. & Henwood, K.L. (2006). Risk, environment and technology. In: P. Taylor-Gooby & J. Zinn (eds) *Risk in Social Science* (pp. 94–116). Oxford: Oxford University Press.

Rosa, E. A. & Freudenburg, W. R. (1993). The historical development of public reactions to nuclear power: Implications for nuclear waste policy. in: R.E. Dunlap, M.E. Kraft & E.A. Rosa (eds) *Public Reactions to Nuclear Waste: Citizens' views of Repository Siting* (pp. 32–63). Durham NC: Duke University Press.

Satterfield, T. (2001). In search of value literacy: Suggestions for the elicitation of environmental values. *Environmental Values*, 10(3) 331–359.

Satterfield, T. (2002). *Anatomy of a Conflict: Identity, Knowledge, and Emotion in Old-Growth Forests*. Vancouver: UBC Press.

Satterfield, T., Mertz, C.K. & Slovic, P. (2004). Discrimination, vulnerability and justice in the face of risk. *Risk Analysis*, 24,115–129.

Scheman, N. (1993). Introduction: The unavoidability of gender. In: N. Scheman (ed) *Engenderings: Constructions of Knowledge, Authority and Privilege* (pp. 1–8). London: Routledge.

Slovic, P. (2000). *The Perception of risk*. London: Earthscan.

Taylor, S. & McEvoy, J. (2015). Researching the psychosocial (A Special Issue), *Qualitative Research in Psychology*, 11(1), 1–90.

Tulloch, J. & Lupton, D. (2003). *Risk and Everyday life*. London: Sage.

Venables, D., Pidgeon, N.F., Henwood, K.L., Parkhill, K. & Simmons, P. (2012). Living with nuclear power: Sense of place, proximity and risk perception in local host communities. *Journal of Environmental Psychology*, 32, 371–383.

Wilkinson, I. (2001). *Anxiety in a risk Society*. London: Routledge.

Willott, S. & Griffin, C. (1999). Building your own lifeboat: Working class male offenders talk about economic crime. *British Journal of Social Psychology*, 38, 445–460.

Wynne, B., Waterton, C., & Grove-White, R., ([2007] 1993). *Public Perceptions and the Nuclear Industry in West Cumbria*. Centre for the Study of Environmental Change, Lancaster University.

Zonabend, F. (1993). *The Nuclear Peninsula*. Cambridge: Cambridge University Press.

11
The Communication and Management of Social Risks and Their Relevance to Corporate-Community Relationships

Philippe Hanna, Frank Vanclay, and Jos Arts

Introduction

The relationship between companies and local communities, especially regarding extractive industries and large infrastructure projects, has historically been marked by conflicts and cases of human rights violations (Ruggie 2010, Kemp & Vanclay 2013). Because of the many internal and external pressures to address this problem, different departments, such as Corporate Social Responsibility (CSR) and Community Relations (CR), have been created or empowered in most multinational companies (Kemp 2004, Porter & Kramer 2006). More recently, CR departments are becoming integrated with risk analysis, risk assessment and risk management, especially through the use of community relations management systems (Kemp et al. 2006). This shift to the use of risk analysis can be attributed to several reasons, particularly the increasing importance of companies keeping good relations with the neighbouring communities of their operations in order to reduce the likelihood of community protest actions, blockades of operations, reputational damage, and the consequential loss of shareholder value (Vanclay & Esteves 2011, Vanclay 2014). In this chapter, three of the many discourses associated with the communication of risk are presented: a sociological discourse (Beck 1999, 2000, 2006, 2009, Giddens 1999, 2006); an anthropological discourse (Douglas 1966, 1985, 1992); and a technical or project management discourse (PMI 2013). Towards the end of our chapter, we critique the technical or technocratic perspective from a socio-anthropological approach and from the perspective of social impact assessment (Esteves et al. 2011; Vanclay et al. 2015).

Discourses about risk

The 'sociological approach' to risk is most notable in the works of Ulrich Beck and Anthony Giddens. In this perspective, risk is comprehended as 'the perceptual and cognitive schema in accordance with which a society mobilises itself when it is confronted with the openness, uncertainties, and obstructions of a self-created future' (Beck 2009: 4). Given uncertainties, risk in our society has also become related to the attempt to 'foresee and control the future consequences of human action, the various unintended consequences of radicalized modernization. It is an institutionalized attempt, a cognitive map, to colonize the future ... It is intimately connected with an administrative and technical decision-making process' (Beck 1999: 3). Giddens (2006: 38) also considers risk analysis as a form of creating possible 'future worlds', or scenarios, using risk jargon. Beck (2009) reflects upon the cultural perceptions around risk, which differ from the actual risk itself – a boundary that is becoming increasingly blurred due to the technocratic belief that risks can be rationally understood and completely calculated. Beck (2006) suggests that risk is incalculable at three levels: at spatial level because of transboundary effects; at a temporal scale because of long latency (e.g. the persistence of pollutants over time); and at a social scale because of the complexity of the social.

Most of the anthropological discussion around risk has been written by Mary Douglas (e.g. 1966, 1985, 1992). Her work has been focused on institutional rather than individual perceptions about risk (Douglas 1985: 83). Douglas (1985) discusses how institutions develop mechanisms of accountability and blame allocation when 'misfortunes' happen – who should be held responsible within the institution? Another important contributor to the cultural theory of risk is Wildavsky (a political scientist), who provides an analysis of risk perception at a more individual level – asking the question what leads different individuals to perceive risks differently? Wildavsky & Dake (1990) argued that self-assigned political ideology (i.e. left wing or right wing) can lead to evaluating some risks as being more dangerous than others (e.g. right wingers tend to accept more technological and environmental risks, while left wingers are more adverse to such risks). They also considered whether or not knowledge about a certain topic leads to different opinions about its risks, and to whether these differences are due to reasons other than knowledge. Their conclusion was that knowledge about a topic plays only a minor role in risk perception, but that trust in institutions and belief in the credibility of the information has a major influence. They suggest that 'individuals perceive a variety of risks in a manner that supports their way of life' (Wildavsky & Dake 1990: 57) – as clearly demonstrated by the example: '98.7 percent of nuclear energy experts thinking nuclear power plants are safe compared with only 6.4 percent of public interest officials' (Rothman & Lichter 1987 cited by Wildavsky & Dake 1990: 56).

Similar findings were presented by Cash et al. (2003) who used the concepts of salience, credibility, and legitimacy to evaluate the capacity of scientific data to mobilise efforts towards sustainability. According to their study, these three criteria are fundamental to the way people (i.e. policy makers, but arguably also the general public) form opinions about sustainable practices and scientific knowledge. Credibility relates to the technical evidence of the information, legitimacy to the trust in the production of the information, and salience is about how relevant this information is to decision makers. From a combined socio-anthropological perspective, risk perceptions are shaped by and part of one's overall cultural symbolic system (i.e. belief system), and are not objective representations of reality (Douglas & Wildavsky 1982). Beck provides a concrete example of this in his description of the differences between US and European risk perceptions as being a *clash of risk cultures and risk religions*:

> There are wide divergences between the prevalent risk faith of most Europeans and that of the current US government. For Europeans, risk (faith) issues, such as climate change and the threat posed by global financial movements for individual countries, are much more important than the threat posed by terrorism. Whereas many Americans think that Europeans are suffering from environmental hysteria ... in the eyes of many Europeans many Americans are afflicted with terrorism hysteria. (Beck 2009: 73)

The third perspective is the technical approach. While there are several technical ways in which project risks are managed, e.g. PRINCE2 (Morris et al. 2011), the Project Management Body of Knowledge (PMBOK) of the Project Management Institute (PMI) guides much of corporate practice and is our reference point for this chapter. The PMI defines project risk as 'an uncertain event or condition that, if it occurs, has a positive or negative effect on one or more project objectives such as scope, schedule, cost, and quality' (PMI 2013: 310). Risk Management is one of the ten areas of the PMBOK, and risk management should be performed proactively during every project stage. Project risk management is the process of dealing with risks in general throughout a project, comprising every activity related to project risks, such as risk impact assessment, risk analysis, risk response and risk control (based on PMI 2013). It is important to clarify the differences between each of these processes, thus the definitions of each concept are given in Box 11.1. When risk analysis, management or response planning in the technical approach is performed, two dimensions are usually deployed: the likelihood of a risk occurring and the severity of the consequences if the risk does occur – in other words, risk probability and impact or consequence. All possible risks are listed in a probability and impact matrix, rating both the probability and impact of each risk as being either very low, low, moderate, high or very high. In order for this process to be

---

BOX 11.1 Components of risk management according the technical discourse

Risk impact assessment 'investigates the potential effect on a project objective such as schedule, cost, quality, or performance, including both negative effects for threats and positive effects for opportunities' (PMI 2013: 330).

Risk response planning 'is the process of developing options and actions to enhance opportunities and to reduce threats to project objectives. The key benefit of this process is that it addresses the risks by their priority, inserting resources and activities into the budget, schedule and project management plan as needed' (PMI 2013: 342).

Control Risks 'is the process of implementing risk response plans, tracking identified risks, monitoring residual risks, identifying new risks, and evaluating risk process effectiveness throughout the project. The key benefit of this process is that it improves efficiency of the risk approach throughout the project life cycle to continuously optimize risk responses' (PMI 2013: 349).

Qualitative Risk Analysis 'is the process of prioritising risks for further analysis or action by assessing and combining their probability of occurrence and impact. The key benefit of this process is that it enables project managers to reduce the level of uncertainty and to focus on high-priority risks' (PMI 2013: 440).

Quantitative Risk Analysis 'is the process of numerically analyzing the effect of identified risks on overall project objectives. The key benefit of this process is that it produces quantitative risk information to support decision making in order to reduce project uncertainty' (PMI 2013: 441).

accurate, it is argued that the risk identification must be conducted properly using the methods suggested by the PMBOK, which include brainstorming, Delphi technique, interviewing experts and root cause analysis. Risk management plans are elaborated as a way to control and monitor project risks. Contingency plans (sometimes called Plan B) are elaborated for each identified risk of high relevance. A contingency plan is only triggered if a given scenario occurs, and is part of the overall risk management plan. A proactive approach to risk management during every project stage is recommended (PMI 2013).

One of the biggest difficulties faced by risk analysts is acknowledging that not all risks are known (Ward & Chapman 2003, PMI 2013). Risks that cannot be identified are called unknown risks. PMBOK recommends that a contingency fund be established as a separate item in the budget to be available to deal with these 'known unknowns' should they occur. Known unknowns are those risks that are qualitatively identifiable but for which it is impossible to determine either the exact consequences or probability. In addition to this, PMBOK also recommends that a management fund be established to deal with the 'unknown unknowns', i.e. potential risks that were not even conceived. Beck (2006) emphasises this issue by referring to the 'incalculability of risks' and he reiterates the importance of 'not knowing' to risk management in general.

Due to the increasing recognition of the importance of external stakeholders to projects, in the fifth edition of PMBOK, stakeholder management has now become one of its knowledge areas (PMI 2013). Project stakeholder management 'includes the processes required to identify the people, groups, or organisations that could impact or be impacted by the project, to analyze stakeholder expectations and their impact on the project, and to develop appropriate management strategies for effectively engaging stakeholders in project decisions and execution' (PMI 2013: 391). Its definition and process are not much different from Project Risk Management – as both try to identify and manage factors that might impact the project. Stakeholder Management matrixes are also based on two axes (power vs. interest). By combining these two criteria, it is possible to determine which stakeholders are more relevant to a project. It is considered that those rated as both high in power and interest should be 'managed closely', while those low on these criteria need only to be monitored or kept informed about the project development. The important differences between stakeholder management and risk management are social and political. It would be a mistake for companies to consider that stakeholders with low power and high interest (as in the case of most impacted communities) don't need to be managed closely, as will be further discussed below.

Risk management, community relations and corporate culture

It is important to appreciate that in business practice taking risks was considered to be an essential part of creating wealth, and the encouragement of risk taking by companies was considered necessary (Dake 1992). A sign that business attitude towards risk is changing is that, since about the mid-1990s, companies have been incorporating management systems into their whole organisation, including project management (Kerzner 2004), risk management (Power 2007), and environmental management (Carruthers & Vanclay 2007, 2012). As a consequence, business now is, in general, more cautious in relation to taking risks (Dake 1992). PMBOK has criteria to assist companies in judging the level of risk they are willing to take, as this varies across different companies and sectors, and is reflected in the terms, 'risk appetite', 'risk tolerance' and 'risk threshold'.

Risk management practitioners, especially in the extractive industries, usually make the distinction between technical and non-technical risks (Joyce & Thomson 2000, Brewer & McKeeman 2011). Although there is a lack of proper definition, literature usually considers technical risks to be the physical, structural, engineering, and environmental aspects of the project, while non-technical risks relate to the managerial, legal, social, and political aspects (e.g. Brewer & McKeeman 2011, Davis & Franks 2011). In industry thinking, non-technical risks are sometimes regarded as being 'external' risks, as they are considered to occur as a result of circumstances outside the control of the project managers, and therefore might strike unexpectedly. Arguably, this distinction between external and internal is inappropriate, as many of these non-technical risks are actually internal, or at least are directly related to corporate activities (and/or the lack of attention to the issues). However, as risk identification is a cultural process (Douglas & Wildavsky 1982), when risks are selected by a homogeneous team (e.g. engineers), risks that are not related to their areas of specialty tend to be ignored or externalised as something beyond control (Kutsch & Hall 2010). For example, conflicts between companies and communities happen because of the company's operations near those communities, and therefore they should not be considered to be 'external'.

Although the distinction between technical and non-technical risks is not made in the PMBOK, it fits well with the differentiation between known and unknown risks – as many technical risks are readily known and predictable, while non-technical risks are usually more difficult to identify and quantify. While Flyvberg (2003) affirms that risks, in general, are systematically overlooked in the development of megaprojects, in the case of mines, the literature indicates that risks related to social issues are the most overlooked, because the 'analysis of risk in mine feasibility studies most frequently focuses on technical and market parameters' (Schafrik & Kazakidis 2011: 87). This is partly due to the lack of involvement of social experts in the planning and feasibility studies of such projects. It is also due to the marginalisation of the 'social', even when it is considered (Baines et al. 2013).

In a study of 190 international oil projects conducted by Goldman Sachs (cited by Ruggie 2010: 15), non-technical risks represented more than half of all project risks, with stakeholder-related risks being the largest single category amongst them. Similarly, Brewer & McKeeman (2011: 1) indicate that non-technical risks 'can account for up to 70–75% of cost and schedule failures in major oil and gas projects in the form of project delays and cost overruns, lost deal opportunities, and host of stakeholder-related issues'. Public dissatisfaction in general is also mentioned as a very relevant non-technical risk, especially in the case of megaprojects, and is usually related to a lack of proper stakeholder engagement and public participation (Flyvberg 2003: 88). Such conflict with local communities can be highly costly for companies if project implementation is not conducted properly. In an interview with Connor (2011: online) and using data collected as part of the research for his work as the United Nations Special Representative on business and human rights, Professor John Ruggie stated that:

> for a world-class mining operation, which requires about $3–5 billion capital cost to get started, there's a cost somewhere between $20 million and $30 million a week for operational disruptions by communities. Another estimate used by the mining industry is that an asset manager [i.e. head person at the mine site] is supposed to spend between 5% and 10% of his or her time on community engagement issues. We found that it can be anywhere from a one-third to 50%, and in some cases 80% of their time [in other words, a significant cost in management time]. So there are opportunity costs, financial costs, legal costs, and reputational costs. All this has escalated tremendously, which is why companies themselves have been so interested in the UN mandate I've led.

Due to the complex nature of social aspects, the high risks involved and how costly community conflicts can be, community relations has become more relevant to companies. Many companies have started implementing social performance and CR systems (Kemp et al. 2006), a feature that has brought the practices of CR and risk management closer together. This has occurred in order to assist companies in avoiding conflicts with communities, protests against the company, blockades of operations, and other 'crisis situations' (Rees et al. 2012). Consultancy firms offer services to avoid conflict with communities, and focus on the benefits to companies of incorporating CR with risk management, as demonstrated in the statement: 'Our community investment strategy planning aims to not only reduce the risk of project delays, but ensure that funds spent in the community are aligned with your business vision, are a part of your overall risk management strategy and deliver tangible and sustainable results to the wider community' (PWC 2012: 3). Kytle & Ruggie (2005) position CSR as an important strategy in managing risks in multinational corporations, referring to stakeholder-related issues as social risk:

> From a company's perspective, social risk occurs when an empowered stakeholder takes up a social issue area and applies pressure on a corporation (exploiting a vulnerability in the earnings drivers – e.g., reputation, corporate image), so that the company will change policies or approaches in the marketplace. (Kytle & Ruggie 2005: 6)

When Indigenous communities are involved, social risks can be even higher, basically due to three reasons: (1) there is increasing prominence being given to Indigenous rights in international law, and there is increasing pressure to ensure that such rights are fully respected; (2) mineral and oil extraction activities are becoming more frequent near Indigenous lands; and (3) the increased use of social media brings more visibility to Indigenous advocacy campaigns (FPW 2013). Additionally, risk communication in cross-cultural contexts can be quite challenging, as the technical aspects of projects are sometimes incomprehensible to Indigenous peoples and, conversely, the social and cultural impacts on communities are not understood or appreciated by project developers. For example, the public consultation meetings for implementing a hydro-electric power plant and dam (Belo Monte) in Brazil have been much criticised by Indigenous organisations and external observers, who have argued that the consultations consisted of highly technical content, making the possible risks inconceivable to the local community

(Anaya 2010). Ideally, consultation meetings with Indigenous peoples should deploy diverse communication methods in order to be effective, such as the use of native interpreters, videos, and letting communities take their own time to consider the risks that were communicated, and determine if they are acceptable or not. Potentially, site visits to similar projects elsewhere could be appropriate. A benefit of doing this is that local knowledge may be harnessed that could potentially improve the technical design of the project bringing benefits to local communities as well as the proponent.

Risks, which are not relevant to project developers but are highly risky to the community's livelihoods, could be communicated back to the engineers to enable them to develop and implement mitigation measures. For example, a nickel plant in Brazil, which neighbours the Xikrin do Cateté Indigenous territory, originally planned to draw its water supply from a river also used by the Xikrin people. After community consultations, planners realised that this alternative was strongly opposed by the community, especially the women. The project design was therefore altered and a water supply dam was built so that water from the river would not be utilised for the operation. In response to requests from the local community, the company agreed to manage the flow rates from the dam so that the local streams would have water available throughout the year, including in the dry season – thus bringing a positive benefit to local communities instead of a negative impact (ICMM 2010).

Indigenous rights have been strengthened with the rise of the concept of Free, Prior, and Informed Consent (FPIC), which states that Indigenous peoples have the right to be consulted on any decision that may affect their lives. This right is becoming established in various international declarations, and Indigenous peoples are becoming highly aware of this right (Hanna & Vanclay 2013). However, many companies are still reluctant to recognise this right, despite a growing international acceptance as reflected in recent changes in the International Finance Corporation (IFC) performance standards (IFC 2012). There is also a growing acceptance of the value of FPIC being applicable to all communities, not only Indigenous (Goodland 2004, Hill et al. 2010, Vanclay & Esteves 2011, Langbroek & Vanclay 2012).

In cases of the occurrence of high impact social risks, such as severe community conflicts resulting in the blockade of operations, companies usually have a crisis management mechanism called a 'crisis room' or 'war room' (Shaker & Rice 1995). In such circumstances, responsible managers from different departments (security, CR, CSR, communications, logistics, legal etc.) are summonsed by the top-management (CEO, Directors) to consider how to solve the problems as soon as possible so as to avoid any further damage or financial loss to the company. Considering that risk in contemporary society is a basis for decision-making, it is necessary to take in to account the opposition between *decision makers* and those *affected* by the decisions (Beck 2009: 112). In discussing how decision makers come to decisions rationally, Douglas (1985: 84) proposes that 'the big choices reach them in the form of questions of whether to reinforce authority or to subvert it, whether to block or to enable action. This is where rationality is exerted'. This is exactly the choice that CEOs, Security or CR managers face in crisis situations of conflicts with communities, whether to resort to the use of police force, judicial action, or direct negotiation with the community.

Douglas (1985) discusses the institutionalising effects on the perception of risk of individuals. Regarding the mining industry, for example, there continues to exist a 'deeply ingrained instrumental logic that continues to underpin management decisions' (Kemp et al. 2011: 106). At an individual level, employees are expected to defend the company/institution they work for. At an institutional level, decisions are expected to be market driven, not necessarily focused on mitigating the impacts on the affected community, but on reducing any further (financial and legal) risks to the company. Such decisions can be related to how to: deal with the press, avoid reputational damage, or apply a legal response. In fact, conflict with communities should be seen as a sign that there might be something wrong with the company's approach towards a given community or towards communities in general, and thus as an opportunity for the company to improve its practices (Prenzel & Vanclay 2014); or in other words, a tipping point for institutional change (Gunderson & Holling 2002). Despite increasing societal pressure for good CSR practices, being loyal and defending the company's interests is expected, even when it clashes with local communities' priorities and rights. Because crises of community relations are intrinsically anthropogenic – and are only rarely due to a technical problem – there is always someone to blame, often the community relations department or the communities themselves (Rees et al. 2013: 7). Blame allocation is intrinsically a political process:

> Blaming is a way of manning the gates through which all information has to pass. Blaming is a way of manning the gates and at the same time arming the guard. News that is going to be accepted as

true information has to be wearing a badge of loyalty to the particular political regime which the person supports; the rest is suspect, deliberately censored or unconsciously ignored. (Douglas 1992: 19)

In the event of a crisis, no one wants to be (held) responsible for the great losses that might occur. Most of the time, conflicts with communities are related to broader corporate practice (or a lack of good practice) and not to a mistake of a given department, or to the actions of local communities. However, as demonstrated in the previous quote from Douglas, blame allocation is political, thus those who are weaker or more marginalised within the institutional context are to be blamed – in this case the CR department or the communities themselves. The process of blame allocation, as explained by Coleman (1982), is also useful in understanding why CR departments get the blame when conflicts with communities happen. Coleman describes how the manufacturing industry divided the production line into different responsible areas, so when a defective product was manufactured, it was possible to identify who, or which department, was responsible for the defect. A similar blaming mechanism still operates in contemporary corporations. When conflict with communities is not avoided, it is rational in industry logic to blame the department perceived to be responsible.

Risks of conducting community relations as risk management

There are risks, for both companies and communities, of conducting CR as if it was simply the management of social risks. Despite being difficult to identify in conventional risk analysis, it is important to be aware that community conflict does not happen out of the blue. It is usually related to the relationship history between a company and a community. Various authors point out that proactive measures, such as early engagement, external stakeholder involvement in planning and evaluation, and a value-based approach to CR, are key to avoiding conflicts (Kemp et al. 2006). After all, social and environmental impacts are also serious risks to impacted communities. Just as communities are a risk for companies, company operations can be a great risk to people's livelihoods and wellbeing. When communities protest, in most cases the protest actions can be considered to be a strategy to communicate to the competent authorities the risks of environmental and social impacts from company operations (Hanna et al. 2014). In fact, different communities have different levels of risk tolerance (O'Faircheallaigh &

Gibson 2012) and it might be argued that political mobilisation is often a community's only way (or last resort) of performing risk management, as mobilised communities may be able to achieve better mitigation measures and compensation (O'Faircheallaigh 2010), thus reducing the risks they experience from company operations.

Corporate methodologies of risk analysis are based on predicting the probability of occurrence and the likely consequences of each risk. More complex analyses are based on software modelling, which may simulate diverse scenarios and involve many stakeholders in the process. A limitation of such risk management systems, especially regarding CR, is that community conflict and protests are almost always of high impact and many times of high probability as well. It is always of high impact because it can be costly in time, resources and even reputational damage to companies. Beck (2000: 215) dismisses the risk statements that are produced by quantitative risk analysis and considers that the 'risk statements are neither purely factual claims nor exclusive value claims. Instead, they are either both at the same time or something in between, a "mathematised morality" as it were. As mathematical calculations (probability computations or accident scenarios) risks are related directly or indirectly to cultural definitions'.

In order to avoid cultural biases in risk analysis, there is the need for a multidisciplinary group comprising engineers, risk specialists, social scientists, anthropologists, economists, and so on. Bringing together professionals from different risk cultures, or 'risk religions' (Beck 2006: 337), can contribute to a more comprehensive and efficient risk analysis (Mahmoudi et al. 2013). Regarding CR-related risks, desktop risk analysis is not sufficient to identify likely risks; there is always a need to include expert advice for specific content and to include representatives from the impacted communities themselves. In the same way as there are requirements for community liaison committees to take part in EIAs and EIA follow-ups (Ross 2004), communities should take part in the risk assessment of all projects, as many company operations represent a direct risk to communities and peoples' lives and livelihoods. A multi-stakeholder risk analysis can be considered an efficient way to identify risks for the different sides involved, as well as being a transparent way for risk communication and conflict avoidance. Also recommended is a shift in corporate culture (especially in the extractive industries) towards greater valorisation and empowerment of CR departments and staff, considering them to be a strategic and essential component of the business, on par with the technical departments in the company (Kemp & Owen 2013). Conducting conflict-sensitive social impact assessments

and discussing the results and potential risks in a transparent way contributes to avoiding conflicts between companies and communities (Prenzel & Vanclay 2014) and lowers the risks for both sides.
  Mahmoudi et al. (2013) recommend integration of social impact assessment and social risk assessment. Should that also be the case for community relations and risk management? We suggest that there are many positive aspects of this integration, such as the empowerment of community relations practice inside companies and the valorisation of community issues in general. Due to potential high risks, CR is becoming an important issue for companies. However, if the goal is to mitigate risks in the long term for all involved stakeholders, other strategies should also be deployed. O'Faircheallaigh (2010: 400) also warns about this issue, as companies are often primarily worried about 'cost minimisation, and by an emphasis on risk management which is short-term and focused on securing initial project approvals rather than on building positive, long-term relationships with affected communities'.
  Kemp et al. (2006: 398) recommend that 'conventional management systems thinking, with its internal focus and rational approach, needs to be balanced with value-based decision-making, a supportive organisational culture, and [an] externally focused stakeholder-driven orientation'. To effectively mitigate risks to companies (and arguably also to communities), a company needs to achieve and maintain a social licence to operate (Prno 2013, Jijelava & Vanclay 2014, Moffat & Zhang 2014). In addition to obtaining any required formal legal licences from governmental authorities, it is necessary to have the community's approval for proceeding with the project. However, O'Faircheallaigh (2010) highlights the problem of companies simply pursuing a social licence to operate as a way to mitigate risks to its operations, or simply focusing on mitigating immediate risks in order to achieve a social licence in the short-term. A genuine attempt to obtain and maintain a social licence should be based on using several long-term strategies, such as implementing internal policies (e.g. on human rights, Indigenous peoples, etc.), conducting cross-cultural training for staff, conducting licensing processes with procedural fairness, conducting conflict-sensitive social impact assessments, clearly communicating potential risks to communities, operationalising social impact management plans, developing community-based grievance mechanisms, and fully observing FPIC principles (Kemp et al. 2006, O'Faircheallaigh 2010, Franks & Vanclay 2013, Moffat & Zhang 2014, Vanclay et al. 2015).

Conclusion

Decisions regarding communities can't be relegated to quantitative mathematical analysis or simulation and modelling. Ball and Watt (2013) point to the limits of risk analysis matrixes, highlighting the several subjective factors that influence how risks are identified and categorised. These issues also apply to the software analysis tools, which can be considered as a complex version of risk matrixes. As highlighted by Beck (2006), the problems of risk assessment are even greater when considering social risks, due to their complexity and their consequential incalculability. As it is not possible to identify all social risks, in practice companies tend to act reactively, what is often called in industry jargon 'putting out the brushfires', with CR staff spending most of their time addressing risks that were not previously identified. However, over time companies and individual staff do learn both from their experience with communities and through the global practice and discourse of social performance and social impact assessment.

Risk analyses are improved when performed by interdisciplinary and multi-stakeholder teams, and through transparent and effective communication of identified risks. However, even when done well, risk management decisions should not be narrowly focused only on mitigating specific risks in particular cases, but should also address broader corporate practices and corporate culture. Risk and community management systems need to become more externally focused, values-based, and stakeholder-driven rather than shareholder-driven (Kemp et al. 2006: 401). Conducting risk management by itself will not lower the risk; actions need to be implemented. There is need for proactive mitigation measures, value-driven early engagement, and early impact assessment studies conducted in the spirit of FPIC. Such respect for local communities and their rights is crucial to avoid conflicts, especially when likely impacts are understood as being high risks for communities. When communities protest, this is usually related to their strategies of attempting to communicate risks to key decision makers and seeking mitigation or avoidance of the impacts from operations, and thus to lowering the risks they will experience.

The good practices mentioned above should help establish and maintain a genuine social licence to operate, minimise the risks of conflict and provide better opportunities for communities. It is also recommended that companies adopt appropriate policies and procedures for risk management (including human rights due diligence), conduct cross-cultural training for company staff, increase the profile of the CR functions, and abide by the spirit of FPIC with respect to all communities, not only Indigenous. All these active measures will lower the risk of conflict, benefiting communities and the company. Despite the fact that the risk management literature provides so much focus on risk analysis and classification, and recommends that such activities are performed regularly, we consider that risk analysis, without the necessary good practice actions to properly address the social issues, will not provide a safeguard against the risks to companies and communities alike. Conversely, considering stakeholders not as 'risks' but as partners, and engaging in good-faith and respectful relations with them proactively lowers the social risks.

References

Anaya, J. (2010). *Report by the special rapporteur on the situation of human rights and fundamental freedoms of indigenous people. Addendum: Cases examined by the special rapporteur.* (June 2009–July 2010). (No. A/HRC/15/37/Add.1). Geneva: United Nations.
Baines, J., Taylor, C. N., & Vanclay, F. (2013). Social Impact Assessment and ethical social research principles: Ethical professional practice in impact assessment Part II. *Impact Assessment & Project Appraisal, 31*(4), 254–260.
Ball, D. J., & Watt, J. (2013). Further thoughts on the utility of risk matrices. *Risk Analysis, 33*(11), 2068–2078.
Beck, U. (1999). *World risk society.* Cambridge: Polity Press.
Beck, U. (2000). Risk society revisited: Theory, politics and research programmes. In J. van Loon, U. Beck & B. Adam (Eds.) *The risk society and beyond: Critical issues for social theory* (pp. 211–229). London: Sage.
Beck, U. (2006). Living in the world risk society. *Economy and Society, 35*(3), 329–345.
Beck, U. (2009). *World at risk.* Cambridge: Polity Press.
Brewer, L., & McKeeman, R. (2011). Deepwater Gulf of Mexico development in a post-Macondo world. OTC 21945, *Offshore Technology Conference Proceedings.* May 2–5, Houston, USA. (pp. 2854–2860).
Carruthers, G., & Vanclay, F. (2012). The intrinsic features of Environmental Management Systems that facilitate adoption and encourage innovation in primary industries. *Journal of Environmental Management 110,* 125–134.
Carruthers, G., & Vanclay, F. (2007). Enhancing the social content of environmental management systems in Australian agriculture. *International Journal of Agricultural Resources, Governance & Ecology, 6*(3), 326–340.
Cash, D. W., Clark, W. C., Alcock, F., Dickson, N. M., Eckley, N., Guston, D. H., ... Mitchell, R. B. (2003). Knowledge systems for sustainable development. *Proceedings of the National Academy of Sciences, 100*(14), 8086–8091.
Coleman, J. S. (1982). *The asymmetric society.* Syracuse: Syracuse University Press.
Connor, J. (2011). *Business and human rights: Interview with John Ruggie.* Retrieved February 04, 2014, from http://business-ethics.com/2011/10/30/8127-un-principles-on-business-and-human-rights-interview-with-john-ruggie/

Dake, K. (1992). Myths of nature: Culture and the social construction of risk. *Journal of Social Issues, 48*(4), 21–37.
Davis, R., & Franks, D. M. (2011). The costs of conflict with local communities in the extractive industry. *Proceedings of the First International Seminar on Social Responsibility in Mining*, October 19–21, Santiago, Chile. 1–13.
Douglas, M. (1966). *Purity and danger: An analysis of concepts of pollution and taboo.* New York: Routledge.
Douglas, M. (1985). *Risk acceptability according to the social sciences.* New York: Russell Sage Foundation.
Douglas, M. (1992). *Risk and blame: Essays in cultural theory.* London: New York: Routledge.
Douglas, M., & Wildavsky, A. (1982). How can we know the risks we face? Why risk selection is a social process. *Risk Analysis, 2*(2), 49–58.
Esteves, A. M., Franks, D., & Vanclay, F. (2012). Social impact assessment: The state of the art. *Impact Assessment and Project Appraisal, 30*(1), 35.
Flyvbjerg, B., Bruzelius, N., & Rothengatter, W. (2003). *Megaprojects and risk: An anatomy of ambition.* Cambridge: Cambridge University Press.
Franks, D. M., & Vanclay, F. (2013). Social impact management plans: Innovation in corporate and public policy. *Environmental Impact Assessment Review, 43*, 40–48.
FPW. (2013). Indigenous rights risk report for the extractive industry (U.S.). Available: http://www.firstpeoples.org/images/uploads/R1KReport2.pdf.
Giddens, A. (1999). Risk and responsibility. *The Modern Law Review, 62*(1), 1–10.
Giddens, A. (2006). Fate, risk and security. In J. Cosgrave (Ed.) *The sociology of risk and gambling reader* (pp. 29–59). New York: Routledge.
Goodland, R. (2004). Free, prior and informed consent and the World Bank Group. *Sustainable Development Law & Policy, 4*(2), 66–74.
Gunderson, L. H., & Holling, C. S. (2002). *Panarchy: Understanding transformations in human and natural systems.* Washington, DC: Island Press.
Hanna, P., & Vanclay, F. (2013). Human rights, Indigenous peoples and the concept of free, prior and informed consent. *Impact Assessment and Project Appraisal, 31*(2), 146–157.
Hanna, P., Vanclay, F., Langdon, E. J., & Arts, J. (2014). Improving the effectiveness of impact assessment pertaining to indigenous peoples in the Brazilian environmental licensing procedure. *Environmental Impact Assessment Review, 46*, 58–67
Hill, C., Lillywhite, S., & Simon, M. (2010). *Guide to free prior and informed consent.* Carlton: Oxfam Australia.
ICMM. (2010). *Indigenous peoples and mining: Good practice guide.* London: International Council on Mining and Metals.
IFC. (2012). *Performance standard 7 – Indigenous peoples.* Washington: International Finance Corporation.
Jasanoff, S. (1993). Bridging the two cultures of risk analysis. *Risk Analysis, 13*(2), 123–129.
Jijelava, D., & Vanclay, F. (2014). Assessing the social licence to operate of development cooperation organizations: The case of Mercy Corps in Samtskhe-Javakheti, Georgia. *Social Epistemology, 28*(3–4), 297–317.
Joyce, S., & Thomson, I. (2000), Earning a social licence to operate: Social acceptability and resource development in Latin America, *Canadian Institute of Mining Bulletin, 93*(1037), 49–53.

Kemp, D. (2004). The emerging field of community relations: Profiling the practitioner perspective. *Minerals Council of Australia Sustainable Development Conference, November 2004,* Melbourne.
Kemp, D., Boele, R., & Brereton, D. (2006). Community relations management systems in the minerals industry: Combining conventional and stakeholder-driven approaches. *International Journal of Sustainable Development, 9*(4), 390–403.
Kemp, D., & Owen, J. R. (2013). Community relations and mining: Core to business but not 'core business'. *Resources Policy, 38*(4), 523–531.
Kemp, D., Owen, J. R., Gotzmann, N., & Bond, C. J. (2011). Just relations and company–community conflict in mining. *Journal of Business Ethics, 101*(1), 93–109.
Kemp, D., & Vanclay, F. (2013). Human rights and impact assessment: Clarifying the connections in practice, *Impact Assessment & Project Appraisal, 31*(2), 86–96.
Kerzner, H. (2004). *Advanced project management: Best practices on implementation.* 2nd ed. Hoboken NJ: Wiley.
Kutsch, E., & Hall, M. (2010). Deliberate ignorance in project risk management. *International Journal of Project Management, 28*(3), 245–255.
Kytle, B., & Ruggie J. (2005). *Corporate social responsibility as risk management: A model for multinationals.* Corporate Social Responsibility Initiative Working Paper No. 10. Cambridge: John F. Kennedy School of Government, Harvard University.
Langbroek, M., & Vanclay, F. (2012). Learning from the social impacts associated with initiating a windfarm near the former island of Urk, The Netherlands. *Impact Assessment & Project Appraisal, 30*(3), 167–178.
Mahmoudi, H., Renn, O., Vanclay, F., Hoffmann, V., & Karami, E. (2013). A framework for combining social impact assessment and risk assessment. *Environmental Impact Assessment Review, 43,* 1–8.
Moffat, K., & Zhang, A. (2014). The paths to social licence to operate: An integrative model explaining community acceptance of mining. *Resources Policy, 39,* 61–70.
Morris, P. W., Pinto, J. K., & Söderlund, J. (2011). *The Oxford handbook of project management.* Oxford: Oxford University Press.
O'Faircheallaigh, C., & Gibson, G. (2012). Economic risk and mineral taxation on indigenous lands. *Resources Policy, 37*(1), 10–1.
O'Faircheallaigh, C. (2010). CSR, the mining industry and Indigenous peoples in Australia and Canada: From cost and risk minimisation to value creation and sustainable development. In C Louche, S. O. Idowu & W. L. Filho (Eds.) *Innovative CSR: From risk management to value creation* (pp. 398–418). Sheffield: Greenleaf.
PMI. (2013). *A guide to the project management body of knowledge (PMBOK guide)* (5th edn). Newtown Square: Project Management Institute.
Porter, M. E., & Kramer, M. R. (2006). Strategy and society. *Harvard Business Review, 84*(12), 78–92.
Power, M. (2007). *Organized uncertainty: Designing a world of risk management.* Oxford: Oxford University Press.
Prenzel, P. V., & Vanclay, F. (2014). How social impact assessment can contribute to conflict management. *Environmental Impact Assessment Review, 45,* 30–37.
Prno, J. (2013). An analysis of factors leading to the establishment of a social licence to operate in the mining industry. *Resources Policy, 38*(4), 577–590.

PWC. (2012). Retrieved January 06, 2014, from http://www.pwc.com.au/industry/energy-utilities-mining/assets/Social-Impact-Assessment-May12.pdf

Rees, C., Kemp, D., & Davis, R. (2012). *Conflict management and corporate culture in the extractive industries: A study in Peru*. Corporate Social Responsibility Initiative Report No. 50. Cambridge, MA: John F. Kennedy School of Government, Harvard University http://www.hks.harvard.edu/mrcbg/CSRI/CSRI_report_50_Rees_Kemp_Davis.pdf

Ross, W. A. (2004). The independent environmental watchdog: A Canadian experiment in EIA follow-up. In A. Morrison-Saunders & J. Arts (Eds.) *Assessing impact: Handbook of EIA and SEA follow-up* (pp. 178–192). London: Earthscan Sterling.

Rothman, S., & Lichter, S. R. (1987). Elite ideology and risk perception in nuclear energy policy. *The American Political Science Review, 81*(2), 383–404.

Ruggie, J. (2010) Report of the Special Representative of the UN Secretary-General on the issue of human rights, and transnational corporations and other business enterprises – Business and human rights: further steps towards the operationalisation of the 'protect, respect and remedy' framework. A/HRC/14/27.

Shaker, S. M., & Rice, T. L. (1995). Beating the competition: From war room to board room. *Competitive Intelligence Review, 6*(1), 43–48.

Vanclay, F. (2014). Developments in Social Impact Assessment: An introduction to a collection of seminal research papers. In F. Vanclay (Ed.) *Developments in social impact assessment* (pp. xv–xxxix). Cheltenham: Edward Elgar.

Vanclay, F., & Esteves, A.M. (2011). Current issues and trends in social impact assessment. In F. Vanclay & A. M. Esteves (Eds.) *New directions in social impact assessment: Conceptual and methodological advances* (pp. 3–19). Cheltenham: Edward Elgar.

Vanclay, F., Esteves, A. M., Aucamp, I., & Franks, D. (2015). *Social impact assessment: Guidance for assessing and managing the social impacts of projects*. Fargo ND: International Association for Impact Assessment.

Ward, S., & Chapman, C. (2003). Transforming project risk management into project uncertainty management. *International Journal of Project Management, 21*(2), 97–105.

Wildavsky, A., & Dake, K. (1990). Theories of risk perception: Who fears what and why? *Daedalus, 119*(4), 41–60.

12
How Structured Dialectical Discourse of Risk Eased Tension in North American LNG Siting Conflicts

Susan Mello

Introduction

As domestic supplies of natural resources wane and energy demand increases, federal governments and local communities face difficult decisions regarding proposed solutions. The siting of wind farms, nuclear power plants, wells for hydraulic fracturing, and natural gas import terminals may help meet rising demand, but all raise serious questions about the risks and benefits associated with energy development and diversification. How communities perceive these risks and benefits and, perhaps more importantly, whether they consider them acceptable depend greatly on effective risk communication.

Because federal energy priorities naturally differ from those of local communities, dialogue concerning development is inherently dualistic and therefore, intractable if not properly mediated (Gould, Schnaiberg, & Weinberg, 1996; Lidskog, 2005). Communication is often complicated by risk amplification, divergent scientific assessments, and competing social value systems. As siting conflicts raise a host of environmental, political, and economic concerns, governments must anticipate greater complexity and potential discord within risk debates. Accordingly, there is a pressing need for new and improved models of risk communication governance. In the context of energy development, the necessity lies in moving risk communication away from the pedagogical and toward the dialectical.

This chapter presents results from a comparative case study of risk communication efforts surrounding liquefied natural gas (LNG) development in the northeast regions of Canada and the United States. In the early 2000s, the town of Lévis, Québec, and the city of Fall River,

Massachusetts, were each marked by the natural gas industry as ideal sites for new LNG import terminals. In both regions, corporate entities promoting the projects invested significant time and resources into risk communication. Although both projects were ultimately abandoned for economic reasons in 2013 and 2011, respectively, the events surrounding these proposals offer a unique opportunity for transnational comparisons of stakeholder dialogue, civic participation in decision-making processes, and policy outcomes.

Using a case study approach (Yin, 2009), this investigation combined in-depth stakeholder interviews with analysis of secondary source documents including official reports, risk assessments, and mass media coverage. Following an introduction to the issue of LNG in North America, this chapter reviews theory and research related to two key frameworks in risk communication: dialectical discourse (Juanillo & Scherer, 1995) and the social arena concept of risk (Renn, 1992). How the strategic coupling of these theoretical models in Québec fostered productive public discourse on risk will be described. Finally, a discussion of how divergent communication strategies likely contributed to acceptance in Québec and rejection in Massachusetts offers practical advice for future siting conflicts.

LNG in North America

Natural gas is used mainly for space or water heating and electricity generation in the industrial, commercial, and residential sectors, and as a raw material in the chemical and petrochemical industries. When natural gas is chilled to minus 260°F (minus 162°C), it converts into a liquid state one 600th of its original volume that is non-toxic and odourless (Shell Global, 2014). The liquification of natural gas has transformed the industry by making it possible for efficient and economical exportation of natural gas in larger quantities overseas, where it is re-gasified, stored, and delivered to in-land markets via pipeline networks and ground transportation.

In the early 2000s, the market demand for LNG experienced the highest growth rate among all energy sectors in the world (U.S. Energy Information Administration, 2004). As the largest market for natural gas, North America saw an unprecedented number of proposals from the private energy sector for additional LNG receiving capacity in both Canada and the U.S. In April 2004, natural gas conglomerates Gaz Métro, Enbridge Inc., and Gaz de France (also known as the Rabaska Consortium) announced their plan to construct an LNG import terminal on the southern banks of the St. Lawrence River in Lévis, Québec. Located across the river from Québec City, the receiving terminal would be sited near a mixed residential-farming area, amid 700 immediate residents and a greater population of 620,000 inhabitants of the capital region (Brinkhoff, 2007).

Risks, benefits and assessment

Compared to other fossil fuels, natural gas is relatively clean burning and less expensive. Most industry analysts, capitalists, and economists viewed the growing investment in LNG as a means of reinforcing the strength of local and national economies. For these particular stakeholders, the benefits of job creation, increased municipal and federal tax revenues, as well as lower energy costs for businesses and residents outweighed potential risks. In order to maximise economic efficiencies, however, sites were proposed along coastlines or rivers and near existing pipeline networks-areas which also tend to been densely populated and of ecological importance. The acceptability of potential risks associated with terminal construction and operation, thus, became a primary concern of host municipalities and their citizens.

Since the 1940s when LNG was first introduced to North America, only four major LNG-related accidents that resulted in one or more fatalities had been recorded. Fires and explosions caused by industrial malfunctions in terminals located in Cleveland, OH (1944), Staten Island, NY (1973), Lusby, MD (1979), and Skikda, Algeria (2004), blemished LNG's safety record (Foss, 2006; Hamutuk, 2008). The worst of the accidents in Cleveland occurred when a vapour cloud of natural gas filled surrounding streets and sewers before it ignited, killing 128 people. The most recent Algerian accident involved an ignited spill that injured 56 and killed 27 people. With the exception of Cleveland, there have been no civilian deaths or injuries and no large-scale maritime incidents to date. In addition to technological risks, LNG development may also have adverse ecological impacts (e.g., ground pollution, carbon emissions, degradation of riverbanks, and endangerment of indigenous species) and sociological implications (e.g., expropriation, property devaluation, disruption of tourism, and local transportation).

The evaluation of LNG terminal proposals in the U.S. and Canada falls under the jurisdiction, respectively, of the Federal Energy Regulatory Commission (FERC) and the Minister of Sustainable Development. To ensure proper siting and security, these agencies require corporate promoters to engage in a long, bureaucratic approval process beginning with an environmental impact statement. This statement explains potential risks associated with the project's construction and operation and presents options for risk attenuation and plan modification to prevent technical malfunctions and minimise environmental harm. In both the U.S. and Canada, these reports are made available to the public during approval proceedings.

In their impact statement, the Rabaska Consortium cited numerous precautionary measures, including the use of double-hulled tankers, retention basins, permanent facility monitoring, and disaster response plans (Rabaska, 2005). An international risk assessment firm also examined 238 potential accident scenarios and estimated the risk of an on-site accident to be once every 1,000 years. An accident extending beyond the site's perimeter was estimated to occur once every 10 million years.

Civic engagement

Technical risk assessments by government agencies often fail to consider factors beyond the likelihood and severity of a hazard, such as dread, culture, politics, and values that influence the social construction of risk (Gutteling & Kuttschreuter, 2002; Slovic, 1999). While the use of LNG in North America is not novel per se, these recent proposals became a major point of contention. In Québec, a dozen citizens groups ranging in size and scope emerged after the proposal's announcement. Using multiple strategies and tactics, ranging from polls and petitions to rallies and media relations, these social groups all voiced opinions in the public forum intent on influencing public opinion and policy. How the discourse on risk unfolded in Québec is of theoretical and practical significance.

Public discourse on risk: Theory and research

The social experience of risk is made up of several interdependent elements, ranging from risk assessment and the exchange of information to institutional procedure and broader social and cultural influences (Renn, 1992). The inherent complexities of risk make it nearly impossible to examine a case study through the lens of a singular theoretical model. In most cases, several concepts and frameworks need to be identified and their interplay with one another further defined in order to capture the full essence of public discourse on risk. Experiential knowledge of the practice of risk communication in everyday life is best reflected on theoretically and infused with formal purpose and structure (Webler, 1999).

Ryfe (2003) argues that the postmodern theoretical conception of productive argumentation and good public discourse consists of six prime factors: reflexivity, radical difference, reciprocity, formalisation, grounded rationality, and moderation. It is the main argument of this chapter that such a prescription for good public discourse on risk can be met by strategically coupling two theoretical models of communication: dialectical discourse and the social arena concept of risk debates (see Table 12.1).

Dialectical discourse on risk

Risk communication models are moving away from traditional one-way, linear models of information transmission toward a more dialectical construction. The dialectical discourse model (Juanillo & Scherer, 1995) is rooted in a type of argumentation that 'uses a series a questions and answers to probe through possibilities and weigh contradictory facts and opinions with a view to their resolution' (McComas, 2004, p. S64). The dialectical model cannot only empower stakeholders to better understand technical assessments, but also to appreciate different perspectives on risk, scrutinise opinions and perceptions, and make informed judgments (Juanillo & Scherer, 1995). The characteristics of the model align well with three of Ryfe's (2003) primary building blocks of good public discourse: reflexivity, radical difference, and reciprocity.

Reflexivity. Reflexivity is a principle based on the postmodern condition, which presupposes the intrinsic and unavoidable complexities of public thought. It assumes truth to be relative, a product of multiple influences and processes. To the domain of risk and its communication,

Table 12.1 Alignment of Ryfe's (2003) factors for good public discourse and theoretical models of risk communication

Theory	Factor	Description
Dialectical discourse (*Juanillo & Scherer, 1995*)	Reflexivity	Iterative consideration of objective and subjective components of risk
	Radical difference	Universal integration of stakeholders in discourse
	Reciprocity	Collaborative dialogue
Social arena concept of risk debates (*Renn, 1992*)	Formalisation	Institutional support of citizen rights and equitable resource allocation
	Grounded rationality	Relevant information and logical argumentation
	Moderation	Structured facilitation of debate

reflexivity imparts a comprehensive vision of and validation to both scientific risk assessments and personal value judgments.

Traditionally, the proper rhetorical construction of risk messages was thought to be grounded exclusively in technological risk assessments (Gutteling & Kuttschreuter, 2002). Objective estimates of the likelihood and severity of a hazard's consequences were the only variables in the equation calculating the acceptability of a risk. As such, risk assessments were the sole responsibility of the technocratic elite. Unavoidable epistemological variations of truth in science, however, leave gaping loopholes in calculations of acceptable risk. Whatever consensus exists among scientists can always be proven wrong by a slight change of either context or method (Collingridge & Reeve, 1986). This vulnerability and uncertainty leaves policymakers, legislative officials, and the public on unsteady ground.

The simultaneous growth of both knowledge and uncertainty in modern society creates much opportunity for competing interpretations. For example, in the face of scientific and technological advancements, citizens tend to view themselves as victims rather than beneficiaries (Fischhoff, Slovic, & Lichtenstein, 1982). Because knowledge, understanding, values, belief systems, and fears vary dramatically among stakeholders, the presumption that science alone could effectively communicate risk to public audiences is antiquated (Sauer, 2003). Such an approach fails to recognise that citizens employ very different analytical thought processes than experts when deciphering and elaborating on risk messages (Margolis, 1996; Vlek & Cvetkovich, 1989). In either context – the scientific or the social – risk is intrinsically subjective because of its origins in human assumption (Slovic, 1992).

Ryfe (2003) argues that successful communication is a function of reflexive conversation and deliberation. In the context of risk communication, this means the integration of both objective and subjective judgments in the form of expert and non-expert stakeholder input is essential. Much academic research suggests the incorporation of public participation in environmental risk assessment is key to improved risk management and communication (e.g., Jardine, Predy, & MacKenzie, 2007; Lidskog, 2005; Stern & Fineberg, 1996). In fact, a review of 239 published case studies of civic participation in environmental decision-making processes (Beierle, 2002) showed that inclusive risk evaluations were more likely to result in higher-quality decisions.

Radical difference. According to Ryfe (2003), good public discourse is also contingent on radical difference, or the notion that all individuals have the right to speak in the voice of any of the groups to which they belong. Heterogeneity and plurality may complicate matters in the short term, but serve to improve discussion and outcomes in the long run.

English and colleagues (1993) support this principle of inclusiveness and radical difference, particularly with regard to its ability to resolve siting conflicts. They argue that the sustainability of outcomes achieved through inclusive decision-making processes is directly proportionate to the degree of diversity among stakeholder participants. Often, problems that tend to be overlooked by elite technocratic groups far removed from immediate threats – such as construction debris or neighbourhood traffic control – are invaluable insights from public participant pools (Fiorino, 1990). Opening access to additional resources better equips participants to consider alternative solutions within a more collaborative context. Furthermore, when interested parties are left out of the risk dialogue, feelings of disenfranchisement spawn alternative means of interfering in the process, leading to heightened public risk perceptions, institutional distrust, financial inefficiencies, and legislative abeyance (Stern & Fineberg, 1996).

Reciprocity. The inclusion of differing views would be fruitless if stakeholders did not actively engage in empathetic consideration of and conversation with one another. Ryfe's concept of reciprocity draws attention to the significance of such collaboration and interaction. The dialogic, or reciprocal, function of risk communication is wholly generative. This means dialogical communication not only transmits established messages, but generates new ones (Lotman, 1990). Such a duality expands the function of risk communication by encouraging, rather than deterring, critical thinking – an invaluable asset to mass deliberations on public policy. Participants must listen and respond to a continuum of arguments, which can limit inertia and illogical reasoning.

Past dependence on a technocratic approach in decision-making processes has triggered disaffection among stakeholders and limited the success of more democratic procedures (Tuler, 2000). Ryfe (2003) argues that reciprocal communication between engaged stakeholders can help overcome limitations of otherwise ritualistic public forums. In the dialectical discourse of risk, it levels the playing field by demanding a certain degree of consideration for all parties. The theoretical delineation between senders and receivers of risk messages in conventional risk communication is considered immaterial since all stakeholders are engaged in creating and collecting messages (Juanillo & Scherer, 1995).

Dialectical risk communication offers all stakeholders the opportunity to participate in sharing information, educating themselves, actively understanding components of an issue, and weighing the feasibility of proposed solutions (Juanillo & Scherer, 1995). During situations of conflict resolution, it enables society to collaboratively carry out the most fundamental purpose of democracy: government by and for the people. It thickens communication and deliberation with the addition of subjective and conversational civic participation. The generative nature of collaborative and dialogic forms of discourse does not alone guarantee its success in resolving disputes. Rather, effective application is dependent on the actual context to which it is applied. This context or setting must have an open orientation – a framework built to encourage interaction and the free flow of information while ensuring fair and formal mediation. These conditions underscore the need for coupling the dialectical approach to discourse with Renn's social arena concept of risk debates.

The social arena concept of risk debates

The social arena concept of risk debates (SACRD; Renn, 1992) is a metaphor rooted conceptually in political arena policies (Hilgartner & Bosk, 1998; Kitschelt, 1980) and used in risk communication to describe how the context of stakeholder actions influences collective policy decisions. From a practical perspective, SACRD provides a structured framework for analysing risk and environmental policy-making processes by defining stakeholder roles and mapping the flow of communication. The application of the social arena concept also helps satisfy Ryfe's three remaining factors for good public discourse – formalisation, rational argumentation, and moderation – in the context of risk debates.

Formalisation. According to Ryfe, formalisation provides government support for citizen rights. In the context of risk debates, formalisation is enacted by the rule enforcement agency who sets rules and conditions by which actors must abide (e.g., the number and location of hearings, accessibility of documents to the general public). Stakeholder tactics are not determined by the rule enforcer, but rather monitored and guided toward greater productivity. Essentially, the rule enforcement agency serves as a system of checks and balances that is dedicated to ensuring fair, democratic discourse. The rules and procedures set forth by the enforcement agency also aim to remove structural inequalities by addressing imbalanced resource allocation among stakeholders.

Renn (1992) highlights five key resources in social arenas: monetary strength, legal authority, social influence, value commitment, and evidence. Corporate enterprises often depend on financial strength, while municipal governments use legislative means (legal authority) to accept or deny siting proposals. Community coalitions with limited funds and authority usually take advantage of more intangible social resources like trust to garner support. In modern democratic and pluralistic societies, however, actors must capitalise on more than one resource in order to be successful (Renn, 1992). The rule enforcement agency supplements diversified resource procurement among actors. This assistance may come in many forms, such as underwriting an independent scientific investigation or public speaking seminars for citizens.

Grounded rationality. Grounded rationality involves first advancing claims, providing evidence and finally, developing relevant counterarguments. Within SACRD, the rule enforcement agency's central position in the arena is to impose order on the exchange of information, as well as on the rationality of arguments presented. Renn (1992) claims that actors in a social arena, while working to maximise their opportunities to influence the outcome of the debate, naturally adhere to a rational approach. Should rationality waver or cease to exist, the rule enforcer's position permits it to demand actors to follow a more logical communication pattern.

Critics of stakeholder engagement in a civic context argue that most laypeople are incapable of making sound contributions to complex risk assessments. Contrary to widespread assumption, citizens can gauge risks sensibly. Disagreements arising between the public and experts should not be qualified as solely based on public irrationality (Fischhoff, Watson, & Hope, 1984; Slovic, 1992). Structure and communicative guidance provided by a rule enforcement agency can facilitate rational dialogue and ultimately assist risk regulators in rendering better decisions (Renn, 1998).

Moderation. Moderation overcomes the inherent fragmentation of modern society and provides structured facilitation to public debate (Ryfe, 2003). Within the social arena, moderation is most often the responsibility of the rule enforcement agency. While in most cases, the rule enforcer is the ultimate decision maker, policy style often characterises the rule enforcement agency more as a moderator than sovereign administer (Renn, 1992). Asymmetric models of environmental risk communication that prevent civic participation in decision-making and lack effective moderation, such as public meetings, can lead to dissatisfaction among participants, decrease perceived source credibility, and even increase risk perceptions (McComas, 2003a, 2003b, 2003c). Facilitated deliberation (e.g., citizens juries), on the other hand, can provide equal opportunities for inquiry and argumentation (Lundgren & McMakin, 2009).

Structured dialectical discourse in Québec

While dialectical discourse is not a panacea for all debates on environmental risk, it played an effective role in Québec's LNG siting conflict. The successful resolution was brought about by the Bureau d'audiences publiques sur l'environnement (BAPE), whose risk assessment process was characterised by a coupling of dialectical discourse and social arena theory.

BAPE: Québec's rule enforcement agency

In 1978, the provincial government of Québec established BAPE, a quasi-judicial government organisation whose primary function is to inquire into any question related to the quality of the environment submitted to it by the Minister of Sustainable Development, including private sector energy development proposals (Gouvernement du Québec, 2008). Official public hearings, or *audiences publiques*, are organised using a distinctive method of stakeholder education and subsequent consultation. BAPE is then required to submit a project report and analysis to the Minister for further consideration and authorisation.

BAPE considers itself a 'gateway' for civic engagement in the project authorisation process, while claiming complete objectivity, transparency, and impartiality (BAPE, 2008). Over the past 30 years, the agency has conducted official risk assessments and mediated public participation in 250 different environmental inquires. During that time, more than 100,000 stakeholders (i.e. citizens, municipalities, businesses) have presented opinions and submitted over 15,000 memoirs to the organisation.

In late 2006, BAPE was commissioned to review the Rabaska proposal. The evaluation began with a primary information and public consultation period (10 October–24 November 2006). During that time, BAPE publicised its involvement and opened consultation centres in the region for public education on the topic. The Minister of Sustainable Development received 50 petitions as a result of the primary consultation – a relatively large number that confirmed the necessity for BAPE's involvement and its creation of the Joint Review Panel dedicated solely to the examination of Rabaska. The considerable level of civic participation in BAPE's proceedings was without precedent in the province. To accommodate the high volume of engagement, BAPE coordinated two additional periods of in-depth public consultation (6–15 December 2006; 29 February–9 January 2007).

The first 14 séances, or hearings, offered citizens the opportunity to pose questions in an effort to increase understanding of the proposal and its potential implications. The second portion of public hearings, consisting of 20 additional séances, permitted citizens to express their concerns and opinions. Both hearings took place before 14 members of the BAPE Joint Review Panel, Rabaska executives and their consultants, resource authorities, government officials, various social groups, hundreds of neighbouring citizens, and the media. A total of 699 memoirs and 15 oral testimonies were given to the Joint Review Panel during these two periods. An average BAPE inquiry consists of 30 memoirs. Prior to the LNG terminal siting, the largest number of memoirs ever received from stakeholders was 340 – half that of Rabaska. During the process, the memoirs and testimonies, along with other documents submitted to BAPE for review, remained open and available to the public both online and at eight official consultation centres in the province.

Dialectical discourse

The Canadian Environmental Assessment Agency (CEAA, 2007), which oversees BAPE, ensures that 'public participation in the federal environmental assessment process [is] an open, balanced process ... [in which] local and traditional knowledge about a project's physical site can help to identify and address potential environmental effects at an early stage'. In accordance with CEAA legislation, BAPE considers citizens to be competent stakeholders and experts in their communities. Throughout the Rabaska review, BAPE encouraged public inquiry, education, and participation, even distributing pamphlets throughout the province entitled *How to Participate* (BAPE, 2008). The pamphlets encouraged locally based 'resource people' to offer not only scientific and technical expertise, but also practical information about what the province needed and could support.

Expert analysis was also solicited from sources beyond the Consortium, more specifically by citizens groups who sought expert advice beyond the technical and political circles within Québec and Canada. Relevant analyses from scholars and officials abroad were incorporated into the memoirs and presentations of citizen stakeholders, and thus considered by BAPE. Opening the dialogue in these ways gave citizens a sense of empowerment that eclipsed any budding feelings of disenfranchisement or victimisation. Contributing to the analysis with independent research infused laypeople with a greater sense of control and a realistic, rational purpose in the matter without undermining the functional authorities of science and government. In the end, Qussai Samak (BAPE,

2007), chair of the panel that oversaw the *audiences publiques* on Rabaska, underscored the success of such collaborative civic participation:

> On behalf of the Panel, allow me to acknowledge the extraordinary participation, unprecedented in Québec, by the people of Lévis, Beaumont, and l'Île d'Orléans ... as well as the scope of their contribution ... Their thoroughness, discipline, and courtesy bear eloquent testimony to their civic engagement and sense of citizenship, and are a tribute to their communities and to the country's vibrant democratic traditions.

André Stainier, president of the environmental group Les Amis de La Vallée du Saint-Laurent, noted that when BAPE is involved in a project inquiry, 'tout le monde est considéré. Le public n'est jamais encadré. C'est vraiment un instrument de la démocratie'. (Stainier, personal communication, 14 February 2008). That is to say, everyone is considered, the public is never shut out of the discussion, and BAPE is truly an instrument of democracy.

In Stainier's opinion, two major schools of risk communication emerged from social group strategies during the Rabaska debate, one based on environmental politics and another primarily on scare tactics. Anti-LNG groups like Rabat-Joie, he noted, sought to create a psychosocial public crisis in order to achieve their objective of blocking the terminal's construction. Stainier admitted that he and his group purposely refrained from garnering excessive media attention as a result of such inflated propaganda. Stainier and fellow environmental protectionists considered radical groups like Rabat-Joie detrimental to the greater cause of preserving the St. Lawrence River Valley, which was his main objective (Stainier, 2008). In the absence of dialectical discourse structures, less provocative stakeholders, like Stainer and Les Amis de La Vallée du Saint-Laurent, would run the risk of being overshadowed in the media or even silenced by extremist protestors. Fortunately, the dialectical discourse encouraged and supported by BAPE during the *audiences publiques* allowed for different groups to openly and independently voice their concerns.

Québec's social arena

The exchange of information and opinions between stakeholders in the Québec debate followed a clear, social arena pattern, as exhibited in Figure 12.1. Primary actors supplied both evidence and opinions to the forum, which were moderated by the rule enforcement agency and

How Structured Dialectical Discourse of Risk Eased Tension   201

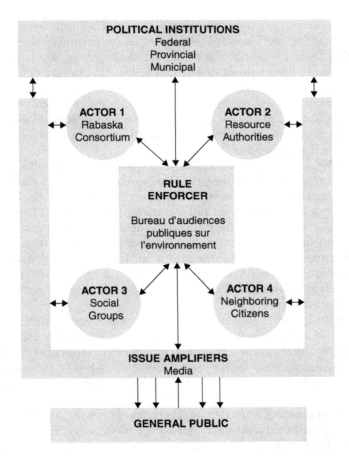

Figure 12.1 Graphical representation of social arena of risk in Québec – LNG debate

ultimately used by the province to arrive at a negotiated decision on its LNG policy. From federal and provincial authorities to grassroots citizen coalitions, the distinct roles played by each were integral to the risk assessment process and governed by formalisation, grounded rationality, and moderation.

Gaston Cadrin, president of the social activist group GIRAM, criticised the Rabaska Consortium for employing an expensive public relations firm – or in his own words *'une machine de communication payée'* – to defraud the public (Cadrin, personal communication, 12 March 2008). But, formalisation by BAPE purposefully off-set such resource advantages on both sides and forced all interested parties onto one stage with a defined set of rules for communication. For civic participants at a financial and evidential disadvantage, BAPE financed a number of projects aimed to generate additional LNG safety information. Groups of engaged citizens (i.e., professors, environmentalists), for instance, received grants to conduct field research on current LNG safety regulations in France. Their findings were considered during the hearings as a means of counterbalancing the expert testimony commissioned by the Consortium.

In order to uphold its mission of impartiality, BAPE also helped Rabaska off-set a common resource disadvantage for private companies with financial motives: value commitment. By bringing the Rabaska Consortium into a formalised and regulated social arena, BAPE created a situation in which the Consortium could demonstrate cooperation and dedication to the community. The Rabaska Consortium's compliance and investment of time and resources in the hearings, as well as its willing acceptance of all BAPE recommendations, increased its credibility with other key stakeholders and positioned it as an invested actor rather than external aggressor.

According to BAPE's formal constitution of procedure (BAPE, 2004), the goal of mediation is to help stakeholders arrive at a consensus. In addition to the *audiences publiques,* mediation is a clearly defined supplemental method employed by the BAPE when structuring the proceedings of stakeholder dialogue. Cadrin criticised BAPE, stating that 'leur procèssus n'était pas de mediation ... ils ont justement collecté des opinions opposantes' (Translation: 'Their process was not a mediation ... they only collected opposing opinions.') (Cadrin, 2008). In BAPE's defence, Jean-Sébastien Filion, spokesman for the organisation, argued that full mediation was not a viable option for Rabaska given the inquiry's unprecedented scale. Thus, the most effective alternative for handling the Rabaska siting conflict involved 'heavily moderating' the hearings as a means of synthesising and organising facts and opinions (Filion, personal communication, 4 April 2008).

Following the completion of the *audiences publiques* in early 2007, stakeholder groups including municipal officials and the promoter of the project made their satisfaction with the public consultation and risk assessment processes known. Even the most staunchly opposed critics of Rabaska could not deny the operational success of the social arena concept as applied to this particular risk debate: 'Le déroulement, il fonctionne bien' (Translation: 'The process, it functions well.') (Cadrin, 2008).

Problems in Massachusetts

The waterfront installation of an LNG terminal in the industrial city of Fall River, Massachusetts, seemed an attractive and feasible option. In early 2003, Amerada Hess submitted its environmental impact assessment to the Federal Energy Regulatory Commission (FERC). The project, sited in an urban neighbourhood within one mile of 9,000 people, would bring $3 million in taxes annually (Keteyian & Hirschkorn, 2007). In 2005, the FERC approved the proposal in a 3–1 vote; however, the decision was not well-received by locals. Bureaucratic barriers were erected by city officials who 'vowed along the way to bleed the project dry through 1,000 paper cuts' (Richmond, 2011). Ad-hoc efforts from stakeholders including government officials (i.e., then U.S. Senators John Kerry and Edward Kennedy), the U.S. Coast Guard, and grassroots environmental groups contributed to repeated delays in the project until Hess eventually withdrew its proposal eight years later.

Arguably, the lack of a framework for dialectical discourse in Fall River was a major drawback to effective risk communication. Consistent with prior research (e.g., McComas, 2003a), traditional public meetings infrequently scheduled frustrated stakeholders and damaged institutional credibility, creating what scholars refer to as a 'crisis in confidence' in Massachusetts (Slovic, Flynn, & Layman, 1991, p. 1606). Although the FERC conducted what it calls 'scoping meetings' to solicit information and statements from experts and the public, true reciprocity was lacking. Stakeholders believed the federal government possessed ultimate authority and public input was meaningless. In an interview with a local radio station, the CEO of Weaver's Cove Energy, Gordon Schearer, even undermined the authenticity of political opposition to the project at the time, stating that 'it's clearly not a problem for a local politician to take an opposing view [because it's a federal decision]. And, of course, you have a convenient excuse if the federal government approves it' (Emery, 2005).

Without an agency like BAPE focused on generating productive two-way discussions and offsetting stakeholder advantages, uncontrolled public conversation evolved into *ecospeak*, a form rhetoric involving reducing public debate about the natural environment into two opposing forces – environmentalist versus developer (Killingsworth & Palmer, 1992). The battle served as a prime example of the degenerative effects of the social amplification of risk (Kasperson et al., 1988). Much of the opposition expressed concerns about the safety risk posed by tankers filled with explosive cargoes, which overshadowed any rational discussion of other risks that posed a more certain threat (King, 2005). For instance, the Coalition for the Responsible Siting of LNG Facilities (CRSLF) often suggested that the proposed terminal would be 'like living near the world's largest Roman candle' (CRSLF, 2006).

Reflexive and reciprocal dialogue among diverse stakeholders formally moderated by an agency with institutional credibility was nonexistent. Instead, wars were waged in local media, online, and on front lawns. Lacking trust, Hess leveraged evidence and financial resources to campaign locally and fight legal attacks. Lacking financial resources, anti-LNG groups leveraged citizen trust and value commitment. Had there been a framework in place for structured dialectical discourse in Massachusetts, it is likely that a great deal of time, resources, and credibility could have been saved.

Conclusion

Although both terminal proposals were ultimately abandoned due to market shifts, comparative analysis of LNG siting conflicts can be useful for identifying critical missteps and highlighting functional strategies in risk communication. The *audiences publiques* process in Québec serves as a prime model for how structured dialectical discourse can facilitate effective decision-making in siting conflicts. BAPE's approach judiciously combines two risk communication models: dialectical discourse and the social arena concept of risk debates. This combination achieved success because it engaged divergent socio-political constituencies, balanced resources, limited delays, and moderated a rational and even exchange of information between stakeholders. Moving forward, institutions engaged in evaluating siting proposals should consider establishing a similar framework for civic engagement in environmental decision-making processes.

Acknowledgements

The author gratefully acknowledges the U.S.-Canada Fulbright Program for its financial support of this project. She also thanks Dr. Jacques de Guise and l'Université Laval in Québec City, Québec, for their guidance and support on the ground.

References

BAPE. (2008). What is BAPE?. Retrieved March 15, 2015, from http://www.bape.gouv.qc.ca/sections/faq/eng_faq_ind.htm#qcq

———. (2007). Rabaska Project – Implementation of an LNG Terminal and Related Infrastructure Inquiry and Public Hearing Report (Report 241). *Bibliothèque et Archives Nationales du Québec*. ISBN: 978-2.550-50148-0. Québec, QC.

———. (2004). Rules of procedure relating to the conduct of environmental mediation. Retrieved October 1, 2014, from http://www.bape.gouv.qc.ca/sections/documentation/proc-mediation-anglais.pdf

Beierle, T. C. (2002). The quality of stakeholder-based decisions. *Risk Analysis, 22*(4), 739–749.

Brinkhoff, T. (2007). Canada: Québec City Population. Retrieved May 10, 2008, from http://www.citypopulation.de/Canada-Quebec.html

Canadian Environmental Assessment Agency (CEAA) (2007). Public Participation. *Government of Canada*. Retrieved March 15, 2015, from http://www.ceaa-acee.gc.ca/default.asp?lang=en&n=8A52D8E4-1

Coalition for the Responsible Siting of LNG Facilities (CRSLF) (2006). Why take the risk? *NoLNG.org*. Retrieved March 15, 2015, from http://nolng.org

Collingridge, D., & Reeve, C. (1986). *Science speaks to power*. New York: St. Martin's.

Emery, C. E. (2005) CEO defends weaver's cove on radio. *The Providence Journal*, January 12, C2.

English, M. A., Gibson, A., Feldman, D., & Tonn, B. (1993). *Stakeholder involvement: Open processes for reaching decisions about the future uses of contaminated sites. Final report to the US Department of Energy*. Knoxville, TN: University of Tennessee Waste Management Research and Education Institute.

Fiorino, D. (1990). Public participation and environmental risk: A survey of institutional mechanisms. *Science, Technology, & Human Values, 152*, 226–243.

Fischhoff, B., Slovic, P., & Lichtenstein, S. (1982). Lay foibles and expert fables in judgments about risk. *The American Statistician, 36*(3), 240–255.

Fischhoff, B., Watson, S., & Hope, C. (1984). Defining risk. *Policy Sciences, 17*, 123–139.

Foss, M. M. (2006). LNG safety and security. Center for Energy Economics. Retrieved October 1, 2014, from http://www.beg.utexas.edu/energyecon/lng/documents/CEE_LNG_Safety_and_Security.pdf

Gould, K., Schnaiberg, A., & Weinberg, A. (1996). *Local environmental struggles: Citizen activism in the treadmill of production*. New York: Cambridge University Press.

Gouvernement du Québec. (2008). Division II.1 – 6.3. Function – The Bureau d'audiences publiques sur l'environnement. Environment Quality Act R.S.Q. Q-2. Retrieved October 1, 2014, from http://www2.publicationsduquebec.gouv.qc.ca/dynamicSearch/telecharge.php?type=2&file=/Q_2/Q2_A.htm

Gutteling, J. M., & Kuttschreuter, M. (2002). The role of expertise in risk communication: Laypeople's and expert's perception of the millennium bug in The Netherlands. *Journal of Risk Research, 5*(1), 35–47.

Hamutuk, L. (2008). Timor-Leste Institute for reconstruction monitoring and analysis: History of accidents in the LNG industry. Retrieved October 1, 2014, from http://www.laohamutuk.org/Oil/LNG/app4.htm

Hilgartner, S., & Bosk, C. L. (1998). The rise and fall of social problems: A public arenas model. *American Journal of Sociology, 94*, 53–78.

Jardine, C. G., Predy, G., & MacKenzie, A. (2007). Stakeholder participation in investigating the health impacts from coal-fired power generating stations in Alberta, Canada. *Journal of Risk Research, 10*(5), 693–714.

Juanillo, N., & Scherer, C. (1995). Attaining a state of informed judgments: Toward a dialectical discourse on risk. *Communication Yearbook, 18,* 278–299.
Kasperson, R. E., Renn, R., Slovic, P., Brown, H. S., Emel, J., & Goble, R. (1988). The social amplification of risk: A conceptual framework. *Risk Analysis, 8,* 177.
Keteyian, A., & Hirschkorn, P. (2007). Safety concerns tie up LNG development. Retrieved October 1, 2014, from http://www.cbsnews.com/news/safety-concerns-tie-up-lng-development/
Killingsworth, J. M., & Palmer, J. S. (1992). *Ecospeak: Rhetoric and environmental politics in America.* Carbondale: Southern Illinois University Press.
King, W. (2005, July 24). As you were saying: LNG siting dispute fuels misguided solutions. *The Boston Herald,* p. 26.
Kitschelt, H. P. (1980). *Kernerenergiepolitik: Arena eines gesellschaftlichen Konflikts.* Frankfurt: Campus Verlag.
Lidskog, R. (2005). Siting Conflicts – Democratic perspectives and political implications. *Journal of Risk Research, 8*(3), 187–206.
Lotman, Y. (1990). *Universe of the mind: A semiotic theory of culture.* Bloomington: Indiana University Press.
Lundgren, R., & McMakin, A. (2009). *Risk communication: A handbook for communicating environmental, safety, and health risks.* Columbus, OH: Battelle Press.
Margolis, H. (1996). *Dealing with risk: Why the public and experts disagree on environmental issues.* Chicago: University of Chicago Press.
McComas, K. A. (2003a). Citizen satisfaction with public meetings used for risk communication. *Journal of Applied Communication Research, 31*(2), 164–184.
McComas, K. A. (2003b). Public meetings and risk amplification: A longitudinal study. *Risk Analysis, 23*(6), 1257–1270.
McComas, K. A. (2003c). Trivial pursuits: Participant views of public meetings. *Journal of Public Relations Research, 15*(2), 91–115.
McComas, K. A. (2004). When even the 'best-laid' plans go wrong. *EMBO Reports, 5*(Suppl 1), S61–S65.
Rabaska. (2005). The World of LNG. Retrieved October 1, 2014, from http://www.rabaska.net/lng
Reed, M. (2008). Stakeholder participation for environmental management: A literature review. *Biological Conservation, 10,* 2417–2431.
Renn, O. (1992). The social arena concept of risk debates. In S. Krimsky & D. Golding (Eds.), *Social theories of risk* (pp. 179–196). Westport, CT: Praeger.
Renn, O. (1998). The role of risk perception for risk management. *Reliability Engineering and System Safety, 59,* 49–62.
Richmond, W. (2011, June 13). Hess LNG withdraws Fall River Weaver's Cove proposal. *The Herald News.* Retrieved from http://www.heraldnews.com/article/20110613/NEWS/306139388
Ryfe, D. (2003). The principles of public discourse: What is good public discourse? In J. Rodin & S. Steinberg (Eds.), *Public Discourse in America* (pp. 163–177). Philadelphia: University of Pennsylvania Press.
Sauer, B. (2003). *The rhetoric of risk: Technical documentation in hazardous environments.* New Jersey: Lawrence Erlbaum.
Shell Global. (2014). What is LNG? Retrieved October 1, 2014, from http://www.shell.com/global/future-energy/natural-gas/liquefied-natural-gas/what-is-lng.html

Slovic, P. (1992). Perception of risk: Reflections on the psychometric paradigm. In S. Krimsky & D. Golding (Eds.), *Social theories of risk* (pp. 117–152). Westport, CT: Praeger.

Slovic, P. (1999). Trust, emotion, sex, politics, and science: Surveying the risk assessment battlefield. *Risk Analysis, 19*(4), 689–701.

Slovic, P., Flynn, J. H., & Layman, M. (1991). Perceived risk, trust, and the politics of nuclear waste. *Science, 254,* 1603–1607.

Stern, P., & Fineberg, H. (1996). *Understanding risk: Informing decisions in a democratic society.* Washington, DC: National Academy Press.

Tuler, S. (2000). Forms of talk in policy dialogue: Distinguishing between adversarial and collaborative discourse. *Journal of Risk Research, 3*(1), 1–17.

U.S. Energy Information Administration. (2004). Country Analysis Briefs: Canada Retrieved October 1, 2014, from http://www.eia.doe.gov/countries/country-data.cfm?fips=CA&trk=m

Vlek, C. A., & Cvetkovich, G. (1989). *Social decision methodology for technological projects.* Dordrecht: Kluwer Academic Publishers.

Webler, T. (1999). The craft and theory of public participation: A dialectical process. *Journal of Risk Research, 2*(1), 55–71.

Yin, R. K. (2009). *Case study research: Design and methods* (4 ed.). Los Angeles, CA: Sage.

13
Framing Risk and Uncertainty in Social Science Articles on Climate Change, 1995–2012

Christopher Shaw, Iina Hellsten, and Brigitte Nerlich

> This research has been funded by the ORA-NWO grant number 464-10-077 and by ESRC ORA grant number RES-360-25-0068. (Climate Change as a Complex Social Issue).

Introduction

The issue of climate change is intimately linked to notions of risk and uncertainty, concepts that pose challenges to climate science, climate change communication, and science-society interactions. While a large majority of climate scientists are increasingly certain about the causes of climate change and the risks posed by its impacts (see IPCC, 2013 and 2014), public perception of climate change is still largely framed by uncertainty, especially regarding impacts (Poortinga et al., 2011). Social scientists and communication researchers have begun to advocate moving from a framing of climate change in terms of uncertainty to one that focuses on risk (Painter, 2013; Silverman, 2013) and they hope that this shift in framing may generate greater public support for climate mitigation policies.

Individual and social responses to climate change risks and uncertainties are, of course, always context-dependent and mediated not only by an understanding of climate science but also political and cultural factors. And while some of the main risks associated with climate change are physical risks, such as flooding and droughts, climate change also poses economic, political and social risks, such as loss of trust and marginalisation of certain constituencies.

Social scientists have studied issues of risk perception, public understanding of risk and policy responses to risk for a long time.

In the context of climate change in particular, public and political understandings of risks and uncertainties interact with values, attitudes, social influences, and cultural identity (Renn, 2001; Hulme, 2009). To understand how framing climate change as a risk issue may play out in society, it is important to reflect back on the wealth of research on risk in the social science literature. How are risk and uncertainty frames discussed in the social sciences and how are risk and uncertainty framed in the social sciences specifically in relation to climate change? Finding answers to these questions will be the focus of our chapter.

Frames act as organising ideas, or cognitive windows which relate a particular version of the topic being reported (Olausson, 2009: 423). Frames highlight specific aspects of the debate while hiding other plausible interpretations (Nisbet, 2009). An uncertainty frame might focus on areas of ignorance in the climate science and stress what is unknown. An example of this frame is found in discussions of the speed at which the planet will warm as a result of increased atmospheric concentrations of greenhouse gas, commonly referred to as climate sensitivity. Such a projection of future changes in a complex system is of course not devoid of uncertainties. Some social commentators focus on these uncertainties to suggest there is insufficient evidence to justify making radical and expensive changes to the production and use of energy in order to reduce carbon dioxide emissions. A risk frame would highlight what is knowable about the relationship between increases in greenhouse gas concentrations and future temperature rises, and from that extrapolate to provide risk assessments which can inform precautionary measures (Painter, 2013: 11). Painter argues the use of risk frames allows decision makers and the public to start thinking about the costs of climate change mitigation policy in terms of paying insurance premiums; we are all familiar with paying money to limit the impact of uncertain future events (Painter, 2013: 26).

In this chapter, our aim is to analyse how risk and uncertainty were framed in social scientific articles on climate change over three decades. We examine the abstracts of 155 social science journal papers addressing risk and uncertainty from the period 1995–2012. In particular, we focus on articles that were co-authored by at least one author from the UK or the Netherlands, as these two countries are part of the top five country affiliations publishing on climate change risks or uncertainties. They also share similar risks arising from climate change, especially risks related to sea level rise and flooding (Bernstein et al., 2007; Veraart and Bakker, 2009; DEFRA, 2010). The UK and the Netherlands also share broadly similar cultures as northern European Annex 1 countries, including the adoption of the precautionary principle which the Netherlands recognises as a guiding framework for environmental policy and which, one might argue, is also the foundation for the UK's Climate Change Act of 2008, making 80% reductions in greenhouse gas emissions by 2050 legally binding. The UK and the Netherlands therefore offer interesting case studies for examining how social science scholars, working within nations at the forefront of efforts to build an international climate regime, discuss the framing of risk and uncertainty in the context of climate change.

Risk, uncertainty and the social sciences

The topics of risk and uncertainty have been discussed for many decades in the social sciences especially following a series of environmental, agricultural, and health crises at the end of the 1980s (see for example Beck, 1986, 1995; for an overview of risk in social science, see Taylor-Gooby & Zinn, 2006). Uncertainty too has attracted much attention from social scientists, especially within Science and Technology Studies. Wynne and Jasanoff (1988: 27–28) identify the media, the law, regulatory agencies, advisory bodies, and advocacy groups as key institutional and political factors in the process of defining uncertainty. These institutions tend to assume (or require) that all uncertainty be quantifiable, leaving qualitative questions such as 'what counts as uncertainty?' unasked (Tickner, 2003: 6, 17). The quantification of uncertainty is seen in some quarters as an ideological act – providing the impression of objectivity while denying the culturally determined components of the knowledge production process (Mulkay, 1991: 8; Wynne & Jasanoff, 1998; Kline, 2010: 9). McKenzie's well known 'uncertainty trough' demonstrates clearly how uncertainty is a cultural artefact. In this model, perception of uncertainty varies with distance from the point of knowledge production – perception of uncertainty is high for those closest to the production of knowledge and higher still for those most alienated, with a greater degree of certainty characterising the attitude of those actors in between these two extremes (MacKenzie, 1990, cited in Wynne & Jasanoff, 1998: 13 and 301–31).

Risk assessment seeks to accommodate uncertainty through statistical modelling (Tickner, 2003: 6), the desired outcome of which is a credible, probabilistic assessment of the likelihood of a particular event or events arising. Persistent uncertainties surrounding how increases in atmospheric concentrations of greenhouse gases will impact the physical properties of the climate system means it is necessary for climate projections to employ both probabilistic assessments and frequency distributions, techniques which rely, to varying degrees, on subjective judgments. Belief about the likelihood of any particular event arising is, from a realist perspective, the product of a rational, consistent thought process which is directly translatable into a probability statement (Baer, 2005: 56–58). Thus realist risk assessment assumes that subjective judgments can be rationally incorporated into science through adequate probabilistic calculations (Gigerenzer, 2001: 22; Nowotny et al., 2001: 34; Kline, 2010).

This hybrid of data and opinion is described as Bayesian probability. Employment of Bayesian techniques is viewed as necessary when dealing with complex problems permeated by uncertainty, as the alternative is to say 'we know nothing at all' (Hulme, 2009: 73). The role of rationality in Bayesian analysis has been questioned by Baer who, in discussing the study of climate sensitivity, argues that faced with huge uncertainties 'what you choose to act as if you believe is fundamentally an ethical choice' (Baer, 2005: 14). Such choices might be influenced not only by quantifications of risk and uncertainty in the science but also by the framing of risk and uncertainty in the context of media, politics, and public discourse more generally. Improving the communication of scientific risks and uncertainties (see Budescu et al., 2009) may therefore not be enough to change individual and social perceptions of risks and uncertainties. The belief that the failure of people to act rationally is the result of insufficient knowledge has become known as the information deficit model of science communication. It is a model which assumes understanding emerges from a one-way process of experts telling non-experts the science, with the result that the recipients of this information will then begin acting in accordance with that information. In this sense it is very different from a dialogue between experts and lay audiences which has as its goal the co-production of knowledge (Timmer, 2013). The co-production process has been identified as particularly important in the area of climate change policy. By allowing citizens to become both 'creators and critics in the knowledge production process' (Rayner, 1987: 8) the chances of finding socially acceptable solutions is improved (Rachlinski, 2000; Cash et al., 2003; Lowe et al., 2006). These are very different assumptions about how to build strong climate engagement than those evident in claims that communicating climate impacts more in terms of risk than uncertainty will result in significant changes in public attitudes to climate change.

Data collection and methods

In order to achieve our research aims, that is, to understand how risk and uncertainty are framed in social science articles about climate change, we collected a longitudinal data set of social scientific articles dealing with climate change between 1995 (when the first article on risk and/or uncertainty appeared in the Web of Science database) and 2012. We used the Web of Science and, in particular, the social sciences and the humanities indices, as these documents reflect social science debates of risk and uncertainty. To further limit our focus we analysed articles that were co-authored by either Netherlands or UK based author in order to compare social construction of risks/ uncertainties by authors working in countries that share similar climates and climate change risks, in particular flooding risk that one would expect to be relevant to researchers in these two countries.

We restricted our analysis to the abstracts as they provide the main content of the research in a condensed manner. The abstracts were coded depending on whether the title of the paper referred to either risk or uncertainty. Twelve papers mentioned both risk and uncertainty in the title.

We manually coded the abstracts into six main frames: adaptation, insurance, perception, governance, assessment and economics (see Table 13.1). We have further divided these main frames according to the overall focus of the article on risks or uncertainties.

Our coding proceeded as follows. Those abstracts which discussed adaptation and made explicit use of that word were put into the 'adaptation' category. The abstracts discussing climate change in terms of flood insurance but which did not necessarily use the word 'adaptation' were also coded into that category The perception code refers to the manner in which risk and uncertainty are framed for the purposes of communication, not just to the public but also amongst the relevant expert and policy communities. Assessment is the frame used for papers which address the role of modelling in turning uncertainty into risk. Whilst it would have been legitimate to place the papers employing economic modelling to define climate policy under uncertainty, it was

Table 13.1 Risk and uncertainty frames

	Adaptation	Insurance	Perception	Governance	Assessment	Economics	Total
Risk	19	6	40	20	17	n/a	100
Uncertainty	4	n/a	17	15	9	11	55

more consistent to place those alongside papers drawing on ideas of cost-benefit analysis to prescribe certain policy trajectories. The governance frame is for those papers which discuss how risk and uncertainty affect decision-making. The coding of these main frames was discussed with all the authors to improve the reliability of the procedure. Since our aim is not to provide statistically significant results, we did not calculate inter-coder reliability. Instead we supplement the quantitative results with qualitative analysis of the main frames.

Results

We will first discuss the results of the quantitative coding that shows the trends over time, and in the second part the results of the qualitative analysis of the main frames.

Overall trends

Overall, uncertainty abstracts were not concerned with uncertainties about the reality of anthropogenic climate change, but the challenges of identifying optimal responses in the face of uncertainties in projections of future climate impacts. From 2007 risk became a more salient theme than uncertainty. It is not clear whether this was a natural progression as research reduced the unknowns in climate science, or the result of a conscious decision to move away from discussing uncertainty in favour of risk frames.

In the abstracts we surveyed, the main focus was on perception of climate change, followed by governance, climate change assessment, and adaptation to global warming (Table 13.1 and Figure 13.1).

The distribution of codes across the risk and uncertainty frames reflects the overall trend of a larger number of risk papers than uncertainty papers. All the economics papers sit within the category of uncertainty, partly because of generally discussing cost-benefit deliberations as a means for overcoming uncertainty.

The UK is more strongly represented in this sample than the Netherlands (Figure 13.2). This difference is to some extent a response to the prominence of sceptic voices in UK public life identified by Painter (2013), and the UK's role as a leading player in international climate negotiations.

The longitudinal distribution of the risk and uncertainty categories and the codes in Table 13.1 are laid out in Figures 13.3–13.5.

Discussion of risk and uncertainty increases from 2005. In the UK, from whence most of the papers originate, 2005 was the year the then

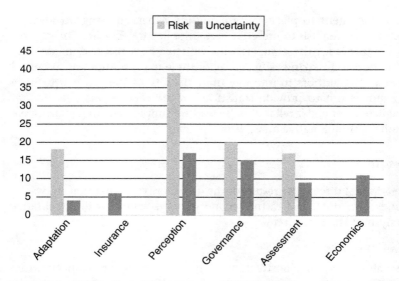

Figure 13.1 Distribution of frames across the 'risk' and 'uncertainty' categories

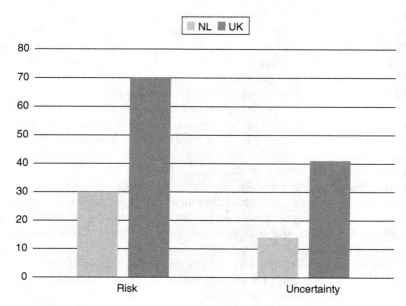

Figure 13.2 Distribution of papers by author country

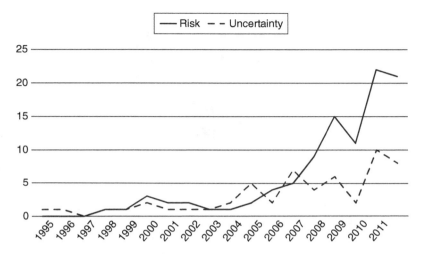

Figure 13.3 Distribution of articles across the time period 1988–2012

Prime Minister called a conference on defining dangerous climate change. It was also the year when the European Emissions Trading scheme was established. In 2007 the IPCC released its 4th Assessment Report (IPCC, 2007) and in 2008 the Climate Change Act introduced legally binding emission reduction targets (IPCC, 2007; Climate Change Committee, 2008). These events are all part of a growing realisation of the seriousness of the climate change problem, and the recognition of the need for the state to fulfil its primary goal of protecting its citizens from harm. Climate change was framed as a dangerous (risk) issue and reducing emissions was the main frame through which climate change mitigation came to be discussed at that time.

The surge in papers addressing uncertainty from the perspective of governance following the doomed Copenhagen climate summit in 2009, which was overshadowed by the 'climategate' affair (Nerlich, 2010) may reflect a move from a purely symbolic politics of climate change (Blühdorn, 2007) to a more substantive policy process. At that time, the risk/danger framing which shaped some of the climate change debates before the summit and before the recession gave way to framing climate change as uncertain and climate science as unsettled.

The perception frame shows an surge in coverage around 2009–2010 as was apparent for the uncertainty category. Whereas for uncertainty codes there was a marked drop off in perception papers from 2011, in the risk category this drop off came for papers coded under

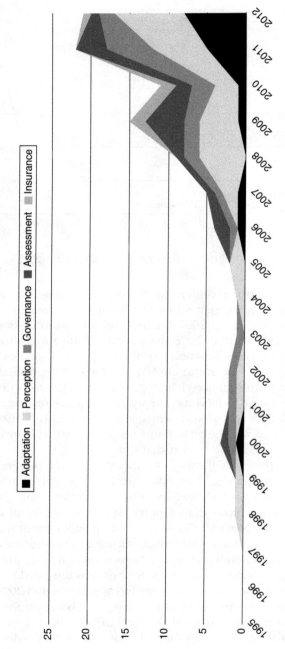

Figure 13.4 Timeline of uncertainty frames, 1995–2012

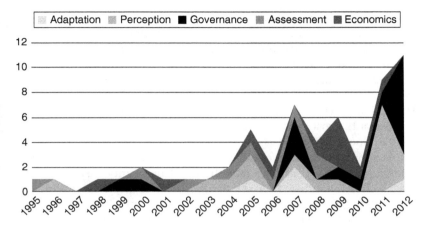

Figure 13.5 Timeline of risk frames, 1995–2012

governance. With climate change impacts becoming increasingly apparent, it is perhaps not surprising that the 'risk' category highlights an increasing level of interest in adaptation from 2010 onwards.

Main frames

In this section, we take a closer look at the six main frames of adaptation, perception, governance, assessment, insurance, and economics.

Adaptation

Risk management was the most salient theme in abstracts concerned with adaptation. Both British and Dutch authored papers were predominantly concerned with flooding risk. Urban environments were of greater concern than agricultural and rural areas. This may reflect the largely urban nature of UK and Dutch societies, or highlight the priority of protecting people from immediate harm. This latter point is clearly illustrated in a paper from Runhaar et al. (2012: 777) which identifies flooding as a higher priority issue for urban planners than climate risks such as heatwaves because lines of responsibility for protecting the public from flood risk are clearly demarcated.

> This paper focuses on the governance of two climate change-related risks in urban areas in the Netherlands, namely heat stress and

flooding from rainfall and rivers. Heat stress hardly seems to be perceived as an urgent problem, mainly because there is no clear 'problem owner'. Because municipalities are responsible for rain and sewage water management and partly for river flooding, increased flood risk is more often recognised as a (potential) problem.

This urban focus also reflects a concern with protecting property because of the cost implications of flood damage that is reflected in the scholarly publications.

Uncertainty did not negate the risk management goals, but was instead framed as a surmountable barrier. Uncertainty was a problem insomuch as it hampered the search for optimal levels of adaptation, i.e. rather than base adaptation on the precautionary principle, the research was directed at identifying just the right amount of adaptation, just where it was needed. Hence, it was argued that 'failure to incorporate uncertainty and flexibility in the economic analysis of flood risk and coastal management strategies can result in maladaptive decisions' (Hall et al., 2010).

Four of the papers in this category were notable for the greater saliency they afforded to uncertainty than was the norm. These papers were trying to identify mechanisms for adaptation planning under conditions of uncertainty. They contrasted with the other papers in the adaptation category by arguing for robust adaptation, i.e. a more precautionary approach.

Insurance

The insurance frame was more prominent in articles authored by Dutch scholars (five out of six articles authored by the Dutch). The abstracts project to a future of increased climate impacts on the economy, and hence the research attempts to understand what form of insurance is required to make these economic risks manageable. Opportunity was a strong frame used in these abstracts; climate change was likely to raise demand for insurance products. Currently there is no private flood insurance market in the Netherlands. As one paper stated: 'The results indicate that there are opportunities for the development of a flood insurance market' (Botzen & van den Bergh, 2012: 1005), while another claimed: 'climate change can also bring new business opportunities for insurers' (Botzen & Bouwer, 2010: 577).

One can reasonably attribute the flood focus to the particular geographic and topographical characteristics of the Netherlands, and the enhanced flood risk resulting from these attributes. One Dutch authored article addressing flood risk in Bangladesh has implications for Dutch policy, given the high flood risk and the question posed about the willingness to pay for insurance addressed in the paper, a question asked of Dutch households in two of the other papers.

Perception

The frame of perception was evident in relatively recent publications (27 out of 57 were published since 2008), and most focused on risk perception rather than perception of uncertainty (40 out of 57). In the earlier papers climate change is framed as a future risk and the problem is the communication challenges posed by this absent risk to lay, business, and expert communities. In 2012, the growing scientific consensus on the anthropogenic causes of climate change '...points to a need for a fundamental revision of our conceptualisation of what it is to do climate risk communication' (Pidgeon, 2012: 951).

The idea that this communication challenge could be overcome through the anchoring of novel climate risks in terms of more familiar framings was apparent in a paper from 2008, which compares climate change tipping point discourses with ideas of social tipping points and states that 'the prospect of runaway or "irreversible" global warming [...] has also revived an interest in its original sociological sense – i.e. tipping points in social and political movement. How do we relate the two?' (Stefan, 2008: 9).

In papers focusing on uncertainty, it is treated as a problem to be removed from the realm of climate science, discourse, and politics. One paper states that '[w]hile scientific consensus and political and media messages appear to be increasingly certain, public attitudes and action towards the issue do not appear to be following suit' (Whitmarsh, 2011: 690).

These papers hint at an ongoing adherence to the information deficit model, and offer the suggestion that the desired actions will result from the correct communication and understanding of risk and uncertainty. For example: 'A major challenge facing climate scientists is explaining to non-specialists the risks and uncertainties surrounding potential changes over the coming years, decades and centuries' (Pidgeon & Fischoff, 2011: 35).

Science is framed as a form of knowledge which, it is implicitly argued, is generally held by lay and policy audiences to be free of uncertainty. Therefore the question posed in these papers is how to get people to trust a science that cannot remove these uncertainties. The problem posed in the oldest paper identified, from 1996 remains essentially the same as that in the most recent paper from 2012, 16 years later, namely uncertainty as a frame-barrier to public engagement.

Governance

The abstracts coded as governance were characterised by a search for consensus under conditions of uncertainty, which included interest in understanding how the uncertainties regarding the linear or non-linear nature of climate change can be reduced to narrow the consensus range. Some authors questioned the possibility of being able to accommodate uncertainty within the political process through quantification, arguing instead that the debate should recognise the role of culture, values, and social relations in determining how people respond to scientific uncertainty:

> The sheer number of attempts to define and classify uncertainty reveals an awareness of its importance in environmental science for policy, though the nature of uncertainty is often misunderstood. The interdisciplinary field of uncertainty analysis is unstable; there are currently several incomplete notions of uncertainty leading to different and incompatible uncertainty classifications. One of the most salient shortcomings of present-day practice is that most of these classifications focus on quantifying uncertainty while ignoring the qualitative aspects that tend to be decisive in the interface between science and policy. Consequently, the current practices of uncertainty analysis contribute to increasing the perceived precision of scientific knowledge, but do not adequately address its lack of sociopolitical relevance. The 'positivistic' uncertainty analysis models (like those that dominate the fields of climate change modelling and nuclear or chemical risk assessment) have little social relevance, as they do not influence negotiations between stakeholders (Maxim & van der Sluijs, 2011: 482).

Economic modelling, with its quantitative framing of uncertainty, was offered as a corrective to this excess of subjectivity. Hence several papers suggested problems in governance generated by uncertainties could be overcome by improved modelling, often with a strong economic component.

Other papers are less concerned with reducing uncertainty and instead look to coping with it through different forms of social interaction, scenario-building, and communication which can co-exist with uncertainty. Floods are the predominant risk discussed from a governance perspective. The social dimensions of risk governance come to the fore, with questions being asked about whose attitudes should count in risk governance frameworks and how these different perspectives are mediated. There was a general sense across the abstracts of trying to close down the space for alternative perspectives opened up by too great an amount of uncertainty. There follows just one example of this kind of discourse:

> For flood risk management in England and Wales, we show that futures are actively constituted, and so imagined, through 'suites of practices' entwining policy, management and scientific analysis. Management has to constrain analysis because of the many ways in which flood futures can be constructed, but also because of commitment to an accounting calculus, which requires risk to be expressed in monetary terms. (Lane et al., 2011: 1748)

This constraining of the analysis to make the issue governable then leads to questions about what kinds of impacts should be accounted for, what sorts of risks should be considered and what sort of temporal and geographical scales are most appropriate for risk management given these uncertainties.

The other important theme in this section was the promise that economic modelling can help policy makers in the decision-making process by quantifying the risks and benefits of varying degrees of precaution in policy making. Again, this was not about specific climate risks, but rather addressed issues such as fixing a global carbon price, or understanding how safe is safe enough. 'There is an increasing demand for putting a shadow price on the environment to guide public policy and incentivise private behaviour. In practice, setting that price can be extremely difficult as uncertainties abound' (Dietz & Fankhauser, 2010: 270).

This longstanding economic framing of climate change (risk) still pervades climate change governance today (Shaw & Nerlich, 2015).

Assessment

Several papers (seven out of 26 assessment papers) address assessment of climate risks and uncertainties through economic modelling, or are concerned to assess the nature of the climate change problem through the financial implications. A prominent issue in these economics papers is the question of how to better assess future flood risk, though heatwaves, hurricanes, and health impacts are also addressed. There is also an interest in trying to better assess the likelihood of high-impact, low-probability events, in a sense trying to constrain the challenges posed by the potential for sudden catastrophic events and retain a role for discounting risk, or more crudely, to justify putting off the changes needed to address the risk to some future date:

> To what extent does economic analysis of climate change depend on low-probability, high-impact events? This question has received a great deal of attention lately, with the contention increasingly made that climate damage could be so large that societal willingness to pay to avoid extreme outcomes should overwhelm other seemingly important assumptions, notably on time preference. This paper provides an empirical examination of some key theoretical points, using a probabilistic integrated assessment model. New, fat-tailed distributions are inputted for key parameters representing climate sensitivity and economic costs. It is found that welfare estimates do strongly depend on tail risks, but for a set of plausible assumptions time preference can still matter. (Dietz, 2011: 519)

The papers in this category have a much stronger focus on place than other papers; whether that is coastal regions, urban heat islands, residential zones, and rivers or West Antarctic ice sheets. This is indicative of attempts to frame and manage risk not through a holistic whole earth view, but through addressing individual, though still large-scale, earth surface and socio-economic processes.

Strikingly, the abstracts that focus on uncertainty are all from pre-2009. This is in stark contrast to the risk assessment codes. The assessment codes are thus the only ones to show the pattern one would anticipate from standard definitions of risk and uncertainty – the early papers address the reduction of uncertainty so that later papers deal with the risk management that emerges from this constrained set of future possibilities.

Overall there is a sense of not trying to reduce the uncertainties so much as understand what the uncertainties are, where the epistemological boundaries of the climate change problem lie and what should count as uncertainty. There are significant doubts as to whether it will prove possible to constrain the uncertainties regarding future climate impacts, a topic addressed in this extract from an abstract for example

> This paper presents a broadly based up-to-date analysis of the main doubts and uncertainties in climate change science. The primary focus is on the different types and scales of climate change uncertainty including data bases, system complexity, the human fingerprint and climate modelling. (O'Hare, 2000: 357)

Economics

In abstracts dealing with economics, climate change is framed as a problem only in so much as there is the potential for negative impacts on economic growth. Having framed the problem as one of weighing costs and benefits, discount rates etc. it then becomes possible to model optimal (optimal in this sense being the most cost-effective) responses which account for a range of uncertainties in climate impact projections. An exemplar of this approach offers a methodology for identifying the optimal rate of emission reductions required to prevent the 'catastrophic' collapse of the West Antarctic Ice Sheet

> A collapse of the West-Antarctic Ice Sheet (WAIS) would cause a sea level rise of 5–6 m, perhaps even within 100 years, with catastrophic consequences. The probability of such a collapse is small but increasing with the rise of the atmospheric concentrations of greenhouse gas and the resulting climate change. This paper investigates how the potential collapse of the WAIS affects the optimal rate of greenhouse gas emission control. We design a decision and learning tree in which decisions are made about emission reduction at regular intervals: the decision makers (who act as social planners) have to decide whether to implement the environmental or not (keeping then the flexibility to act later). By investing in the environmental policy, they determine optimally the date of the optimal emission reduction. (Guillerminet & Tol, 2008:193)

Two papers deal with specific economic transactions – uncertainties about whether green innovations will succeed and whether people will pay a voluntary carbon tax on flight tickets. Otherwise the focus is on cost-benefit analysis, social costs of carbon, discount rates, marginal damage costs, and the other epistemologies which characterise economic analysis. The anomaly was the oldest paper in this category, from 1998, which included issues of fairness and income distribution as of equal importance as economic models. Explicit references to issues of equity were absent from the other abstracts.

Discussion and conclusion

The epistemological themes addressed in the scholarly articles on risks related to climate change were predominantly perception, adaptation, and assessment. Adaptation, or coping with the risks generated by climate change impacts is a theme one would have anticipated being prominent in this category. Assessment and perception are to some extent related categories, though assessment deals with much more applied approaches to risk management, especially in so much as the goal is to find the most optimal (i.e. cheapest or most cost-effective) level of intervention.

Perception dealt more closely with understanding which risks were deemed significant enough to require a response and how perception affected the willingness of actors to change their behaviour. The perception of the risks associated with activities such as nuclear power generation and carbon capture and storage technologies were also explored. The risk addressed most frequently related to water, with flooding receiving greater attention than drought. Insurance was frequently identified as an economic instrument which was, respectively, an underdeveloped market, a means of limiting financial risk and a mechanism for communicating price signals about climate impacts which could motivate public and private behaviour change. The other risks discussed were heatwaves, food production, and impacts on tourism (for example from increased hurricane activity).

Uncertainty, in turn, was, broadly, approached as a problem for effective governance. Most of the papers were addressing uncertainties relating to the identification of the most cost-effective mitigation and adaptation responses, not with any uncertainty relating to the fundamentals of climate change.

Risk, from 2007 onwards, is a much more salient theme in the papers analysed than uncertainty. The trend may reflect changes in funding priorities, with more money being directed to risk analysis than more abstract assessments of uncertainty. Or the pattern may be the result of research reducing uncertainties, turning unknowns into more constrained scenarios which could be assessed as risks.

Perception, governance, and assessment became increasingly prominent themes in the later papers, for both risk and uncertainty themes. This is to be expected if the discourse is moving towards the language of risk, and away from uncertainty. The questions about the gravity of the issue are largely settled. It is now a case of overcoming the challenges to delivering emission reduction targets.

Qualitatively the differences between papers from the UK and the Netherlands were not marked. Dutch papers were more concerned with risk; the UK papers gave greater attention to uncertainty. Recent papers from the Netherlands showed an interest in the development of a private flood insurance market. Governance issues were more prevalent in the UK, perhaps because delivery of effective mitigation is given greater urgency in the UK because of the legally binding targets set out in the Climate Change Act.

Looking across the whole data set, it is apparent that risk and uncertainty are points on a continuum, rather than distinct categories. There are no standard criteria in this data for distinguishing between uncertainty and risk. That is to say, the points on the continuum occupied by uncertainty and risk are fluid, according to the nature of the risk, the policy issue, the scale employed, or the ontological assumptions upon which the analyses are built.

Whilst reference to non-linear impacts were on occasion made, these were not seen to be significant enough risks to put climate change beyond management through risk frameworks.

These findings highlight a core problem with the assumption that one can simply substitute uncertainty for risk in climate communications. Some climate policy and climate science issues might be more amenable to a risk framing, for example flooding. But that does not represent a break with historic framings of flood – the insurance market in the UK has always treated it as a risk issue. If risk language is the means by which the costs and benefits of responding to climate change are made communicable to a wider audience then it is possible to raise doubts about the appropriateness of trying to extend that risk language into other areas of the policy debate. Asides from the charge of economic imperialism, such cost-benefit considerations are not a break with subjectivity, but merely a means of masking value choices and further extending the technocratic construction of climate change which has to date not only failed to engage the public, but actively excluded them from any meaningful stake in the choices being made.

The overall framing of communication accompanying the 2013–2014 IPCC reports has been one of cost-benefits analysis. We can afford to fight climate change and it is cheaper to fight it than do nothing. But fighting in this instance rests on the unexamined value choice of defining two degrees of warming as an acceptable risk, of emission cuts which give only a 66% chance of achieving that limit as sufficient and seeking to address climate risks via the generation of novel risks such as those associated with yet to be realised carbon capture and storage programmes and a massive expansion of extremely damaging biofuel production (Rogers, 2010).

This analysis raises questions over the rather simplistic claims that a move from the language of uncertainty to risk will overcome the very factors which have to date hampered communication, around trust, culture, values, and uncertainty. Putting these issues into a risk language does not magic away the social and subjective dimensions of the policy debate. There are some areas where a move from the framing of uncertainty to risk may help some actors advance a mitigation strategy. How much hope should be placed in such a shift in language remains uncertain. Assumptions that the climate debate can be reframed in terms of risk, and in so doing overcome obstacles to the delivery of effective mitigation policies may have some value for market actors and institutions who already operate with such frames. One should be more cautious in assuming such a reframing will have great traction with a cynical public audience.

References

Baer, P. (2005). *Anchors Away: Why it's Time for the IPCC to Get Rid of the 1.5–4.5° Range and Replace it with an Evaluation of PDFs* (Unpublished doctoral dissertation, UC Berkely).

Beck, U. (1986). *Risk Society: Towards a New Modernity*. London: Sage Publications.

Beck, U. (1995). *Ecological Politics in an Age of Risk*. Cambridge: Polity Press.

Bernstein, L., Bosch, P., Canziani, O., Chen, Z., Christ, R., Davidson, O. and Yohe, G. (2007). IPCC Climate Change 2007: Synthesis Report. Retrieved from http://www.ipcc.ch/pdf/assessment-report/ar4/syr/ar4_syr.pdf

Blüdhorn, I. (2007). Sustaining the unsustainable: Symbolic politics and the politics of simulation. *Environmental Politics, Vol. 16*, 2, 251–275.

Botzen, W., Bergh, J. and Bouwer, L. (2010). Climate change and increased risk for the insurance sector: A global perspective and an assessment for the Netherlands. *Natural Hazards, Vol. 52*, 577–598.

Botzen, W.J. and Jeroen, C.J.M. van den Bergh (2012). Monetary valuation of insurance against flood risk under climate change. *International Economic Review, Vol. 53*, 3, 1005–1026.

Budescu, D.V., Broomell, S.B. and Por, H. (2009). Improving communication of uncertainty in the reports of the intergovernmental panel on climate change. *Psychological Science, Vol. 20*, 299–308.

Carrington, D. (2014, May 19th). Taming the floods, Dutch-style. *The Guardian*. Retrieved from http://www.theguardian.com/environment/2014/may/19/floods-dutch-britain-netherlands

Cash, D.W., Clark, W.C., Alcock, F., Dickson, N.M., Eckley, N., Guston, D.H., Jaeger, J. and Mitchell, R.B. (2003). Knowledge systems for sustainable development. *Procedures of the National Academy of Sciences, Vol. 100*, 14, 8086–8091.

Climate Change Committee (2008). The Climate Change Act and UK regulations. Retrieved from http://www.theccc.org.uk/tackling-climate-change/the-legal-landscape/global-action-on-climate-change/

DEFRA. (2010). *Flood Management Act*. Retrieved from http://www.defra.gov.uk/environment/

Dietz, S. (2011). High impact, low probability? An empirical analysis of risk in the economics of climate change. *Climatic Change, Vol. 108*, 519–541.

Dietz, S. and Fankhauser, S. (2010). Environmental prices, uncertainty, and learning. *Oxford Review of Economic Policy, Vol. 26*, 2, 270–284.

Gigerenzer, G. (2002). *Reckoning with Risk: Learning to Live with Uncertainty*. London: Penguin.

Guillerment, M.L. and Tol, R. (2008) Decision making under catastrophic risk and learning: the case of the possible collapse of the West Antarctic Ice Sheet. *Climatic Change, Vol. 91*, 193–209.

Hall, J.W., Lempert, R.J., Keller, K., Hackbarth, A., Mijere, C. and McInerney, D.J. (2012). Robust climate policies under uncertainty: a comparison of robust decision making and info-gap methods. *Risk Analysis, Vol. 10*, 1657–1672.

Hulme, M. (2009). *Why we Disagree about Climate Change*. Cambridge, UK: Cambridge University Press.

IPCC (2013). Summary for policymakers, in *Climate Change 2013: The Physical Science Basis. Contribution of Working Group I to the Fifth Assessment Report of the Intergovernmental Panel on Climate Change*, Stocker, T.F, Plattner, G.-K, Tignor, M. and Allen, S.K. et al. (Eds.) Cambridge, United Kingdom and New York: Cambridge University Press. Retrieved from http://www.climatechange2013.org/images/uploads/WGI_AR5_SPM_brochure.pdf.

Kline, R.B. (2010). *Principles and Practice of Structural Equation Modelling* (3rd ed.). New York: Guilford Press.

Lane, S.N., Landstrom, C. and Whatmore, S. (2011). Imagining flood futures: Risk assessment and management in practice. *Philosophical Transactions of the Royal Society A., Vol. 369*, 1942, 1784–1806.

Lowe, T., Brown, K., Dessai, S., Franca Doria, M., Hayes, K. and Vincent, K. (2006). Does tomorrow ever come? Disaster narrative and public perceptions of climate change. *Public Understanding of Science, Vol. 15*, 435–457.

Mackenzie, D. (1990). *Inventing Accuracy: A Historical Sociology of Nuclear Missile Guidance*. Cambridge, MA: MIT Press.

Maxim, L. and van der Sluijs, J.P. (2011) Quality in environmental science for policy: Assessing uncertainty as a component of policy analysis. *Environmental Science & Policy, Vol. 14*, 482–492

Mulkay, M. (1991). *Sociology of Science: A Sociological Pilgrimage*. Milton Keynes: Open University Press.

Nerlich, B. (2010). 'Climategate': Paradoxical metaphors and political paralysis. *Environmental Values, Vol. 19*, 419–442. doi: 10.3197/096327110X531543

Nisbet, M. (2009). Communicating climate change: Why frames matter for public engagement. *Environment Magazine*. Retrieved from http://www.environmentmagazine.org/Archives/Back%20Issues/March-April%202009/Nisbet-full.html

Nowotny, H., Scott, P. and Gibbons, M. (2001). *Re-Thinking Science. Knowledge and the Public in an Age of Uncertainty*. Cambridge: Polity.

O'Hare, G. (2000). Reviewing the uncertainties in climate change science. *Area, Vol. 32*, 4, 357–368.

Olausson, U. (2009). Global warming – Global responsibility? Media frames of collective action and scientific certainty. *Public Understanding of Science, Vol. 18*, 4, 421–436.

Painter, J. (2013). *Climate Change in the Media: Reporting risk and Uncertainty*. London: I.B Tauris.

Pidgeon, N. and Fischoff, B. (2011). The role of social and decision sciences in communicating uncertain climate risks. *Nature Climate Change, Vol. 1*, 35–41.

Pidgeon, N. (2012). Climate change risk perception and communication: Addressing a critical moment? *Risk Analysis, Vol. 32*, 6, 951–956.

Poortinga, W., Spence, A., Whitmarsh, L., Capstick, S. and Pidgeon, N. (2011). Uncertain climate: An investigation into public scepticism about anthropogenic climate change. *Global Environmental Change, Vol. 21*, 3, 1015–1024.

Rachlinski, J.J. (2000). The psychology of global climate change. *University of Illinois Law Review, Vol. 1*, 299–231.

Renn, O. (2011). A comment to Ragnar Lofstedt: Risks versus hazards. *European Journal of Risk Regulation, Vol. 2*, 3, 197–202.

Rogers, H. (2010). *Green Gone Wrong. Dispatches from the Front Lines of Eco-Capitalism.* London: Verso.

Runhaar, H., Mees, H., Wardekker, A., van der Sluijs, J and Driessen, P. (2012) Adaptation to climate change-related risks in Dutch urban areas: Stimuli and barriers, *Regional Environmental Change, Vol. 12*, 777–790.

Shaw, C. and Nerlich, B. (2015). Metaphor as a mechanism of global climate change governance: A study of international policies, 1992–2012. *Ecological Economics, Vol. 109*, 34–40.

Silverman, H. (2013, October 4th). *Amidst Uncertainty, Perceiving risk.* (Web log comment) Retrieved from http://www.solvingforpattern.org/2013/10/18/amidst-uncertainty-perceiving-risk/

Stefan, S. (2008) Approaching the tipping point – Climate risks, faith and political action. *European Journal of Science and Theology, Vol.4*, 2, 9–22.

Taylor-Gooby, P. and Zinn, J.O., Eds. (2006). *Risk in Social Science.* Oxford: Oxford University Press.

Tickner, J. (2003). Introduction, in *Precaution, Environmental Science and Preventative Public Policy.* Tickner, J. (Ed) (pp. xiii–xvii). Washington DC: Island Press.

Timmer, J. (2013). Applying science to communicate science, ARS Technica http://arstechnica.com/staff/2013/08/applying-science-to-communicate-science/. Accessed 2nd March 2015.

Veraart, J. A. and Bakker, M. (2009). Climateproofing The Netherlands, in *Climate Change Adaptation in the Water Sector.* F. Ludwig, B. Kabat, H. van Schaik, and M. van der Valk (Eds.) (pp. 110–122). London: Earthscan.

Whitmarsh, L. (2011). Scepticism and uncertainty about climate change: Dimensions, determinants and change over time. *Global Environmental Change, Vol. 21*, 690–700.

Wynne, B. and Jasanoff, F. (1998). Science and decision making, in *Human Choices and Climate Change vol. 1. The Societal Framework.* Rayner, S. and Malone, E. (Eds.) (pp. 1–112) Ohio: Battelle Press.

14
Between Two Absolutes Lies Risk: Risk Communication in Biosecurity Discourse

Sue McKell and Paul De Barro

Introduction

Biosecurity and quarantine laws operate internationally to protect national and state borders from potential threats to human and animal health, the environment, agriculture, and trade. This chapter explores how risk is communicated within biosecurity contexts through a discourse analysis of public communication texts from a federal government agency responsible for biosecurity in Australia. An overview of the relatively new policy concept of biosecurity is provided, suggesting some complexity in terms of how the concept is communicated by public agencies and, in turn, how it might be understood by the public. The analysis provided by this study points to similarities in the way that biosecurity discourse conflates risk with activities that are associated with (but are not actually the 'real') risk source, while adopting fundamentally different conceptualisations of risk itself, and different levels of agency onto members of the public in negotiating those risks. The difference in the ways that risk is communicated by these texts suggests that the task of gaining community awareness of biosecurity issues may be complicated by apparently contradictory public messages relating to biosecurity risks, which has implications for how messages that target personal responses to manage public risks are communicated by agencies internationally.

Background

Biosecurity and quarantine are not new concepts within Australia, but there is a growing awareness of their roles within international trade negotiations. An increasing national and global awareness of threats to biodiversity and the economic wellbeing of the agricultural sector posed by alien invasive species has seen biosecurity achieve new levels of importance. While there is no universally agreed definition for the term 'biosecurity' in Australia and New Zealand, 'biosecurity' is taken to mean protection of the economy, environment, and community from negative impacts of pests (animal and plant) and diseases. At its core, biosecurity is a concept that involves a complex mix of policy, regulation, planning and actions with the overall aim of reducing the threat and impact of external biological threats. Within the framework of biosecurity is the concept, first devised by Nairn et al. (1996) of a biosecurity continuum, entailing pre-border, border, and post-border protection. Quarantine, on the other hand, is one instrument used within a biosecurity framework at the level of border protection, most notably to prevent pests and diseases entering Australia's borders (such as at airports), but has existed as a formal legal framework in Australia since 1908. At each stage of protection, biosecurity policy is interpreted in the form of measures and actions that are taken to reduce the risk of a threat occurring.

National biosecurity policy and implementation within the agriculture sector in Australia is primarily the role of the Australian Government Department of Agriculture. The Department of Agriculture manages quarantine controls at the border to minimise the risk of exotic pests and diseases entering Australia and to protect Australia's agriculture export industries as well as the natural environment and community. The Department works closely with other Australian, state and territory government agencies (such as the Department of Immigration and Border Protection, the Department of Health, Food Standards Australia and New Zealand and state and territory governments), to support their management of post-border detections and incursions of pests and diseases, and to support Australia's verification and certification activities for the export/import of agriculture and food products. As part of this process, the Department provides import risk analyses and policy advice that protects Australia's relatively low pest and disease status and Australia's access to international markets. These analyses and advice need to be consistent with international agreements such as the International Plant Protection Convention, WTO Sanitary and Phytosanitary (SPS) Agreement, Technical Barriers to Trade Agreement, and International Standards for Phytosanitary Measures. A key issue under these arrangements is the need for a strong level of scientific evidence for any measures imposed to manage risks presented by import and, by implication, the exclusion of other factors (including commercial and social) that are not directly associated with those risks.

The Department of Agriculture also provides a national and international focal point for protecting Australia's plant health and for coordinating responses to plant pest, disease, and weeds incidents. Their work helps open new, and maintain existing, domestic and international markets. As part of this effort, the Department of Agriculture coordinates and develops national technical and operational plant health policy, leads and coordinates national responses to plant pest incursions that affect Australia's plant industries, encourages greater competence and capacity in plant health management in Australia, collects and manages information on Australia's plant health status, builds capacity within countries in the region to assist Australia's biosecurity efforts, and coordinates proposals and policies for improving international plant protection policy and standards so that they are consistent with Australia's trading policy.

One of the inherent difficulties in communicating biosecurity issues is to gain an understanding of the concept of biosecurity itself and, with it, the activities used to implement the concept. A further complication is that while the use of quarantine is not new, in real terms the use of biosecurity is. Further, while the use of quarantine as a term has not changed, the way that it is implemented has, as it has had to meet the new requirements of managed international trade such as trade governed through the auspices of agreements under the World Trade Organisation. The relative newness of the term 'biosecurity' compared to an old, but well understood term 'quarantine' is an ongoing challenge for the Department in communicating complex biosecurity activities to the Australian public.

The work of the Department as it relates to preventative biosecurity activities can be broadly divided into policy and operational roles. While the Department of Agriculture's operations role has been plauded for its communication public awareness activities, the policy side has been often criticised, particularly in relation to import risk analysis and transparency of the process. The most recent example of this is the Australian Senate Rural and Regional Affairs and Transport References Committee report into the risk assessments for Malaysian pineapples, Fijian ginger and New Zealand potatoes, released in March 2014. In the Senate Committee's Report, it was noted that one of the intentions of proposed new legislation in Australia (the Biosecurity Bill 2012 and the Inspector-General of Biosecurity Bill 2012), were to increase the transparency of the biosecurity system for clients and stakeholders (including trading partners) particularly in relation to the assessment and management of biosecurity risks.

While both operations and policy undertake public communication activities relating to biosecurity issues, the role of the communications functions within each organisation differs to the extent that they aim to increase public awareness of biosecurity issues. The main communication programme within operations provides 'public awareness and education activities', reflecting a relatively (pro-)active goal of influencing quarantine understanding within the community (and by overseas travellers, in particular). In other words, operations focus on selling an action. The communication unit within policy, on the other hand, aims to '[support] engagement with stakeholders and provides public information', suggesting that its activities are concerned less with increasing public awareness of biosecurity issues in a general sense, as they are with providing information on specific activities undertaken by the Department. This distinction is an important one, as it indicates that the organisation responsible for ensuring aspects of Australia's national biosecurity carries two different, and possibly competing, objectives in terms of communicating biosecurity risks to the public.

Risk communication

Discourse analyses of risk have traditionally focused on the potential for adverse outcomes affecting individuals to be realised, most frequently in relation to medical and interpersonal issues (Hamilton, Adolph and Nerlich 2007). While a growing number of discourse analysis studies have been undertaken within specific risk-related discourses such as terrorism and disease control (see, for example, Ditrych 2014; Garoon and Duggan 2008), biosecurity discourse remains a relatively unexplored topic in this field. An investigation of biosecurity risk discourse offers the potential to explore how the concept of risk is instantiated and potentially (mis)understood by the public in a context that intersects a range of commercial, environmental, social, and personal risks.

Risk itself refers to the (potential positive or negative) effect of uncertainty on objectives (ISO 31000:2009), and is often represented using the following formula:

Risk = [Probability of an event occurring] × [Consequences of that event]

It follows that an assessment of risk is the product of an evaluative process at the individual or organisation level to determine the chances of an event occurring and the likely consequences in the event of its occurrence. While the term 'risk' itself is narrowly defined in a legal or scientific sense in the context of government risk analyses, it remains an inherently complex concept, and the communication of 'risk' in public communication can carry much broader meanings. Risk can exist simultaneously as mathematically correct, and subjectively wrong, making communication of risk inherently complex and potentially problematic.

Risk communication can be defined as the exchange of information between stakeholders about risk, involving: the level of risk, the significance or meaning of risk, or decisions, actions or policies aimed at managing or controlling risks (Covello, von Winterfeldt and Slovic 1986). Risk communication involves an implicit objective of motivating people or organisations to make a decision to act (or stop acting) in a certain way, thereby changing the status quo to avoid or capitalise on opportunities presented by a perceived risk. Lupton has argued that the meaning of risk has changed over time, and is increasingly the result of scientific-based assessment within a discourse of modernity that emphasises rationality, science and technology (1999). Power roles in the modern 'risk society' are aligned within the official risk assessment processes and risks are determined by scientists and perceived by the public (Beck 1992), leading to complexities between institutional, professional, and personal modes involved in communicating risk. Jasanoff (1998) describes how professional languages and formal analytic practices (such as quantitative risk assessment and cost-benefit analysis) can operate to privilege authoritative knowledge and constrain personal perceptions of risk and, thus, how 'risk discourses may systematically exclude valid, but powerless, viewpoints'.

The role of government can be problematic if risk is conceived as a tool for bureaucracy to seek to order and control individual freedom (Bauman 1991). In the current context, the process of biosecurity risk assessments aims to advise on the 'acceptability' of risks presented within complex systems, and thus involves reaching a qualitative evaluation based on a range of incomplete quantitative measures within complex legal and scientific frameworks. In doing so, risk assessment operates within a paradigm of incomplete knowledge, both in terms of understanding the risk context itself as well as in relation to excluding other (scientific and non-scientific) knowledge. In this way, risk assessment is a science-based evaluative process that is also inherently ideological in nature.

In addition to what actual information is communicated, the way in which it is communicated is known to play a key role in influencing how that information is perceived or used by individuals (Tverskey and Kahnemann 1981). Moreover, presenting low-probability risks as either absolute rates or relative expressions also plays a role in how people perceive levels of risk, with low incidence rates more likely to be interpreted as 'essentially nil risk' and relative expressions more likely to be understood in inflated terms (Stone, Yates, and Parker 1994). Of particular relevance to the current study, concepts such as the 'appropriate level of protection' as used in biosecurity risk analyses work to present 'a bright line, dividing safe from unsafe' (Margolis 1996), and can thus 'mislead as the general public sees the level of protection selected as ... safe' (Botterill and Mazur 2004).

Method: Critical discourse analysis

Critical Discourse Analysis (CDA) provides the methodological basis for the current study, in its aim to 'reveal connections between language, power and ideology, and ... how power and dominance are produced and reproduced in social practice through the discourse structures of generally unremarkable interactions (Stubbe et al. 2003)'. While CDA has traditionally been used as a tool with which to reveal communication strategies and, in doing so, critique existing (particularly political) power structures (van Dijk 2001), it brings two other important methodological perspectives to this study. On the one hand, CDA offers a tool with which to better understand how biosecurity risks might be both understood and potentially misunderstood by individuals. On the other hand, CDA offers the potential to dissect how concepts are constructed discursively within a 'gatekeeping' context in order to examine complex interplays between institutional, professional, and personal modes of communication (Roberts and Sarangi 1999). Central to both of these perspectives is the aim of gaining a better understanding of what biosecurity risks are presented and why, with a focus on practical issues of how they are communicated.

For this current analysis, we take an approach of 'joint problematisation' (Roberts and Sarangi 1999) between the fields of discourse analysis and biosecurity practice, with a view to identifying areas where public messages on biosecurity risks negotiate, and make trade-offs between, institutional, professional, and personal responsibility. To achieve this, we look at descriptions of specific activities that present apparent biosecurity risks within two official communication texts taken from the public internet websites of the Australian Government's Department of Agriculture, which is the federal agency responsible for quarantine and biosecurity in Australia. The first of these texts is the official Australian Government announcement for incoming passengers on planes and ships travelling to Australia (henceforth 'Passenger Announcement'), which is required to be read out by crew on all international journeys under Section 74AA of the Quarantine Act 1908. The second text is a Departmental 'Biosecurity Advice' notifying stakeholders of the findings of the final import risk analysis report allowing importation of fresh ginger from Fiji into Australia.

These texts were selected for this study as examples of biosecurity communication messages in Australia due to the nature of their biosecurity-related messages, general audience, and public relevance. The Passenger Announcement, while only short in length (390 words in total), is clearly a significant example of public biosecurity communication given the fact that it is required by law to be read to all passengers on international planes and vessels entering Australia, including over 6.4 million visitors in 2013 (Australian Bureau of Statistics). By contrast, the Biosecurity Advice targets a much smaller audience, that is, stakeholders with an interest in fresh ginger imports, but is also representative of Departmental 'Biosecurity Advices' of this kind, as it follows a distinctly formulaic design in terms of its content (often repeating key sections of text exactly) and structure as other similar advice issued by the Department. While Plant Import Risk Analyses (IRAs) are relatively few in number, with only 24 completed since 1998, IRAs that have allowed plant imports have been the subject of at times intense public debate in Australia, leading to a federal government review of the Import Risk Analysis process in 2014. Both of these texts contain messages about managing risks that are aimed at a general public audience, and so are useful to examine how successfully the concept of risk is negotiated within biosecurity discourse in the Australian context and between institutional, professional, and personal modes of communication.

Communicating biosecurity risk

The way in which the concept of risk itself is communicated within each of the texts in question is of specific interest to this study of biosecurity communication. Locating the exact point where the concept of risk lies in each text, however, relies first on locating the object or activity that carries that risk (henceforth, 'risk activity'). The risk activity described in the Passenger Announcement is not, as one might assume, the actual plant and animal materials that might contain a potential quarantine threat, nor is it the person carrying those materials, nor the act of importing those materials into Australia. The risk activity highlighted by the quarantine announcement is in a literal sense associated with the actual act of declaring whether those materials are being carried. The following extract is the only section of the Passenger Announcement that deals explicitly with plant biosecurity issues (with the remainder addressing communicable diseases):

> You must mark 'yes' on your Incoming Passenger Card if you have certain food, plant or animal products, or equipment or shoes used in rivers and lakes or with soil attached. Food supplied onboard must be left onboard. You must answer this card truthfully, this card is a legal document and a false statement may result in a penalty. It will be checked by an official on your arrival.

It is the failure to properly declare, or to make a false statement about the possession of plant and animal materials, that might warrant punishment (by way of an unspecified 'penalty'), rather than the 'real' risk of importing plant and animal materials which might lead to 'real' consequences (such as economic, agricultural, environmental, social impacts). By implication, it is the declaration form itself that constitutes the primary risk activity in the Passenger Announcement, rather than actual plant and animal materials that carry 'real' biosecurity risks.

By conflating the location of risk primarily with the declaration form (rather than with the act of importing plant and animal products), the Passenger Announcement also operates to shift the concept of risk itself away from 'real' biosecurity risks and onto the act of declaring (or failing to declare) those products. Passengers who are in possession of plant and animal materials are in fact given three (not necessarily mutually exclusive) options for action in response to the quarantine announcement: to declare risk-associated products; not to declare them (and risk punishment); and/or to leave food supplied on-board on the plane. These options, combined with a literal interpretation of the words used in the announcement, has the effect of identifying all passengers as potential importers of risk-associated products – a role that can be removed, but only through declaration or disposal of goods.

The implications of this analysis are that risk within the Passenger Announcement is conceptualised in absolute terms: without declaration or disposal, risk exists and negative consequences are possible; with declaration or disposal, risk disappears and, with it, so too the consequences. The act of declaration thereby becomes akin to the act of absolution, even though the actual source of risk still exists. A pattern can be seen in the Passenger Announcement whereby the message of biosecurity is simplified in order to make clear what action is being asked of passengers, to the point where it potentially: compromises the accuracy of the message; characterises risk incorrectly (in absolute terms); and removes both the risk activity and consequences associated with that risk activity away from their 'real' origins. In other words, the message is that if you declare and material is allowed through, then no risk exists.

To compare the Passenger Announcement with how risk is conceptualised within the Biosecurity Advice, one must similarly locate the point or activity where risk is seen to reside. Where the Passenger Announcement was seen to conflate the primary risk activity with the declaration form, the Biosecurity Advice works similarly to conflate the principal risk activity with the act of 'importation of fresh ginger for consumption from Fiji', rather than with the 'real' source of risk, which are specific pests and diseases that are found in Fiji ginger, but which are not currently found in Australian ginger.

> The final IRA report recommends a combination of risk management measures and operational systems that will reduce the risk associated with the importation of fresh ginger for consumption from Fiji into Australia to achieve Australia's ALOP [Appropriate Level of Protection]...

Similar to the Passenger Announcement, the effect of shifting the risk activity onto an essentially administrative activity (in this case, the process of importation) has the effect of distancing risk from its 'real' origins, while also conflating the focus of any remedial action (to remove the source of this risk) onto the importation process itself (rather than the plant material being imported).

Unlike the Passenger Announcement, where the risk consequence was articulated in terms of punishment for passenger's failure to declare plant and animal products, the Biosecurity Advice does not associate any consequences with the risk identified, nor does it attempt to quantify the level of risk, except to say that 'the risk management measures set out in this final report will achieve Australia's appropriate level of protection (ALOP) against identified quarantine pests', without defining what an 'appropriate level of protection' is, nor indeed who determines whether it is appropriate. In the case of the Passenger Announcement, the passengers receiving the communication are positioned as active risk agents: they are the source of risk, and are provided with the option and capacity of accepting or absolving themselves of risk. In the Biosecurity

Advice, however, the stakeholders receiving the communication are positioned as passive risk recipients. The public here is not the source, but the target of risk, and are provided with no options to avoid the risk: the risk they are about to receive, they are told, is 'acceptable' to Australia. Although the latter occurs within a trade context, it is passengers that are provided with a bargain that is open for negotiation, while import stakeholders are offered no such bargain.

While both texts similarly conceptualise risk in absolute terms, they differ in how they project notions of 'good' and 'bad' onto those risks. Risk associated with the Passenger Announcement is implied in absolute terms (as either declared or undeclared), yet with a clear indication that the truthfulness of the declaration is what is being assessed: a truthful declaration is good, even if food is being carried, while a dishonest declaration is bad, irrespective of whether or not food is being carried. In this case, both the risk itself and the consequences of that risk are presented in absolute terms. Risk associated with the importation of ginger from Fiji within the Biosecurity Advice is similarly couched as an absolute (as within an 'appropriate level of protection'); yet, provides a more complex notion of 'good' and 'bad', that is, all risks are inherently bad, but beneath a certain (undisclosed) level, they should be tolerated and therefore considered 'good', or at the very least not 'bad'. Here, risk is presented in absolute terms, but the consequences of that risk are given in relative terms. In these ways, both texts work to obscure the 'real' sources and consequences of risk being addressed by providing contradictory messages on the fundamental nature of risk, and on the role of the public in managing risk.

The similarities between how risk itself is characterised in the two documents is clear: risk is shifted away from its actual source, and is characterised in absolute terms. Conceptualising risk in absolute terms in fact fits quite logically with the historical (and, indeed, literal) interpretation of the concept of 'quarantine', which may be used as a verb to describe something being isolated in absolute, and indeed the more modern concept of 'biosecurity', which refers to protecting plants/animals/humans against pests and diseases (again, in absolute terms). Yet, within these interpretations lies another legal and statistical reality in which risk is not (or very rarely) zero, and where risk factors may be managed in such a way as to minimise the risk associated with them. Somewhere between the absoluteness of these terms lies the reality of risk that is neither objective nor fixed.

The real problem presented by this situation is not so much that either text may be more or less correct in the way that it presents the concept of risk, but that the effect of communicating risk differently is that risk is likely to be understood differently by the audience of these texts. It is reasonable to assume that the simple message being put forward in the Passenger Announcement, in which risk is characterised in absolute terms, with the public occupying an active role in negotiating that risk, is more easily able to be understood by the passengers receiving the message, and that members of the public therefore understand very clearly the concept of risk in absolute terms: as either present (bad) or absent (good). It would not be difficult to also assume that a member of the public, when faced by the information presented in the Biosecurity Advice, in a way that is also absolute, but inherently negative and not able to be avoided by individuals, and therefore acceptable, that they might resist that interpretation.

Conclusion

This chapter highlights the usefulness of critical discourse analysis and joint problematisation to the field of risk communication in general and biosecurity communication in particular, by demonstrating how risk activities described in public messages about biosecurity negotiate, and make trade-offs between, institutional, professional and personal knowledge, and responsibility. The concept of risk is central to the public's understanding of biosecurity, yet this study argues that risk itself presents an inherent logical contradiction to both of these concepts. Moreover, the (incorrect) conceptualisation of risk as either absent (good) or present (bad) gives the appearance of a logical fit with the concept of biosecurity and works further to reinforce the logical opposition to the (correct) conceptualisation of risk as being unavoidable but manageable.

The way in which risk is communicated in these texts points to an area of inconsistency in the way in which the concept of biosecurity may be (mis)understood by the public, and suggests an important area of opportunity for improving public biosecurity communication in the future. Further analysis of descriptions of risk activities in other policy contexts would be helpful to understand how this risk communication process operates in areas that are similarly concerned with explaining risk concepts to the public. These findings have potential relevance to other 'gatekeeping' policy contexts, such as public safety, finance, and immigration, to better understand how risk activity processes shape (and are shaped by) individuals' understandings of risk concepts, and potentially also help understand public responses to risk management policies.

References

Australian Bureau of Statistics (2004). *Overseas Arrivals and Departures in Australia*, cat. no. 3401.0, ABS, Canberra.

Australian Government. Department of Agriculture. (2014). *Guidelines for Airline and Aircraft Operators arriving in Australia*. Retrieved from http://www.agriculture.gov.au

Australian Government. Department of Agriculture. (2013). *Biosecurity Advice 2013/03: Final Import Risk Analysis Report for Fresh Ginger for Consumption from Fiji*. Retrieved from http://www.agriculture.gov.au

Australian Government. Senate Rural and Regional Affairs and Transport References Committee. (2014) *Effect on Australian pineapple growers of importing fresh pineapple from Malaysia; Effect on Australian ginger growers of importing fresh ginger from Fiji; Proposed importation of potatoes from New Zealand*. Retrieved from http://www.aph.gov.au

Bauman, Z. (1991). *Modernity and Ambivalence*. Oxford: Polity.

Beck, Ulrich (1992). *Risk Society: Towards a New Modernity*. New Delhi: Sage.

Botterill, L., & Mazur, N. (2004). *Risk & Risk Perception: Prepared for the Rural Industries Research and Development Corporation*. Retrieved from http://www.rirdc.gov.au

Calderwood, Kathleen, Rural News, Australian Broadcasting Commission (2014, July 15). Horticulture Industry Welcomes Import Risk Review. Retrieved from http://www.abc.net.au

Covello, V.T., von Winterfeldt, D., & Slovic, P. (1986). Risk Communication: A Review of the Literature. *Risk Abstracts*, 3, 171–182.

Ditrych, O. (2014). *Tracing the Discourse of Terrorism: Identity, Genealogy and State*. London: Palgrave Macmillan.

Garoon, J.P., & Duggan, P.S. (2008). Discourses of Disease, Discourses of Disadvantage: A Critical Discourse Analysis of National Pandemic Influenza Preparedness Plans. *Social Science & Medicine*, 67(7), 1133–1142.

Hamilton, C., Adolphs, S., & Nerlich, B. (2007). The Meanings of 'Risk': A View from Corpus Linguistics. *Discourse & Society*, 18(2), 163–181.

Jasanoff, S. (1998), The political science of risk perception. *Reliability Engineering and System Safety*, 59, 91–99.

Lupton, D. (1999). *Risk*. New York: Routledge.

Margolis, H. (1996). *Dealing with Risk: Why the Public and the Experts Disagree on Environmental Issues*. Chicago: University of Chicago Press.

Nairn, M.E., Allen, P.G., Inglis, A.R., & Tanner, C. (1996). *Australian Quarantine: A Shared Responsibility*. Canberra: Department of Primary Industries and Energy.

Roberts, C., & Sarangi, S. (1999). Hybridity in gatekeeping discourse: Issues of practical relevance for the researcher. In S. Sarangi & C. Roberts (Eds.), *Talk, work and institutional order: Discourse in medical, mediation and management settings* (pp. 473–504). Berlin: Mouton de Gruyter.

Standards Australia/Standards New Zealand Standard Committee. (2009). *AS/NZS ISO 31000:2009: Risk Management-Principles and Guidelines*.

Stone, E.R., Yates, J.F., & Parker, A.M. (1994). Risk Communication: Absolute versus Relative Expressions of Low-Probability Risks. *Organizational Behavior and Human Decision Processes*, 60, 387–408.

Stubbe, M., Lane, C., Hilder, J., Vine, E., Vine, B., Marra, M., Holmes, J., & Weatherall, A. (2003) Multiple Discourse Analyses of a Workplace Interaction. *Discourse Studies*, 5(3), 351–388.

Tverskey, A., & Kahnemann, D. (1981). The Framing of Decisions and the Psychology of Choice. *Science*, 211, 453–458.

Van Dijk, T.A. (2001) Multidisciplinary CDA: A plea for diversity. In R. Wodak, & M. Meyer (Eds.), *Methods of critical discourse analysis in the series: Introducing qualitative methods*. London: Sage Publications (pp. 95–120). London: Sage.

Part V
Mediating Risk

15
Negotiating Risk in Chinese and Australian Print Media Hard News Reporting on Food Safety: A Corpus-based Study

Changpeng Huan

Introduction

Risk is defined by Beck as a 'systematic way of dealing with hazards and insecurities induced and introduced by modernization itself' (Beck, 1992, p. 21). Douglas (1992) and Luhmann (1993) maintain that it is essential to distinguish risk from danger, and to sustain a relationship between risk and responsibility. In this sense, risk almost always invokes external attribution (Luhmann, 1993; Sarangi, Bennert, Howell, & Clarke, 2003; Sarangi & Clarke, 2002), and is best understood as manufactured and constructed (Adam, Beck, & Van Loon, 2000) among a nexus of practices (Scollon & Scollon, 2004). Beck (1992, 1999) argues that meanings of risk are primarily constructed and shaped between government and science. This argument however has largely underplayed the importance of media as a crucial site of engagement where meanings of risk are negotiated among different stake-holders and where such negotiation is mediated by journalistic professional practices (see also Mythen, 2004).

In recognition of the research gap, the aim of the study on which this chapter draws was threefold: first, to interrogate the negotiation of risk in Chinese and Australian print media hard news reporting with a particular focus on food safety reports; second, to encourage a discursive approach to risk research; and third, to highlight the usefulness of corpus-based studies in the manifestation and interpretation of risk.

In this following, I will operationalise the concept of hard news reporting in relation to discourses of risk (Section 2) and outline the procedures in constructing and annotating the food-safety corpora (Section 3). Corpus findings will then be presented (Section 4) and discussed in relation to the arising socio-cultural contexts (Section 5).

Operationalising the concept of risk

In the context of Beck's (1992) formulation of risk society, the late twentieth century has been marked by rising hazards associated with risk and uncertainty in many professional practices (also see Sarangi & Clarke, 2002, p. 142). The central issue of risk society concerns 'the ability [of human beings] to define what may happen in the future and to choose among alternatives' (Bernstein, 1996, p. 2). Negotiating risk is therefore linked to the ways 'that lay actors and technical specialists organise the social world' (Giddens, 1991, p. 3). As Candlin and Candlin (2002, p. 128) put it succinctly, it may be not so much the ontological underpinning of risks that is contestable and controversial but rather the epistemological representation or negotiation of discourses of risk.

In the light of this, the study reported in this chapter detached itself from studies of relational and rational nature of risk *per se*. Rather, it took a discursive approach to an understanding of risk communication in news discourse, specifically in the context of food safety in the Chinese and Australian print media hard news. The discursive approach afforded a way of conceiving risk as a negotiable construct demanding sustained negotiation between different voices in discourses. Such a view was premised on the subjectivity (e.g. Lupton, 1999) and uncertainty (e.g. Sarangi & Clarke, 2002) of risk. Negotiating risk through news discourse was inevitably subject to journalistic (inter)subjective stance-taking practices in their mediation of various voices. Each stance-taking practice implied a process of knowledge negotiation of risk 'between different epistemologies and subcultural forms, amongst different discourses' (Beck, 1992, p. 5). In recognition of this, the study described in this chapter highlighted the interconnected relationship between discourses of risk and the mediating role of journalistic stance in the Chinese and Australian press.

Constructing a comparable corpus of Chinese and Australian hard news reporting on food safety risk

The rationale for confining the focus to food safety was twofold. First, there was the methodological consideration that comparing discourses on the same topic could reduce the influence of topic variation on the interpretation of results. The other consideration was the significance of food safety in people's daily life in China and Australia, in that food safety has been deemed one of the 11 top concerns for humanitarian affairs by the United Nations. Although food safety denotes different meanings across cultures, it is hardly deniable that it concerns the whole population worldwide.

By print media hard news reporting on food safety issues, I mean the social issues raised by food safety were reported in hard news with a focus on the societal consequences of the events and in a detached rather than sensational or interpretive fashion.

Data collection

The Chinese data focused on a website that listed almost all food safety reports in China (www.zccw.info) since 2004. I selected those news stories reported in important Chinese newspapers (e.g. the Database for Chinese Key Newspapers). The Australian newspapers were limited to four broadsheet dailies: the *Australian*, the *Age*, the *Canberra Times* and the *Sydney Morning Herald*. Considering that language may change subtly in a relatively short period of time, the corpus was drawn from one decade from 2004 to 2013. The areas where news events occurred were limited to those in Australia reported by newspapers in English and those in mainland China in simplified Chinese. The Australian data were encoded in ASCII characters by default, whereas the Chinese data were encoded in Unicode-8. Eventually, a number of 118 news stories on food safety reported in the selected Australian newspapers were included in the Australian food-safety corpus, amounting to 8,026 word types (49,998 word tokens). The Chinese food-safety corpus incorporated 485 news stories, amounting to 13,197 Chinese word types (215,433 Chinese word tokens). In this sense, rather than achieving the ideally 'balanced' corpora (e.g. Biber, 1990; McEnery, Xiao, & Tono, 2006) by including equal amount of words in each corpus, I have prioritised collecting as many news stories on food safety as possible in each corpus.

Analytical resources

Three analytical resources were used in the study: corpus annotation: POS, semantics and stance annotations.

POS annotation is a process to assign a part-of-speech mnemonic, a POS tag, to an individual word in a corpus (McEnery et al., 2006, p. 34). It is the most common and widely used annotation allowing the annotated corpus to address various linguistic questions. CLAWS 4 (http://ucrel.lancs.ac.uk/claws/) was adopted to POS tag the English data with a reliable accuracy rate of 96% to 98%. The Chinese data were POS tagged with the assistance of the lexical analysis system – ICTCLAS (http://ictclas.org/) that can achieve a precision rate of 97.58% (Zhang & Liu, 2002).

Semantic annotation brings together word senses which are related to one of the many semantic fields (e.g. mental concept) in a corpus. I have deployed the web-based corpus tool W-Matrix (http://ucrel.lancs.ac.uk/wmatrix3.html) to semantically annotate the Australian corpus. This tool uses the semantic tag set of USAS (http://ucrel.lancs.ac.uk/usas/) to assign semantic tags to individual words. It has a multi-tier structure with 21 major discourse fields which can be further subdivided. Since there was no reliable Chinese semantic tagger available, a comparison of semantics of the Australian and Chinese corpus was not conducted.

The negotiation of risk in news discourse was often mediated by journalists' crafted manoeuvre of various voices of different social groups or individuals. When engaging with these voices, journalists may choose to either contract or expand the 'dialogic space' (e.g. Bakhtin, 1981) in news texts by warranting, confirming, or challenging the introduced voices. The engagement subsystem (Figure 15.1 below) under the Appraisal framework (Martin & White, 2005) seeks to capture such dynamic processes of meaning negotiation. In fact, 'contract' and 'expand' constitute the two subsystems of the engagement system.

Under 'contract', journalists narrow down the dialogic space by directly rejecting or replacing other dialogic alternatives (i.e. disclaim)

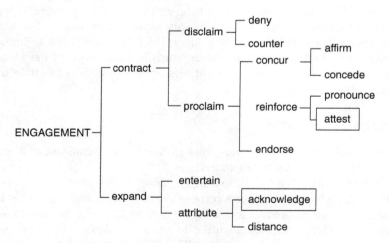

Figure 15.1 The framework of engagement
Source: Modified from Martin & White, 2005.

or by offering a preferable voice over others (i.e. proclaim). 'Disclaim' is divided into 'deny' and 'counter', while 'proclaim' consists of 'concur', 'pronounce', and 'endorse'. Concur involves semantic values which explicitly project the writer as affirming the voice or positioning of the putative readers. 'Pronounce' concerns explicit authorial intrusion into the dialogue so as to de/value one positioning over another. It differs from 'endorse', which 'refers to those formulations by which propositions sourced to external sources are construed by the authorial voice as correct, valid, undeniable or otherwise maximally warrantable' (Martin & White, 2005, p. 126).

Under 'expand', dialogic voices are construed as more negotiable, contingent, contestable, or challengeable, and degree of writer subjectivity moves down towards objectivity. This covers two types of expansive values – 'entertain' and 'attribution'. 'Entertain' mainly concerns those meanings 'by which speaker/writer makes assessment of likelihood via modality' (Martin & White, 2005, pp. 104–105). Attributing one utterance to other sources, writers have the choice of either declining to make a choice/indication of preferred voice, but simply acknowledging the voice as one of a range of possible voices (i.e. acknowledge), or even explicitly detaching/alienating from the position being advanced (i.e. distance).

Direct evidence such as visual, auditory evidence may be considered as marginal in other discourses (e.g. academic discourse), but they constitute a substantial means for journalists to show the vividness of news events, to indicate their presence on site and more importantly to increase reliability of the news reporting in question. The inclusion of 'attest' sources (e.g. eye-witness sources) in news reporting was favourably perceived as true, first-hand, new, and perhaps objective. On this premise, I have added this new semantic feature into the engagement system. Due to space limitations, I will not elaborate on each category. Instead, I will outline how these resources are deployed in Chinese and Australian press, and then will focus exclusively on 'attest' and 'acknowledge' to illustrate the usefulness of engagement in understanding the negotiation of risk.

In line with this framework, I drew on two resources to extract a list of English stance markers. One was Coffin's (2006) *Historical Discourse* and the other was the Collins COBUILD English Grammar (third edition). Chinese stance markers were afforded by Hsieh's (2008) study of evidentiality in Chinese newspapers, and the Dictionary of Chinese Verb Usages. For copyright issues, these two lists were not allowed to be presented here, but they could be made available upon request for academic purposes.

The next step was to categorise the identified stance markers according to their corresponding engagement feature. This was accomplished manually as there was no software that can do this reliably. Although every effort has been spent to achieve the highest possible rate of accuracy, I acknowledge that a marginal portion of error may still exist. However, as my primary focus rested on patterns of journalistic stance in mediating the negotiation of risk rather than exact figures, such marginal errors were minimally relevant.

Findings: What corpus data reveal about negotiating the food-safety risk

Language use is always ideologically invested (e.g. Bourdieu, 1991; Fairclough, 2010, 2014). As such, certain linguistic resources may be adopted more frequently than others in the expression of language users' specific perceptions of the social world. Acknowledging this, a frequency list of words may serve as the starting pointing for subsequent analysis of the interrogated discourse. However, such a frequency list was rather limited in terms of the insight it can provide into an understanding of the saliency of a particular corpus. To address this potential shortcoming a key word analysis was conducted to determine which words occur statistically more often in the node corpus, the corpus in question, when compared with the reference corpus (Baker, 2006). In the following, I will present the frequency (Section 4.1), keyword (Section 4.2), and semantic analyses (Section 4.3) to the food-safety corpus before profiling the selected journalistic stances (Section 4.4).

Negotiating risk through lexical priming

The frequency list of Australian corpus of food safety has been generated with the assistance of W-Matrix. The corpus was dominated by functional words in terms of their frequency because 'with few exceptions, almost all forms of language have a high proportion of grammatical words' (Baker, 2006, p. 53). The general frequency list was not provided here due to space limitations. Instead, I have shown in Table 15.1 below the list of the most frequent lexical words (i.e. verbs, nouns, adjectives, and adverbs). This list immediately suggested chief concerns of the corpus, as illustrated by the high frequency of 'food' and 'safety'. If combining 'said' and 'says' together, the lemma 'SAY' would surpass 'food' as the most frequent word (665 occurrences). This indicated the most prominent feature of news discourse, that of frequent quotation

Table 15.1 Top 20 frequent lexical words in the Australian food-safety corpus

No.	Word	Frequency	Per cent
1	food	565	1.13
2	said	543	1.09
3	safety	181	0.36
4	he	175	0.35
5	they	173	0.35
6	Australia	172	0.34
7	there	164	0.33
8	which	156	0.31
9	this	154	0.31
10	people	143	0.29
11	health	135	0.27
12	their	135	0.27
13	we	133	0.27
14	Australian	127	0.25
15	found	124	0.25
16	says	122	0.24
17	products	112	0.22
18	who	105	0.21
19	NSW	90	0.18
20	she	89	0.18

in news reporting. Pronouns such as 'he, 'they', 'people', 'their', 'we', 'who' and 'she' in the corpus showed those persons involved in the categorisation and negotiation of risk meanings in the Australian press. Based on the frequency of these pronouns, it seemed that that males were more often involved in news reporting than females were since 'he' appeared far more frequently than 'she' did, and that news makers were more often presented in indirect quotation than direct quotation as suggested by the frequent occurrence of 'they' and 'their'. Some place nouns such as 'Australia, Australian, and NSW' revealed the places where the majority of reported news events happened.

Table 15.2 shows that the Chinese news corpus communicated food safety in slightly different ways as it focused on the quality of food, illustrated by frequent occurrence of 'problem', 'safe/safety', 'excessive' and 'quality'. The voices brought in by Chinese journalists in the negotiation of risk tended to highlight two sources: one was that of journalists themselves while the other was the government official voice, evidenced by 'reporter' and 'department', respectively.

Despite these findings, limited knowledge has as yet been obtained concerning the salience of the discursive resources in negotiating risk

Table 15.2 Top 20 frequent lexical words in Chinese food-safety corpus

No.	Word	English translation	Frequency	Per cent
1	食品	food	1950	0.91
2	有	have (main verb)	1565	0.73
3	记者	reporter	1479	0.69
4	生产	produce	1134	0.53
5	产品	products	1032	0.48
6	发现	find	816	0.38
7	人员	people	680	0.32
8	市场	market	667	0.31
9	合格	safe/safety	655	0.30
10	这	this	648	0.30
11	问题	problem	607	0.28
12	说	say	581	0.27
13	进行	conduct	579	0.27
14	销售	sell	530	0.25
15	超标	excessive	526	0.24
16	大	big	517	0.24
17	部门	department	505	0.23
18	表示	show	491	0.23
19	人	people	489	0.23
20	质量	quality	479	0.22

meanings in each corpus. To do so, the node corpus (i.e. my specialised news corpus) was compared against a large reference corpus (i.e. British National Corpus, BNC).

Negotiating risk through projecting key words

When a node corpus was compared against a reference corpus, those words that occurred more often than expected in the node corpus constituted a key word list. I have adopted the W-Matrix (Rayson, 2003) to perform key word analysis of both key POS and semantic domains against the reference corpus of BNC. Unfortunately, there was no corpus tools of parallel function that may generate a key word analysis of the Chinese data. Therefore, the keyness of the Chinese data was unable to be produced here. The same applied to the semantic analysis of the Chinese corpus (see Section 4.3 below). That said, this section (Section 4.2) and Section 4.3 mainly serve to illustrate the methodological usefulness of corpus-based keyword analysis and semantic analysis in understanding the construction and negotiation of risk, rather than to compare risk meanings across two cultures.

Table 15.3 below showed the most salient keywords in the Australian food-safety corpus as compared with the BNC written sampler. Column O1 represents observed frequency in the Australian corpus; O2 is observed frequency in normalised data in BNC written sampler; %1 and %2 values show relative frequencies in the texts. '+' indicates overuse in O1 relative to O2. 'LL' shows the log-likelihood (LL) value, according to which the table is sorted. The LL value expresses how many times more likely a keyword is in the food-safety corpus than it is in the BNC written sampler. The higher the LL value is, the more likely a corresponding keyword is to be found in the food-safety corpus.

Table 15.3 shows that such key words as 'salmonella', 'bacteria', 'poisoning', 'risk', 'hygiene', 'BPA' (Bisphenol A), and 'cancer' brought to the focus a variety of diseases and risks that were prioritised in constructing food risk meanings in Australian press. It also highlighted a close link of risk meanings to the products or places where risks could more probably be identified, such as in eggs, in restaurants, in drugs, in vitamin D, and in imported foods. The voice of FSANZ (Food Standards Australia New Zealand) was more salient in organising the social world involving food safety than it was in the general BNC corpus.

Table 15.3 Top 20 key words in the food-safety corpus against the BNC written sampler

N.	Word	O1	%1	O2	%2	LL
1	says	122	0.24	227	0.02+	306.51
2	eggs	70	0.14	31	0.00+	300.51
3	salmonella	51	0.10	1	0.00+	297.63
4	bacteria	52	0.10	4	0.00+	285.03
5	restaurant	61	0.12	21	0.00+	276.50
6	consumers	59	0.12	19	0.00+	270.94
7	standards	74	0.15	61	0.01+	266.30
8	drug	62	0.12	29	0.00+	262.73
9	poisoning	47	0.09	3	0.00+	260.91
10	imported	54	0.11	15	0.00+	254.75
11	foods	54	0.11	17	0.00+	249.05
12	risk	76	0.15	84	0.01+	245.16
13	hygiene	40	0.08	2	0.00+	225.23
14	BPA	36	0.07	0	0.00+	217.00
15	product	69	0.14	87	0.01+	210.49
16	FSANZ	34	0.07	0	0.00+	204.94
17	vitamin_D	33	0.07	0	0.00+	198.92
18	cancer	43	0.09	15	0.00+	194.40
19	outbreak	42	0.08	15	0.00+	188.97
20	found	124	0.25	474	0.05+	184.69

Negotiating risk through semantic priming

Since the Australian food-safety corpus was semantically tagged, I was able to examine the key meanings of food-safety risk that had been highlighted in my Australian food-safety corpus as compared against the general BNC corpus (see Table 15.4 below).

Each item (e.g. F1) in this table was glossed by its corresponding semantic theme (e.g. food). In terms of the risk meanings, Table 15.4 showed that Australian news prioritised those food-safety risks relating to people's daily meals (e.g. F1 – Food, peanuts, cheese), living creatures (e.g. L2), sold food (e.g. I2.2), substance (e.g. O1), science and technology (e.g. Y1). As some examples, Figure 15.2 showed a snapshot of examples of risk concerning living creatures (e.g. L2).

Table 15.4 indicates that the food-safety risk was also linked to ultraviolet (UV) radiation (e.g. W2: Sunlight/ultraviolet radiation) in

Table 15.4 Top 20 frequent semantic tags in the Australian food-safety corpus against the BNC written

N	Item	O1	%1	O2	%2	LL	Glosses
1	F1	1595	3.19	2974	0.31+	4002.51	Food
2	B2-	815	1.63	1275	0.13+	2245.72	Disease
3	B2	241	0.48	129	0.01+	987.19	Health and disease
4	A15+	241	0.48	166	0.02+	919.08	Safe
5	B3	468	0.94	1711	0.18+	726.16	Medicines and medical treatment
6	L2	583	1.17	3225	0.33+	578.91	Living creatures: animals, birds, etc.
7	Q2.1	885	1.77	7024	0.73+	498.18	Speech: communicative
8	A15-	185	0.37	370	0.04+	445.86	Danger
9	I2.2	458	0.92	2738	0.28+	409.83	Business: selling
10	O1	226	0.45	689	0.07+	408.66	Substances and materials generally
11	G2.1	341	0.68	2418	0.25+	235.06	Law and order
12	X2.2+	327	0.65	2302	0.24+	228.15	Knowledgeable
13	S2	360	0.72	2896	0.30+	197.41	People
14	X2.4	283	0.57	2176	0.22+	169.13	Investigate, examine, test, search
15	A13	28	0.06	0	0.00+	168.78	Degree
16	T3-	79	0.16	197	0.02+	165.52	Time: new and young
17	Y1	147	0.29	778	0.08+	154.35	Science and technology in general
18	G1.1	378	0.76	3542	0.37+	148.58	Government
19	W2	24	0.05	0	0.00+	144.67	Sunlight/ultraviolet radiation
20	A2.2	427	0.85	4362	0.45+	133.97	Cause effect/connection

officials are preoccupied with	swine	flu , Dr Shiv Chopra says there	23 More	Full
link between antibiotic use in	animals	and the spread of antibiotic-re	24 More	Full
spread of antibiotic-resistant	bacteria	in humans , and to the outbreak	25 More	Full
eeding slaughterhouse waste to	animals	. And he raises the warning fla	26 More	Full
ines , is not in a panic about	swine	flu . He says the fear of influ	27 More	Full
He says the fear of influenza	viruses	generating a new mutant capable	28 More	Full
uenza viruses generating a new	mutant	capable of wiping out millions	29 More	Full
n living and evolving with flu	viruses	for millions of years . "It is	30 More	Full
ays . Chopra 's concerns about	bovine	growth hormone and other hormon	31 More	Full
ce in Canada , and in 1999 saw	bovine	growth hormone banned . But he	32 More	Full
uestionable safety , including	bovine	growth hormone . And he tried f	33 More	Full
ones , no genetically modified	organisms	, no slaughterhouse waste and n	34 More	Full
tice of feeding food made with	ruminants	to ruminants , as countries lik	35 More	Full
ng food made with ruminants to	ruminants	, as countries like Canada have	36 More	Full
derived feed is still given to	pigs	or chickens , and they in turn	37 More	Full
feed is still given to pigs or	chickens	, and they in turn are turned i	38 More	Full
turn are turned into feed for	cattle	. So insidious and complete is	39 More	Full
onventional beef . The ears of	pigs	and cows (where the hormones a	40 More	Full
al beef . The ears of pigs and	cows	(where the hormones are inject	41 More	Full

Figure 15.2 Examples of risk in relation to living creatures

believed harmful and approved for	infant	use in Europe , is not approved un	25 More	Full
and urging parents to wean their	infants	off the formula . A chall	26 More	Full
r 10 people , including a newborn	baby	, fell ill with thyroid problems i	27 More	Full
olive oil blend and three canned	infant	foods from Heinz . Heinz smooth cu	28 More	Full
Safety Authority for adults . But	infants	and small children are at greater	29 More	Full
g of the top three BPA-containing	infant	foods tested by Choice delivers ab	30 More	Full
rograms in a single hit so a 10kg	infant	would get 10 per cent of the safe	31 More	Full
food , including that consumed by	infants	from baby bottles , and concluded	32 More	Full
ing that consumed by infants from	baby	bottles , and concluded that level	33 More	Full
vel found by Choice was in canned	baby	custard and our analysis shows a n	34 More	Full
analysis shows a nine- month-old	baby	weighing 9kg would have to eat mor	35 More	Full
phase out use of the chemical in	baby	products within 12 months as new B	36 More	Full
milk , causing the deaths of four	babies	. Food Standards Australi	37 More	Full
powdered milk . The four	infant	deaths occurred after the babies d	38 More	Full
infant deaths occurred after the	babies	drank milk tainted with melamine ,	39 More	Full

Figure 15.3 Examples of risks in relation to babies

Australia. This is because UV constitutes one source of vitamin D for human beings, but long and frequent exposure to UV is detrimental to human health. The food-safety risk could hardly be associated with UV radiation elsewhere (e.g. China) other than in a country as sun-focused as Australia, illustrating the interlocked relationship between the perception of risk and a given culture.

When characterising food-safety risks, Australian journalists highlighted the relevance to various diseases (e.g. B2) and health problems (e.g. B2-) that those risks may incur. Some agents, particularly, young children, senior adults and food allergic people (e.g. S2) and toddlers (see Figure 15.3) were extremely fragile to the exposure of these risks.

To identify these risks foregrounded issues of the measurement of risk and the capacity of participants to judge such risks. Sources of the formulation of risks were attributed to food-safety experts (e.g. X2.2+) and government (G1.1), as shown in Figure 15.4 below.

```
ng in the past year , a    government               food survey reveals . </   14 More | Full
he next 12 months after    Government               health bosses were left    15 More | Full
ur weeks . The NSW Food    Authority                launched a " name and sh   16 More | Full
standards . Department     officials                expected to uncover kitc   17 More | Full
<p> Primary Industries     Minister                 Ian Macdonald said 40,00   18 More | Full
r Mr Macdonald said the    Government               had considered the gelat   19 More | Full
Food Safety Information    Council                  , said . Perishable food   20 More | Full
South Australian health    officials                said they would visit th   21 More | Full
f gastroenteritis . The    state                    's chief medical officer   22 More | Full
t risk . ` ` Department    officials                have returned to the sit   23 More | Full
ontinues to work with a    local council            to control the outbreak    24 More | Full
itamin D. </p> <p> In a    country                  as dazzled by sunlight a   25 More | Full
```

Figure 15.4 Examples of the semantic theme of government (G1.1) in the Australian corpus

```
y inspectors have issued 160    fines         in four weeks . The NSW Food     8 More | Full
ot envisage dishing out 1000    fines         to 600 businesses in 10 month    9 More | Full
ds in Crows Nest , which was    fined         $330 for failing to stop rode   10 More | Full
oad at Ashfield received two    fines         worth $660 for housing live c   11 More | Full
on Restaurant received three    fines         last year for failing to keep   12 More | Full
ons Point both received $660    fines         because there was no temperat   13 More | Full
reet at Bondi Junction , was    fined         $660 for failing to maintain    14 More | Full
recent months by a series of    legislative   changes aimed at protecting t   15 More | Full
ordination . They 're now in    court         arguing unfair dismissal . Bo   16 More | Full
countries and facing strict     regulation    in others , Food Standards Au   17 More | Full
ts labelling will proceed to    trial         . </p> <p> In the NSW Supreme   18 More | Full
```

Figure 15.5 Examples of the semantic theme of law and order (G2.1) in the Australian corpus

The capacity of government officials to evaluate risks was highlighted by their reasoning of cause and effect (e.g. A2.2) coupled with scientific testing and examination (e.g. X2.4). The recognition of risks formed the grounds by which people were judged and directed attention to those individuals and institutions who should hold responsibility for those risks. Normally the consequences for those creating a risk issue would include fines and/or other charges, as illustrated in Figure 15.5 below.

Negotiating risk through mediated voices

In the negotiation of risk, the dialogic space for some voices may be contracted, whereas that for other voices may be expanded. Figure 15.6 showed that when negotiating risk in news discourse, Australian journalists tended to open the dialogic space mainly through acknowledging (77.13%) and entertaining (6.91%) other voices. By contrast, their Chinese peers were reluctant to do so, as evidenced in two ways. One was that the number of attest, endorse, affirm, and pronounce stance markers well exceeded that in the Australian corpus, all these markers

Negotiating Risk in Print Media 257

functioning to contract the dialogic space. The other was that acknowledging other voices (59.67%), a way to open the dialogic space, was adopted far less frequently in the Chinese corpus as compared against that in the Australian corpus. Due to space limitations, this section will focus on 'acknowledge' and 'attest', illustrating how the dialogic space of different voices in risk communication was opened up and closed down, respectively.

The corpus software WordSmith (V5.0) was adopted to investigate the ways food safety was negotiated in the Australian corpus through acknowledging other voices. I have selected the most frequent word in realising acknowledgement – 'said' (528 occurrences) – to show the patterns of the expanded dialogic space in negotiating food risk in the Australian press.

Figure 15.7 was sorted by the frequency of collocation of the searched word, 'said', with the word immediately to its left (L1). It showed that 'said' was most frequently collocated with 'he' to its left, followed by 'she' and others in the 'L1' column. To explicate the persons whom 'he' refers to in the corpus, I went through all collocates of 'he' with 'said' in the co-text. The manual analysis showed that among the 85 collocates, the government official voice accounted for 31 (36%), voices from different associations amounted to 24 (28%), voices from experts

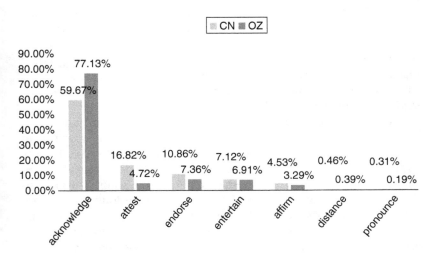

Figure 15.6 Percentage of key engagement features across Chinese and Australian corpora

258 *Changpeng Huan*

N	L5	L4	L3	L2	L1	Centre	R1	R2	R3	R4	R5
1	THE	THE	A	MR	HE	SAID	THE	WAS	HAD	THE	THE
2	TO	AND	SAID	DR	SHE		IT	A	THE	TO	OF
3	A	FOOD	THE	THE	SPOKESWOMAN		IN	THE	STATEMENT	OF	TO
4	OF	OF	MINISTER	MS	FSANZ		THERE	HAD	OF	NOT	FOR
5	AND	FOR	OF	HEALTH	SPOKESMAN		HE	WOULD	TO	WAS	IN
6	IN	A	FOOD	STEVE	MACDONALD		DR	HEALTH	SAID	WERE	HAD
7	HEALTH	TO	DR	PROFESSOR	COUNCIL		THAT	ARE	WAS	A	A
8	FOR	IN	SPOKESMAN	IN	BUCHTMANN		WE	AND	A	NO	THAT
9	PRIMARY	HEALTH	AUSTRALIAN	IAN	WHAN		A	OF	NOT	FOOD	WERE
10	BE	BE	SAFETY	LYDIA	PHILLIPS		THEY	THIS	BEEN	BE	WAS
11	FOOD	HE	AND	SAID	AMES		WHILE	FOOD	WERE	HAD	SAFETY
12	AT	AUSTRALIA	DIRECTOR		KINGMA		YESTERDAY	HAS	NO	ALL	AND
13	WITH	INDUSTRIES	FOR		GALLAGHER		THIS	WERE	ALSO	FOR	IS
14	HAVE	CHIEF	BUT		WOOLLARD		IT'S	COMPANY	IS	NEED	NOT
15			OFFICER		ALSO		FOOD	SAID	WOULD	ON	
16			HEALTH		GUEST		SHE	DRUG	HAVE	THAT	
17			AUSTRALIA		HOUSE		THESE		AND		

Figure 15.7 Patterns of concordances of 'said' in the Australian food-safety corpus

occupied 22 (26%) and voices from individual and institutional risk makers accounted for the rest eight (9.4%).

If we return to and continue reading the collocates of 'said' with those in the 'L1' column (see Figure 15.7 above), it would be clear that the most frequently mediated voice in communicating risk in the Australian corpus was the official voice, as shown by frequent collocation of 'said' with 'spokeswoman, FSANZ, spokesmen, Macdonald (NSW Primary Industries minister), council, Buchtmann (spokeswoman Ms Lydia Buchtmann of FSA), Whan (Mr Steve Whan, Primary Industry minister), Philip (Professor Paddy Phillips, state's chief medical officer)', etc. (Figure 15.8)

To produce acknowledge patterns in the Chinese corpus, I relied on the corpus concordance tool Sysconc designed by Dr Canzhong Wu at Macquarie University. The most frequent 'acknowledge' marker, '说' (shuo, say, 448 occurrences), was taken to illustrate how voices were mediated in negotiating risk in the Chinese press.

Due to the unique nature of Chinese grammar, the available corpus tool disallowed me to gain the information concerning which word collocated more frequently with the searched word. To examine the patterns of Chinese journalists acknowledging different voices, I manually checked all concordances of the searched word 'shuo'. The analysis showed that among the 448 concordances Chinese journalists prioritised the voice of risk makers (183 occurrences, 41%), followed by that of ordinary citizens (106 occurrences, 24%) [government official (86 occurrences, 19%), experts (60 occurrences, 13%), associations (7 occurrences, 1.6%) and journalists themselves (6 occurrences, 1.3%). In comparison with the voices presented in negotiating risk in the Australian

Negotiating Risk in Print Media 259

N	Concordance	Set Tag	Word #	Sent.	#ent. Pos	Para.	#ara. Pos	ead	ead t. Pos	File	%
1	that we pack for lunch these days," he said. "You can reduce your risk of		106	2	100%	0	31%		31%	OZ_FS60.txt	32%
2	fridges and kitchens clean," he said. "They can also supply pens and		303	9	100%	0	88%		88%	OZ_FS60.txt	89%
3	Safety Australia New Zealand, he said. "The claims of several years ago		110	4	100%	0	51%		51%	OZ_FS77.txt	55%
4	25th or Saturday 26th November," he said. "Food safety inspections of the		288	14	100%	0	89%		89%	OZ_FS38.txt	87%
5	of them had already been used. He said there was no need for parents to		232	9	7%	0	52%		52%	OZ_FS12.txt	53%
6	not saying what the percentage is," he said. "These products are safe for use.		206	9	100%	0	52%		52%	OZ_FS66.txt	53%
7	owners still needed to be vigilant, he said. "The vast majority of kebab		124	5	100%	0	70%		70%	OZ_FS56.txt	69%
8	their implantable medical device," he said. Mr Miller said that despite this,		814	27	100%	0	85%		85%	OZ_FS33.txt	84%
9	meat products are still safe to eat, he said. "This is not the strain of		144	5	12%	9	84%		51%	OZ_FS20.txt	47%
10	the products over about 11 days. He said the potential contamination was		379	16	12%	0	84%		84%	OZ_FS12.txt	85%
11	the site boundary (in Japan)" he said. Dr Higson said there was a risk		367	14	100%	0	71%		71%	OZ_FS85.txt	71%
12	diverse that incredibly small," he said. "We have gone hell for leather		383	18	100%	0	71%		71%	OZ_FS6.txt	71%
13	and stop it spreading further. He said members of the public were not		149	6	27%	0	71%		71%	OZ_FS105.txt	70%
14	by Melbourne's better restaurants. He said Australian and New		282	12	25%	0	52%		52%	OZ_FS6.txt	53%
15	the infection and stop it spreading. He said the outbreak remained contained		140	6	11%	0	54%		54%	OZ_FS62.txt	55%
16	there will be no further problems," he said. "These are common issues in		376	12	100%	0	91%		91%	OZ_FS116.txt	91%
17	with the latest attack techniques," he said. "This will ensure that all realistic		868	29	100%	0	91%		91%	OZ_FS33.txt	89%
18	10 packs," Mr Zhou said. He said the company had not imported		348	17	15%	0	92%		92%	OZ_FS88.txt	91%
19	regarded the offences as serious. He said Teng's partner had been		344	13	8%	0	66%		66%	OZ_FS29.txt	65%
20	Department of Primary Industries. He said "every necessary precaution" was		233	9	13%	0	91%		91%	OZ_FS76.txt	90%

Figure 15.8 Examples of collates of 'said' with 'he' in the Australian corpus

corpus, the Chinese corpus highlighted the presence of risk makers and ordinary citizens (e.g. consumers). Due to space limitations, examples of each voice were not provided here, but the general collocation was that 'officials/risk makers/ordinary citizens/experts/associations + 说 (shuo, say)', illustrated in example (1) below.

> X工商所长高峰说，柠檬黄不能用作花椒的添加剂，这是违法添加。[**Official voice**]
> Gao-Feng, the director of the X Administration of Industry and Commerce, **said:** lemon citron could not be used as an additive of pepper and it was an illegal additive.

Let us now turn to how the dialogic space of various voices was contracted in the Chinese and Australian press by the adoption of 'attest' (e.g. eyewitness sources). As previously defined, 'attest' relates to direct witness through human sensory senses (e.g. SEE, HEAR, and SMELL). Table 15.5 below shows that Chinese and Australian journalists mediate news sources in different ways in attesting risk events.

In the Australian corpus, there were considerable cases where sources of eye-witnessed events were unspecified by Australian journalists (34 occurrences, 47%). By unspecified sources, I mean those sources that were not immediately known in the local clause such as example (2) below. This is not to say that all these sources were kept anonymous, but journalists decided to keep them absent in a clause for various reasons.

> Glass has been **found** in a children's snack food … [**Unspecified voice**]

Table 15.5 Distribution of attest voices across Chinese and Australian corpora

N.	Sources	Chinese corpus Freq.	Per cent	Australian corpus Freq.	Per cent
1	authorial	351	40%	0	0%
2	official	261	30%	25	34%
3	risk maker	32	4%	2	3%
4	ordinary citizen	222	25%	6	8%
5	expert	6	1%	6	8%
6	unspecified	9	1%	34	47%
TOTAL		881	100%	73	100%

Australian journalists adopted the source of government officials as one of the main sources of eyewitness of the reported risk events (25 occurrences, 34%), such as example (3) below. The sources of ordinary citizens (6 occurrences, 8%) and those of experts (6 occurrences, 8%), illustrated in examples (4) and (5) respectively, were less frequently mediated by Australian journalists.

> Department of Sustainability and Environment air-attack supervisor Shaun Lawlor told the Bushfires Royal Commission he **saw** 100-metre flames in the mountain ash forest ... [**Official voice**]
>
> 'I was a bit staggered when I came here and **found** eight or nine students who were anaphylactic at the school,' he says. [**Ordinary citizen voice**]
>
> 'Most of the vehicles we **see** associated with outbreaks are foods where the eggs are completely uncooked ...' Mr Kirk said. [**Expert voice**]
>
> The manager in charge of the Villawood KFC in 2005 told the court that ... some staff members also testified that they ... saw chicken fall to the floor and people handling food without gloves on. [**Risk maker voice**]

The voice of risk makers with two occurrences only was rather marginal in negotiating risk meanings, as illustrated in example (6) above. The two cases of risk maker sources found in the Australian corpus were concerned with institutional rather than individual risk maker. The Australian corpus has documented none of journalistic authorial attest voice. The rare occurrences of risk maker voice and no presence of journalistic authorial voice suggested that Australian journalists were hardly on site in the reporting of risk events.

Negotiating Risk in Print Media 261

Figure 15.9 and Figure 15.10 below visualise the standardised distribution of attest sources in the Chinese and Australian food-safety corpus, respectively.

In sharp contrast with the distribution of attest sources in the Australian food-safety corpus, Chinese journalists relied heavily on the journalistic authorial voice (351 occurrences, 40%) in reporting risk events such as 记者见到 ... (reporter **saw** ...), followed by official sources (261 occurrences, 30%) such as 执法人员发现了 ... (law enforcers **found** ...) and sources of ordinary citizens (222 occurrences, 25%). The heavy reliance on authorial voice in the Chinese corpus stood in sharp contrast with that in the Australian corpus in which there was no occurrence of authorial attest source.

There were relatively more sources of risk makers in the Chinese corpus (32 occurrences, 4%) than those in the Australian corpus (2 occurrences) in attesting risk events/issues. Resembling their Australian peers in mediating sources of experts (eight occurrences), Chinese journalists tended to marginalise the voice of experts (six occurrences, 1%) as well. However, unspecified sources (nine occurrences, 1%) were nearly negligible in the Chinese corpus. These suggested that Chinese journalists,

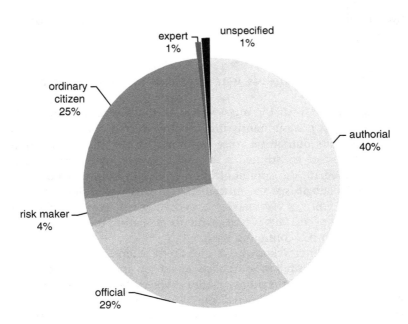

Figure 15.9 Distribution of attest sources in the Chinese food-safety corpus

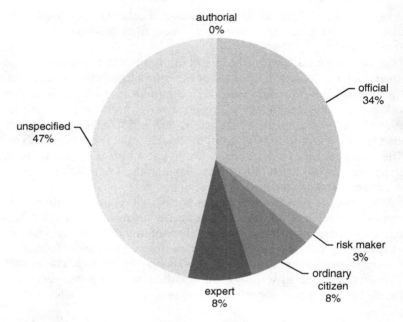

Figure 15.10 Distribution of attest sources in the Australian food-safety corpus

unlike Australian journalists, were more likely to be on site in reporting risk events.

In comparison with the attest voices in the Australian corpus, the Chinese corpus predominantly drew on the voice of journalists themselves while no journalistic voice was found in the Australian corpus. The official voice constituted a significant source for Australian journalists to mediate the communication of risk in both expanded and contracted dialogic spaces. Australian journalists were less likely to mediate the voice of risk makers in reporting risk events/issues than their Chinese peers. This was also true in mediating the voice of ordinary citizens in risk communication.

Bringing together acknowledge and attest sources in both corpora, it seemed that Australian journalists relied too much on official and organisational (e.g. different associations) voices, but the dialogic space of the official voice was opened up (i.e. acknowledge) as much as it was closed down (i.e. attest) in the Australian corpus. In comparison, Chinese journalists were more likely to open the dialogic space for alternative voices when mediating the voices of risk makers and ordinary citizens (e.g. consumers) through acknowledging their voices. This suggested that these voices were more negotiable than others. However, Chinese journalists were less likely to do so for the official voice, since official sources were more frequently adopted to attest risk events than they were to acknowledge risk events. This indicated that the official voice was less negotiable and was not supposed to be further challenged by alternative voices.

Discussion and conclusion

The main focus of this research was on how food safety in Chinese and Australian print media hard news reporting was represented and negotiated through journalists priming different lexico-grammatical and semantic resources and through their mediating different voices in journalistic professional practices. Taking these discursive resources as the analytical point of entry, this research found that the Australian corpus tended to represent food-safety risks in relation to the influence these risks may have on human health, while that the Chinese corpus perceived food-safety risk mainly with a focus on the low quality of food. When categorising the same food-safety risk, each corpus prioritised different risks through privileging certain discursive resources over others. The Chinese corpus prioritised those foods with excessive additives, manufactured without legal licenses, sold to consumers with fake substance, and eaten by consumers with serious medical consequences. In comparison, the Australian corpus prioritised risks concerning those foods containing bacteria, manufactured below food standard, and sold to consumers with hygiene problems. The different priming on discursive resources to categorise the same food-safety risk reflected the diverging situations of food safety in China and in Australia, since China had far more serious food-safety issues. Such a difference mirrored the nature of risk as a social and cultural category, since the meanings of risk were found in this study to differ across cultures.

Recognising risks provoked issues concerning responsibility and blame. In the Australian context, inflicting a food-safety issue often resulted in fines to and/or other legal charges of those responsible. In comparison, Chinese journalists often highlighted the practices of law enforcers on site, often without explicating legal consequences for those who were liable for the concerned risk. This contrast brought to view the capacity of those authorities in dealing with food-safety issues. Based on my participant observation of journalists following law enforcers in their food-safety report in China, law enforcers from the Administration of Industry and Commerce alone were not accorded with the power to issue fines or to legally charge those risk makers. Instead, what they often did was to suspend the business operation license of and confiscated food illegally sold in those shops involved in food-safety issues. The law enforcers were often perplexed by those manufacturers without a legal licence for three reasons. One was that risk makers could easily run away without serious financial loss when they spotted the law enforcers. Another was that law enforcers in China had limited human, material and financial resources to oversee and manage all illegal food firms. Lastly, in some cases there was no food standard for the law enforcers to judge and assess the risk.

Section 4.4 showed that risk meanings were not only constructed and shaped between government and science, as Beck (1992) argued, but the construction and negotiating of risk meanings also involved risk makers, lay persons and journalists. The latter has been found particularly true in the Chinese context. When negotiating risk, voices were not mediated by journalists in the same way. In the Chinese press, the voices of risk makers and consumers were contestable, while the voices of journalists and government officials were warranted. The situation was different in the Australian context, as the official voice was challenged as much as it was warranted, but the voices from different associations were mediated by journalists as contestable. Such differences reflected the different distribution of power represented by each voice in constructing and negotiating risk meanings in news discourse. The Chinese press were affiliated with the Communist Party of China and the government, and as such it was hardly unexpected to find the upholding of the official voice by Chinese journalists in mediating the negotiation of risk. In comparison, although Australia has had a high concentrated press ownership, the institutional practices of news institutions were not directly supervised by the Australian federal government. In this sense, the Australian press were not obliged to attach to or detach from the official voice. However, as a substantial amount of news sources were provided by professional organisations/associations to Australian press in the form of pre-formulated press conference reports, the number of voices from associations far exceeded that in the Chinese context. This is because the main source of daily news in the Chinese press, especially the Party-government affiliated papers was provided by the government who often took all journalists from the affiliated newspapers when there was an official activity to either inspect a food manufacturing factory or to combat illegal food manufacturers. As the Chinese press prioritised social impact over economic profits of news reporting, there were sufficient journalists available to spend days reporting food-safety issues that were relevant to local citizens. In contrast, the Australian press was owned by its proprietors with the chief purpose of profits accumulation. The fierce competition from other media outlets, especially the rising online news, had resulted in massive layoffs in Australian newspapers to reduce the cost of news production. It was therefore impossible for Australian journalists to spend much time to report food-safety issues, and instead they needed to rely heavily on feeds provided by official sources and relevant organisations and associations.

The practical consequences of this research are twofold. On one hand, there is a dire need for Chinese government to establish a more powerful institution to manage food safety, to normalise food production standards and to establish and put in force laws to protect legal food production and penalise illegal food production and transaction. On the other, voices of direct news makers, consumers, and risk makers, need to be heard more often in communicating food-safety risk in the Australian press.

Despite the illuminating findings generated by the newly built comparable corpus on food safety in Chinese and Australian press, the study reported in this chapter is limited by its overemphasis on the negative facet of food safety, and by its insufficient distinction between real and potential food safety risks. Future studies may improve this inadequacy by scrutinising each news report on food safety, rather than by examining general patterns of negotiating risks across cultures. Diachronic change of risk meanings in discourses is worth investigating as well with a view to understanding risks in the ever changing social world.

References

Adam, B., Beck, U., & Van Loon, J. (Eds.). (2000). *The Risk Society and Beyond: Critical Issues for Social Theory*. London: Sage.
Baker, P. (2006). *Using Corpora in Discourse Analysis*. London/New York: Continuum.
Bakhtin, M. M. (1981). *The Dialogic Imagination (translated by C. Emerson & M. Holquist)*. Austin: University of Texas Press.
Beck, U. (1992). *Risk Society: Towards a New Modernity* (Vol. 17). London: Sage.
Beck, U. (1999). *World Risk Society*. Cambridge: Polity Press.
Bernstein, P. L. (1996). *Against the Gods: The Remarkable Story of Risk*. New York: Wiley and Sons.
Biber, D. (1990). Methodological issues regarding corpus-based analyses of linguistic variation. *Literary and Linguistic Computing, 5*(4), 257–269.
Bourdieu, P. (1991). *Language and Symbolic Power*. Cambridge, MA: Harvard University Press.

Candlin, C. N., & Candlin, S. (2002). Discourse, expertise, and the management of risk in health care settings. *Research on Language and Social Interaction, 35*(2), 115–137.
Coffin, C. (2006). *Historical Discourse: The Language of Time, Cause and Evaluation.* London/New York: Continuum.
Douglas, M. (1992). *Risk and Blame: Essays in Cultural Theory.* London/New York: Routledge.
Fairclough, N. (2010). *Critical Discourse Analysis: The Critical Study of Language.* Harlow: Pearson Education Limited.
Fairclough, N. (2014). *Language and Power* (3rd ed.). London: Longman.
Giddens, A. (1991). *Modernity and Self-identity: Self and Society in the Late Modern Age.* Cambridge: Polity Press.
Hsieh, C.-L. (2008). Evidentiality in Chinese newspaper reports: Subjectivity/objectivity as a factor. *Discourse Studies, 10*(2), 205–229.
Luhmann, N. (1993). *Risk: A Sociological Theory.* Berlin/New York: Walter de Gruyter.
Lupton, D. (1999). *Risk: Key Ideas.* London: Routledge.
Martin, J. R., & White, P. R. R. (2005). *The Language of Evaluation.* Basingstoke: Palgrave Macmillan.
McEnery, T., Xiao, R., & Tono, Y. (2006). *Corpus-Based Language Studies.* London: Routledge.
Mythen, G. (2004). *Ulrich Beck: A Critical Introduction to the Risk Society.* London: Pluto Press.
Rayson, P. (2003). *W-Matrix: A Statistical Method and Software Tool for Linguistic Analysis through Corpus Comparison.* (Ph.D. Thesis), Lancaster University, Lancaster.
Sarangi, S., Bennert, K., Howell, L., & Clarke, A. (2003). 'Relatively speaking': Relativisation of genetic risk in counselling for predictive testing. *Health, Risk & Society, 5*(2), 155–170.
Sarangi, S., & Clarke, A. (2002). Zones of expertise and the management of uncertainty in genetics risk communication. *Research on Language and Social Interaction, 35*(2), 139–171.
Scollon, R., & Scollon, S. W. (2004). *Nexus Analysis: Discourse and the Emerging Internet.* London: Routledge.
Zhang, H. P., & Liu, Q. (2002). Model of Chinese words rough segmentation based on NYShortestYPaths method. *Journal of Chinese Information Processing, 16*(5), 1–7.

16
The Uses of Biological Sciences to Justify the Risks of Children's Mental Health and Developmental Disorders in North American News Magazines: 1990–2012

Juanne N. Clarke and Donya Mosleh

Introduction

The rates of diagnosis of children's mental health and developmental (CMHI) issues have been growing in the past decades particularly in Anglophone North America and in other Western neo-liberal democracies. The rates of growth in the United States of three specific diagnoses, attention deficit hyperactivity disorder, bipolar disorder and depression have been especially steep since about 1980 (Whitaker 2010). More recently there has been a dramatic growth in the diagnosis of autism spectrum disorder (Silverman 2008). Today the incidence of diagnosed and community reported CMHI is between 14.3% in Canada (Boyle and Georgiades 2010) and 16.7%, in the United States or 25.6% if addiction disorders are included (Perou et al. 2013). A recent US national survey estimated that approximately 20% of youth had (during their lifetime) experienced some type of mental disorder that limited their functional ability (Merikangas et al. 2010). There has also been a significant increase in the prescription of certain drugs to treat mental illnesses in children and adolescents. For example, there has been an increase in the prescription of anti-psychotics (associated with the escalation in bipolar diagnoses) in the past several decades (Olfson et al. 2006). Clearly, a substantial minority of children and young people are succumbing to the contemporary labelling of 'having' and 'needing treatment' for one or more CMHI.

The context of this dramatic change is contemporary society and the particular focus of this chapter is on its characterisation as a 'risk society' where risk is seen as an inevitable, increasing and virtually ubiquitous consequence of modern life (Beck 1992; Giddens 1999). Risk consciousness and risk management, along with minimal government intervention, privileging of the market, emphasising individual responsibility, and increased inequality are all associated with the contemporaneously dominant political rationality called neo-liberalism (Ayo 2012). The political-economic philosophy of neo-liberalism is also connected with the growth in emphasis on the achievement and maintenance of the healthy body or 'healthism' (Crawford 2006) as a signal and symbol of being a good (neo-liberal) citizen. This emphasis on risk occurs in the context of an expansion of medicalisation (Conrad 2005) and biomedicalisation (Clarke et al. 2003) and illustrates Foucault's notion of biopower (Foucault 1976). What this means for the recognition and treatment of CMHI is that children's health issues are likely considered to be of heightened importance as a part of investing in good and healthy citizens-to-be. CMHI are also, along with other children's health issues a way to demonstrate good citizenship through prescribing good parenting and appropriate social, political, and economic investment in the future through individuals (Lee, Macvarish, and Bristow 2010).

The salience and the risks of different diseases as well as their symptoms and experience vary widely across time. Klawiter (2004) who studied breast cancer experience in two time periods uses the term disease regimes to refer to the complex and deeply internalised ways that diseases are created and 'shaped' by intersecting, structural and cultural forces, and thus involve changed experiences over time. The term regime is derived from Foucault's idea of 'regime of practice' (Foucault 1988), and invokes the notion of governance or management of individual bodies in the population through biopower (Foucault 1976). It refers to the (self)government of, or the 'conduct of conduct' (Foucault 1988) of mundane routine life by individuals. It does not refer to formal governmental structures or power. Instead it is to be understood in the ways that individuals discipline themselves or govern their own lives having integrated (and sometimes resisted) circulating discourses that are pervasive, invisible, and ubiquitous. In other words people police, govern, and manage themselves (and in the case of CMHI, children through 'good' parenting) via the deployment of 'technologies of the self' (Foucault 1988). It appears then that the medical diagnosis of the anomalous behaviours of children is increasingly regarded as a ready possibility as new disease regimes are produced to explain and change undesired behaviours and emotional expression in children. Along with the notion of CMHI, the general category of 'the child at risk' has been expanding (Oulton and Heyman 2009:295). It is to the media contributions to such governing of children at risk for mental health and developmental issues that we turn.

This chapter focuses on the portrayal of children's mental health and developmental issues (CMHI) in mass-print news media produced and widely circulated in North America. It is a part of a larger project that has also examined various specific CMHI such as attention deficit hyperactivity disorder (ADHD), autism, and depression over time in North American magazines. We have taken up this focus because we believe that our social constructions of and actions upon children are fundamental to our ideas of personhood and identity. They are linked inextricably to social/political and economic policies, social programs, and normative behaviours. They are permeated with ethical considerations. Ideas about normal and disordered or diseased childhood and identity reflect diverse and powerful interests in our social structure, culture, and values. In this focus on CMHI we are especially concentrating on powerful forces involved in the expansion of risks enlisted by the medicalisation (Conrad 2005) and biomedicalisation (Clarke et al. 2003) of childhood behaviours. We agree with Rose that childhood today 'is the most intensely governed sector of personal existence' (Rose 1989:123).

Medicalisation or 'defining a problem in medical terms, usually as an illness or disorder, or using medical intervention to treat it' (Conrad 2005:3) has been growing since the middle of the last century. It is linked to the ascendency of medical definitions of reality and the diminution of political, economic, religious, and judicial understandings and remedies. The behaviour of children has come to be conceived within medicalisation discourses (Whitaker 2010) through a process in which a variety of children's behaviours have become pathologised as mental health and developmental diagnoses (Furedi 2008:108). In this growth of diagnoses, Furedi (2008) argues that 'children who deviate from the norm, who are unusually active or shy, or sensitive or naughty will be medicalised and rewarded with the diagnosis of a newly invented syndrome' (15) and parenting too has become imbued with the concerns of medicalisation as parenting has come to be considered a perilous occupation requiring skill and scientific information (Richman and Skidmore 2000; Lee 2008; Oulton and Heyman 2009).

The idea that the portrayal of children's mental health and developmental issues has been growing and changing significantly over the past century in Canada (Clarke 2010a, 2010b, 2011, 2012; Clarke et al. 2014) has been documented through a series of studies of mass high-circulating Anglophone North American magazines as well as Canada's premier magazine for women, Chatelaine (Clarke 2013). The contemporary portrayal in Canadian and US publications of specific developmental and mental disorders such as depression, autism, and ADHD has also been recognised in a series of papers (Clarke 2010a, 2010b, 2011, 2012). Growth and changes in the content and meanings of medicalisation, biomedicalisation, and the risk society along with variations in intensive mothering and parenting expectations and practices have all been noted in these analyses.

Mass media are an influential source of health knowledge, belief, and experience in modern societies. There is no doubt that people turn to mass media to gain a perspective on, or information about, what is going on and what should be going on in the world. Changes in disease regimes are reflected in mass media. As Altheide (2002) argues our everyday lives are mediated. media reflections on how people perform their lives infuse our modern subjectivities showing us what kind of people and citizens we might, and should, want to be. The significance of this study is that the portrayal of CMHI likely filters into and shapes the viewpoints of readers as it sets agenda (McCombs 1972) with ideas about what they need to be aware of as they raise their children, observe others raising their own and work to set social, political, health and economic policy. The powerful circulating discourses evident in the magazines and reflective of different disciplines may also be associated with the development and maintenance of related policies, practices, organisations, and corresponding disease regimes.

This is then a study of the risk governing practices implicated by news stories about the topic of children's behaviours and emotions (mental and developmental health). Disorder or disease is not considered an objective reality experienced differently in different regimes but that the very notion of the existence and characteristics of any physical or mental disorder or disease is socially constructed and thus varies in different periods and places. The purpose of this chapter is to describe, analyse, and theorize about the discourses evident in the content of all of the news stories about children's 'mental' health and developmental issues in the high-circulating news magazines from 1990–2012. It will do this by asking of the magazine portrayal of CMHI the following questions. What is it and of what is said to consist? What causes it, and, what can and should be done about it?

Methods

Sample

This study was based on all available full text English-language articles (without graphics) indexed on the topic of mental health and developmental issues in children in news magazines and one highly circulated newspaper indexed in *The Readers Guide to Periodical Literature* published in the North America from 1990–2012. These magazine stories were selected as data because such news magazines are highly subscribed and then circulated. They are often left for perusal in the waiting rooms of doctors and dentists and other services. Readers can reread articles of interest and share them in a material way with friends and family members. In these ways, print media may at times circulate widely and be more influential than television or radio reports. They may also be more influential than the internet in agenda setting as each magazine sets out for the reader the topics considered worthy of thought and then describes the parameters of the selected topics. However, only once a topic is deemed worthy of further research and investigation is the internet search likely to begin.

All of the 38 articles available that were indexed in *The Reader's Guide* under children's mental health from 1990 to 2012 and were found in news magazines *Time*, *MacLean's* and *Newsweek* were included in this sample as well as the newspaper, *U.S. News* and *World Report*. The following table describes the circulation rates and the audience characteristics. The appendix lists all of the magazine articles included in our analysis. It is important to note that these are considered to be 'gender-neutral' news magazines because the readership is approximately half male and half female and the focus of the stories is news. Further, as can be seen from Table 16.1 the readership would be considered middle and upper class based on their income levels. Further, the average age of the readership is approximately middle aged.

The years 1990–2012 were selected in order to provide enough data over a long enough period of time (approximately 22 years) for in-depth qualitative analysis and to document the contemporary picture of the portrayal of CMHI in the select news magazines. It was also a comparator sample to that of the women's magazines as described by Clarke and colleagues (2014). This multi-year sample also has the advantage of avoiding the bias of the selection of one year only when a particularly newsworthy event relevant to children's mental illness may have occurred.

The search terms for the relevant magazine articles were comprehensive and reflected our belief that the concepts used to reflect mental

Table 16.1 Summary of magazine readerships

Magazine	Circulation	Country	Readership	Demographic
Maclean's	321,275	Canada	2,474,000	53% Male 47% Female Mean age: 47 Income: $92 223
Newsweek	15,000,000	USA	18,000,000	Male: 57% Female: 43% Median age: 49.2 Income: $76 062
Time	3,289,377	USA	17,525,000	Male: 55% Female: 45% Median age: 48 Income: $72 131
U.S. News and World Report	20,000,000	USA	120,000,000	Male: 52% Female 48% Median age: TBD Income: $90,568

health and developmental concerns might well reproduce the probable flux in diagnoses and symptoms, i.e., different disease regimes. They included behaviour, emotion, health, mental health, parenting, psychiatry, psychology, childrearing and associated terminology. As time went on the searches began to include new terms such as attention deficit disorder and other specific diagnoses.

Data analysis

Initially all articles were read by each author to categorise them for manifest content as relevant for the study. Irrelevant articles were eliminated from further analysis. Subsequent readings involved selective coding to identify and label the major (manifest and latent) emerging themes and to choose illustrative quotations in answer to the questions posed This method and these questions are based on earlier work on the portrayal of depression over time (see Clarke and Gawley 2009). Once the specific illustrations were chosen we discussed the findings and reread the articles to ensure that the coding was complete. I (first author) had already read the available sociological literature on the topic of children's mental health issues and was familiar with the theoretical concepts that are pivotal to this particular chapter. This is a type of discourse analysis and is seen as particularly apropos in the study of mental disorders as they are produced almost entirely within language (Harper 1995:348).

Findings

What is it and of what is it said to consist?

The first notable feature of the stories is that they usually entail an emphatic assertion that CMHI are real and growing. Stories stress that CMHI have been overlooked in the past but are coming to light now because of the hard work and the powerful and concerted investigations of biological and medical scientists. Buttressing the idea that they are real and growing is the contention that they (often) begin in infancy and that they reflect innate, biological differences. These ideas are utilised to reinforce the new and material reality of the growing claims-making (Best 1987:106) and risk associated with CMHI. Articles emphasise the expanding incidence of diagnoses well as an increase in untreated children. The characteristics of such children are described as seriously problematic. The linkage of pathologies or diagnoses with ordinary behaviours also supports the notion that CMHI have grown, spread, and are commonplace. Following are a few illustrations of the emphasis on the growth and the extensiveness of CMHI. '1 in 10 children in the US has a mental illness'. 'Fewer than 1 in 5 get the proper attention' (Shute et al. 2000:42). 'Clearly half the patients I see have some kind of serious emotional problem' (Szegedy-Maszak 2002:15) 'Mental illness is a closet problem in this country and it's got to come out' (Wingert and Kantrowitz 2002:52). The following quotation both emphasises that problems begin at birth and that even behaviours that might otherwise be thought of as normal can be indicative of these new and growing diagnoses.

A child with problems, they insist, makes no secret of it from the start, coming into the world timid, moody, jumpy, or worse. Experts often dismiss such claims as hooey at best, blame ducking at worst, but there may be more to it than that. A growing body of research shows that newborns do tip their emotional hand early on, giving parents a chance to take control of behavioural problems and maybe even prevent conditions such as attention deficit hyperactivity disorder (ADHD) or depression from fully taking hold (Klugger 2002:87).

And again emphasising the biological facticity of CMHI because they originate from birth and thus inhere in the body and not social relations is the following, 'Autism, the new findings suggest, is not a sudden calamity that strikes children at the age of 2 or 3 but a developmental problem that can be tracked back to infancy' (Cowley 2003:46).

In spite of the claim that CMHI are growing there is little consistency in the definition of CMHI from article to article. Instead there is wide variation in the classification of what constitutes the developmental or mental health issue under consideration. The articles reflect a prevailing sense of risk through focusing on confusion, contradiction, and uncertainty. Sometimes this state of perplexity is made explicit by the author as in the next four examples, each of which in a different way asserts the confusion and uncertainty involved in diagnosing CMHI. 'What if the apparent explosion in autism numbers is simply the unforeseen result of shifting definitions, policy changes and increased awareness among parents, educators and doctors?' (Wallis 2007:69). 'We need to figure out if the child is experiencing a developmental process versus a developmental delay versus a real illness' (Szegedy-Maszak 2002:13). 'Even adults who make a career of working with kids – teachers, coaches, and pediatricians – can misread symptoms [...] That's why many kids still suffer unnoticed, even though more schools are using screening tools that identify kids who should be referred for a professional evaluation' (Wingert and Kantrowitz 2002:8). Finally, '[...] there is nothing clear about ADHD. The diagnosis is based on a checklist of subjective judgments' (Smith 2000).

The 'symptoms' that the CMHI diagnosis is meant to encapsulate are also disparate and wide-ranging yet they are usually quickly associated with biological causation. Notice in the first quote the problem is described as shyness but the question immediately becomes what sort of a neurological problem is or causes shyness. Readers might wonder why this is not just an 'ordinary' human trait but instead is promptly interpreted as a problem that the 'hard' medical and biological sciences are poised to understand and then treat.

> There may be physical causes of shyness too [...] If blood isn't circulating properly, the brain and other organs don't get enough oxygen for engagement in social interactions [...] A pediatric neurologist can determine if a child's shyness is actually a condition lying on the autism spectrum. (Oz 2012:46)

The following quotations point to another issue in the definition of the problem cited in the news magazines. 'Symptoms' are categorised together and argued to represent particular and specific medical diagnoses even though the 'symptoms' are often not logically linked. Speaking of autism the following reflects this process of association of different 'symptoms' with one cause. 'Researchers still don't know what causes it, nor do they know how best to treat a condition that prompts one child to stop speaking and another to memorize movie scripts' (Kalb 2005:44). Linking any problem to suicide is another technique that may be understood as emphasising the materiality and seriousness of CMHI. 'Approximately half of the teenagers with untreated depression may attempt suicide, which remains the third leading cause of death in this age group [12–18 years]' (Bostic and Miller 2005). In the next quote the association of CMHI with diabetes is made. This notion both justifies the biological certitude and legitimates the use of pharmaceuticals to treat the condition. 'They are ill it's that simple. And the illness is as medical as diabetes' (Szegedy-Maszak 2002).

In conclusion, the answers to the first question are that the stories claim that the problem is a big one and that it is growing. However, what it is that is growing is still quite uncertain and even contradictory at times. The characteristics of the symptoms of each diagnosis are in dispute. Although the symptoms are widely disparate and range from trivial to potentially serious the diagnoses are argued to be as material and concrete as biologically measureable physical diseases such as diabetes and to possibly end in the most serious of outcomes, suicide.

What causes it?

Thus, a paradoxical combination of expansive reification coupled with ambiguity characterises the constructions of the very definition CMHI. There is a pervasive claim that CMHI are real problems that need to be understood. Their causes must be explained and the discipline that is privileged to do the explaining is one or another of the medical or biological sciences, including neurology, biochemistry, genetics, and/or epidemiology among others. This viewpoint is made explicit in many articles. There are myriad hypothesised causes of CMHI. Although the putative causes are numerous, complex, and contradictory, they are written about with a confidence that the answer will definitively come from research using one medial/biological scientific discipline or another. The majority of the discussion of causation focuses on the link to individual biological differences amongst infants and youngsters. Research is described in areas related to hormones, brain structures, genes, and neurology. Regardless of the disparate putative causes invoked by the many different directions of the research, the findings are discussed authoritatively as if they are the singular approach to the problem. Despite the fact that positivist science is based on probabilities not certainties and that the history of science has involved repeated paradigm shifts (Kuhn 1962) including the repudiation of previous findings and reflecting entire new ways of looking at issues, the magazines tend to report the widely varying types of research in a particular and consistent manner. Each study is discussed as if it reflects the truth and absolutely the best scientific approach. Research is described as definitive, objective, and cumulative. Following are a few illustrations of the assertion of varieties of biological causes. One of the strategies used is to underscore and privilege new directions of biological research as in the first quote. The second group of quotes reflects a different assertion: this time that the cause is genetic. The third group attributes CMHI to biochemistry and neurology. Finally other biological causes are put forward including growth patterns, oxygen, and vaccines.

Hormones and brain structure

The hormonal surges of puberty have long been shown to affect moods, but now new research says that changes in brain structure may also play a role. (Wingert and Kantrowitz 2002:4)

Genes

[...]it's likely they share genetic predisposition to the illness. (Szegedy-Maszak 2002:33)

Scientists also believe that there's a genetic predisposition to depression. 'The closer your connection to a depressed family member [...] the greater the individual's likelihood of suffering depression. (Wingert and Kantrowitz 2002:11)

[...] Asperger's appears to be even more strongly genetic than classic autism [...] About a third of the fathers or brothers of children with Asperger's show signs of the disorder. There appear to be maternal roots as well. (Madeline 2002)

Neurological abnormalities

There are three subtypes of the disorder: hyperactive, inattentive, and a combination of the two. Each version is likely to involve deficiencies in dopamine and norepinephrine activity along the pathways between the basal ganglia and the prefrontal cortex, which control attention and short-term memory. The cells that transport both dopamine and norepinephrine may be working overtime in people with the disorder, sucking up the neurotransmitters that are essential for smooth mental functioning. (Szegedy-Maszak 2004:55)

The reigning theory is that attention deficits are related to faulty biochemical communication in the brain. (Szegedy-Maszak 2004:55)

Autism is a neurological disorder characterized by language problems, repetitive behaviours, and difficulty with social interaction. (Kantrowitz and Sceldo 2006:46)

Other biological reasons

The new study, published in *The Journal of the American Medical Association*, links the condition [autism] to abnormally rapid brain growth during infancy – and it raises new hopes for diagnosis and treatment. (Cowley 2003:46)

There may be physical causes of shyness too [...] If blood isn't circulating properly, the brain and other organs don't get enough oxygen for engagement in social interactions [...]. (Oz 2012:46)

The final group of causes is proximately biological but indirectly the result of medical and health policy and practices. Vaccines were implicated in the growing frequency of autism, while surgical/medical interventions in general were associated with other issues.

Vaccines and medical practices

[...] scientists found that children who had had two or more surgeries by the time they were 2 were twice as likely to be diagnosed with ADHD by the time they were 19, compared with youngsters who had had only one surgical procedure. (Park 2012)

Autism stems from a severe immune reaction to something in the vaccine. (Park 2002:53)

What should be done about it?

Notwithstanding the fact that there is no clear problem being discussed under the heading of children's mental health and developmental issues nor are there particular causes that are described as more likely than others, specific solutions are clearly offered. These solutions emphasise the need for greater financial investment in biological and medical science and an expansion of the numbers of psychiatric professionals. The following are a few of the relevant quotations. The first quote emphasises that the problem has been exacerbated by the lack of funding directed to this problem. Budget cuts have contributed to this (Szegedy-Maszak 2002:5). The next quotation is more explicit and more medicalising as it lays the problem clearly at insufficient medical care. 'Fewer hospital beds and long-term care is scarce' (Szegedy-Maszak 2002:5). Finally the next quote asserts that both the lack of funding and medical care are together implicated in the problem of CMHI. 'Richard Harding, former president of the American Psychiatric Association and a child psychiatrist [...], calls the national problem a "perfect storm", where budgets are cut, and inpatient facilities are closing, and more children than ever need help' (Szegedy-Maszak 2002:5). The emphasis on medicine and science is again made in answer to the three questions posed.

Anomalies

There is very little in the news magazines about social, educational, or familial explanations of behavioural differences or of their amelioration. Here is an example of this anomalous where the impact of neighbourhood poverty is noted. 'The shortage is most acute in low-income areas and there are severe consequences in communities with more than enough traumatic circumstances to trigger a major depression' (Wingert and Kantrowitz 2002:9). One other anomaly is noted in the following statement that questions the subjectivity of the diagnosis of ADHD. The first quotation, an anomaly, actually takes issues with the usual assumptions and questions the use of medication in toddlers. '[...] there is nothing clear about ADHD. The diagnosis is based on a checklist of subjective judgments' (K. Smith 2000:84). Finally, and unlike the magazines for women (Clarke et al. 2014) these news magazines tend not to implicate or address the audience as mothers or family members in the problem of CMHI (Table 16.2).

Discussion

This chapter has described the portrayal of CMHI in mass-print news magazines and one newspaper in North America in the contemporary period of slightly more than the past two decades. The focus has been on examining the portrayal of CMHI through answering fundamental questions about each news story. What is it and of what is it said to consist? What causes it, and what can be done about it? As the above table summarises, the story told in the news magazines is consistent. It is a story of the rapid, even epidemic growth in the risk of various children's mental health issues. The definitions or the characterisations of the disorders are often incommensurate from article to article. There is some acknowledgement that definitions and symptom patterns may be shifting but there is no questioning that there is an 'it', a CMHI, to be scrutinised for, understood, explained, and treated. CHMIs are as described above real, biologically rooted, pervasive, and growing. They

Table 16.2 Answers to the three guiding questions

What is it and of what does it consist?	What causes it?	What can be done about it?
It is real.	It is biological.	Funding needs to be increased for scientific research and for treatment.
It is growing.	It could be hormones.	The number of child psychiatrists and medical specialists needs to rise.
It is diverse.	It could be brain structure.	
It is incongruent and contradictory.	It could be brain chemicals.	
It encompasses a wide variety of 'symptoms'.	It could be genes.	
It is biological.	It could be neurological.	
It should be studied objectively, using positivist science.	It could be vaccines or medical interventions.	
	There may be some correlations with some of the social determinants of health such as poverty.	

are portrayed as if they are tangible and objective things that are in the process of being scientifically studied, measured, and then treated. They are based in the body of the child, and, although, in a very few stories, they may be influenced by socialisation, parenting, schooling, racial discrimination, and the like. They occur and can be dealt with in the child's body (virtually) singularly. 'Symptoms' are sometimes dissimilar, idiosyncratic, and range from unwillingness to clean one's room to memorising movie scripts. Nevertheless their reality as an objective facticity is assumed. They are spoken of as if they are inherent in the biology of some children and thus evident from birth. Despite the assurance of their biological substrate a wide variety of putative biological causes is listed from story to story (for e.g. genes, neurological abnormalities, hormones, brain structure, and brain chemicals). Each article advocated for the credibility of one or several of a multitude of various types of objective and positivistic research in the realm of biological sciences. Consequently, the reader is left with a sense of what to do only if they are to adopt the idea of 'positivistic and objective science as truth' and as the way to the truth.

How do the discourses enlisted in the analysed articles contribute to the neo-liberal focus on risk? They underscore the necessity to be on guard for an ever-increasing panoply of potential dangers. They work together to teach the reader that children's mental health and developmental issues are real, serious, and potentially threatening. They do this by explicitly describing them as located in the material reality of the individual body and brain of the child discoverable through the hard work of science. This works implicitly in the service of biopower through medicine as the discipline and jurisdiction under which the body is formally defined, described, evaluated, assisted, and improved. Risk awareness is enhanced by the focus on the wide and expanding dispersion of mental health diagnoses and the assertion of their materiality and threat through the life-course and beginning of infancy. Risk expansion is also evident in the fact that there are many different scientific avenues for research including such directions as neurochemical, brain structures, and genes. This broad focus may also serve to enhance the awareness of imaginable risk in the sense that readers could come away from these stories with a sense of a wide variety of the possible sources of bodily dangers that may await children. Risk is also emphasised by the association of CMHI to a multitude of potential symptoms ranging from the seemingly common and perhaps previously 'innocent' (such as shyness or timidity) to the more serious including suicide. As Gagnon (2010) suggest this may work as a sort of governing through insecurity. This emphasis on CMHI and their material reality is evidently a new disease regime but its contours, causes, and treatments are ambiguous and incommensurate both within and between the many different diagnoses.

The idea that children's health needs governing is emphasised by the dictates of positivistic science with a focus on biology and objectivity. A neo-liberal commodification of children as objects for health improvement and the enactment of 'healthism' (Crawford 2006) is also evident and underscored by the assumption that anyone and everyone would prefer an ideal or normal child. The following quote, for example, audaciously suggests that children are potentially commodified as they may be subjected to a ranking and with the same criteria as are used in competitive sports events. 'Who wouldn't want a golden child? The problem, of course, is that if boldness is golden, reticence must be silver or bronze – or tin. Introverted children are everywhere – you may be raising one' (Oz 2012:46) Representing children as commodities to be governed and fixed in order that they better fit in behaviourally and emotionally reminds us that despite the rhetoric emphasising that the twentieth century was the century of the child (Koops and Zuckerman 2003:43), the evidence here suggests that children are in a sense to be considered a type of pre-person (Mayall 2000). In fact, rather than the century of the child, Mayall (2000) argues that the twentieth century should be called the century of the child professional (p. 244). In particular childhood has grown to be increasingly risky and is now under the jurisdiction of the developmental psy-scientists and practitioners whose purview is the description and maintenance of the 'normal' child and 'normal' stages of child development.

Further, the sciences deployed in legitimating the reality of CMHI neglect utterly the subjectivity of children. The subjectivity of children is thus accorded a private and insignificant place in social and health policy. Children are to obey the dictates of the social order and to be good and healthy citizens-to-be. They are to be acted upon rather than actors in the interests of the neo-liberal, individualising state. In this emphasis children are not accorded human rights but are assumed to live outside of politics and under the authority of the medical and biological sciences. The human rights of children are ignored under the jurisdiction of psychiatriztion (Breggin 2014; La Francois and Coppack 2014). This is despite the fact that many young people with autism, for example, claim they are not disordered but just different. They may call themselves Aspies and those who think differently Neurotypicals (Clarke and Van Amerom 2008).

This portrayal also invokes medicalisation and biomedicalisation (Conrad 2005; Clarke, et al. 2003) implicitly, just as medicalisation and biomedicalisation locate the problem as expanding and at the same time inhering in the individual and a growing number of 'sick' individuals in need of medical intervention. Even though there is not a great deal of discussion of medical treatments, the focus of the search for answers is almost entirely within the body and it is fixing the individual body that is the purview of medical practice. Moreover, there is a call for more child psychiatrists as the medical profession assumed most suited to the problem of CMHI as it is defined in the stories assessed. A medical approach to the issue is also supported by the objectivistic and positivistic type of research that is advocated as the appropriate methodology and subject matter for addressing the claimed problem.

Finally focusing on the individual child reinforces neo-liberalism, individual risk and with it inequality by concentrating on the individual brain or genes or other biological reality. This view reinforces risk as it obscures the powerful inputs of the social determinants of health and social inclusion on the physical and mental well-being of children and adults (Case, Paxon and Vogl 2007; Cordell 2004). There is very little discussion of the fact that mothers, fathers, and their children differ in many ways such as social and economic status, sexual orientation, family structure, education level and type, ethnicity, 'race', gender, occupation, neighbourhood and the like. These distinctions lead to dramatic differences in illness experiences, death rates, and the perceptions of the types, locations, and frequencies of risks faced by the audience.

Works cited

Altheide, D. L. (2002). Children and the discourse of fear. *Symbolic Interaction*, 25(2), 229–250.

Ayo, N. (2012). Understanding health promotion in a neoliberal climate and the making of health conscious citizens. *Critical Public Health*, 22(1), 99-105.

Beck, U. (1992). *Risk society: Towards a new modernity*, translated by Mark Ritter. London: Sage.

Best, J. (1987). Rhetoric in claims-making: Constructing the missing child problem. *Social Problems*, 24(2), 101–121.

Boyle, M. H. & Georgiades, K. (2010). Perspectives on child psychiatric disorders in Canada. In J. Cairney and D.L. Steine (Eds.), *Mental disorder in Canada an epidemiological perspective* (205–206). Toronto: University of Toronto Press.

Breggin, P. R. (2014). The rights of children and parents in regard to children receiving psychiatric diagnoses and drugs. *Children and Society*, 28, 231–241.

Case, A., Paxson, C. & Vogl, T. (2007). Socioeconomic status and health in childhood: A comment on Chen, Martin and Matthews. *Social Science and Medicine*, 64, 757–761.

Clarke, A. E., Shim, J. K., Mamo, L., Fosket, J. R. & Fishman, J. R. (2003). Biomedicalization: technoscientific transformation of health, illness and US biomedicine. *American Sociological Review*, 68, 161–194.

Clarke, J. N. (2010a). Childhood depression and mass print magazines in the USA and Canada: 1983-2008. *Child and Family Social Work*, 16, 52–60.

Clarke, J. N. (2010b). The domestication of health care: Health advise to Canadian mothers 1993–2008. *Today's Parents, Family Relations*, 59, 170–179.

Clarke, J. N. (2011). Magazine portrayal of attention deficit/hyperactivity disorder (ADD/ADHD): A post-modern epidemic in a post-trust society. *Health, Risk and Society*, 13(7–8), 621–636.

Clarke, J. N. (2012). Representations of autism in US magazines for women in comparison to the general audience. *Journal of Children and Media*, 6(2), 182–197.

Clarke, J. N. (2013). Medicalisation and changes in advice to mothers about children's mental health issues 1970–1990 as compared to 1991–2010: Evidence from Chatelaine magazine. *Health, Risk & Society*, 15(5), 416–431.

Clarke, J. N. and Gawley, A. (2009). The triumph of pharmaceuticals the portrayal of depression 1980–2005. *Administration and Policy in Mental Health Service Research, 36*(2), 91–101.

Clarke, J. N. & VanAmerom, G. (2008). Asperger's syndrome: Differences between parents' understanding and those diagnosed. *Social Work in Health Care, 46*(3), 85–106.

Clarke, J. N., Mosleh, D. & Janketic, N. (2014). Discourses about children's mental health and developmental disorders in North American women's magazines, 1990–2012. *Child and Family Social Work, 19*(2), 1–10.

Cordell, A. (2004). Public health and children. *Children and Society, 18,* 243–246.

Conrad, P. (2005). The shifting engines of medicalization. *Journal of Health and Social Behaviour, 46,* 3–14.

Crawford, R. (2006). Health as a meaningful social practice. *Health, 10*(4), 401–420.

Foucault, M. (1976). *The history of sexuality: Volume 1: An introduction.* London: Allen Lane.

Foucault, M. (1988). The political technology of individuals. In Luther H. Martin, Huck Gutman and Patrick H. Hutton (Eds.), *Technologies of the self, A seminar with Michel Foucault* (151–162). London: Tavistock.

Furedi, F. (2008). *Paranoid parenting: Why ignoring the experts may be best for your child.* London, England: Continuum.

Gagnon, M. J. D. J. & Holmes, D. (2010). Governing through insecurity: A critical analysis of a fear-based public health campaign. *Critical Public Health, 20*(2), 245–256.

Giddens, A. (1999). Risk and responsibility. *Modern Law Review, 62,* 1–10.

Harper, D.J. (1995). Discourse analysis and 'mental health'. *Journal of Mental Health, 4*(4), 347–357.

Klawiter, M. (2004). Breast cancer in two regimes: The impact of social movements on illness experience. *Sociology of Heath and Illness, 26*(6), 845–874.

Koops, W. & Zuckerman, M. (2003). *Beyond the century of the child: Cultural history and developmental psychology.* Pennsylvania, U.S: University Press.

Kuhn, T. (1962). *The structure of scientific revolutions.* Chicago: The University of Chicago Press.

Lee, J. (2008). Changes in mothering of Korean women: Based on narrative interview data. *Development and Society, 37*(2), 141–167.

Lee, E., Macvarish, J. & Bristow, J. (2010). Risk, health and parenting culture. *Health, Risk and Society, 12*(4), 293–300.

Le Francois, B. A. & Coppock, V. (2014). Psychiatrized children and their rights: Starting the conversation. *Children & Society, 28,* 165–171.

Mayall, B. (2000). The sociology of childhood in relation to children's rights. *The International Journal of Children's Rights, 8,* 243–259.

McCombs, M. E. (1972). The agenda setting function of the mass media. *Public Opinion Quarterly, 36*(2), 176–182.

Merikangas, K. R., He, J., Burstein, M., Swanson, S. A., Avenevoli, S., Cui, L., Benjet, C., Georgiades, K. & Swendsen, J. (2010). Lifetime prevalence of mental disorders in U.S. adolescents: Results from the National Comorbidity Study-Adolescent Supplement (NCS-A). *Journal of the American Academy of Child and Adolescent Psychiatry, 49*(10), 980–989.

Olfson, M., Bianco, C., Liu, L., Moreno, C. & Laje, G. (2006). National trends in the outpatient treatment of children and adolescents with anti-psychotic drugs. *JAMA Psychiatry, 63*(6), 679–685.

Oulton, K. & Heyman, B. (2009). Devoted protection: How parents of children with severe learning disabilities manage risks. *Health, Risk and Society*, *11*(4), 303–319.
Perou, R., Bitsko, R. H., Blumberg, S. J., Pastor, P., Ghandour, R. M., Gfroerer, J. J. & Huang, L. N. (2013). Mental health surveillance among children – United States, 2005–2011. *Morbidity and Mortality Weekly Report*. (Supplement) May 17, 2013: *62*(2), 1–35.
Richman, J. & Skidmore, D. (2000). Health implication of modern childhood. *Journal of Child Health Care*, *4*, 106–110.
Rose, N. (1989). *Governing the Soul: The shaping of the private self*. London: Routledge.
Silverman, C. (2008). Critical review fieldwork on another planet: Social science perspectives on the autism spectrum. *Biosocieties*, *3*, 325–341.
Whitaker, R. (2010). *Anatomy of an epidemic*. New York: Random House Inc.

Media references

Bostic J. & Miller M., (2005). When should you worry? *Newsweek*, (April 1).
Cowley, G. (2003). Predicting autism. *Newsweek*, (July 1).
Kalb, C. (2005). When does Autism start? *Newsweek*, (February 1).
Kantroqitz, B. & Sceldo J. (2006). What happens when they grow up: Teenagers and young adults. *Newsweek*, (November 1).
Klugger, J. (2002). Preventive parenting. *Time Magazine*, (January 1).
Madeline, J. (2002). Geek syndrome. *Time Magazine*, (May 1).
Oz, M. (2012). Charms of the quiet child. *Time Magazine*, (February 1).
Park, A. (2002). Are the shots safe. *Time Magazine*, (April 1).
Park, A. (2012). Can Anesthesia raise the risk of ADHD? *Time Magazine*, (February 1).
Shute, N., Locy, T., & Pasternak, D. (2000). Perils of pills. *U.S. News & World Report*, (March 1).
Smith K. I. (2000). Ritalin for toddlers. *Time Magazine*, (March 1).
Szegedy-Maszak, M. (2002). Demons of childhood. *U.S. News & World Report*, (November 1).
Szegedy-Maszak, M. (2004). Tune-ups for misfiring neurons. *U.S. News & World Report*, (April 1).
Wallis, C. (2007). What autism epidemic? *Time Magazine*, (January 1).
Wingert, P. & Kantrowitz B. (2002). Young and depressed. *Newsweek*, (October 1).

17
'It's just statistics ... I'm kind of a glass half-full sort of guy': The Challenge of Differing Doctor-Patient Perspectives in the Context of Electronically Mediated Cardiovascular Risk Management

Catherine O'Grady, Bindu Patel, Sally Candlin, Christopher N. Candlin, David Peiris and Tim Usherwood

Introduction

Best practice guidelines for the prevention and management of cardiovascular disease recommend that management decisions be informed by estimation of a patient's 'absolute risk' of a cardiovascular event over time. Such a risk calculation is based on a combination of non-modifiable factors such as age and gender and modifiable factors that include blood pressure, lipids, diabetes, body mass index (BMI), tobacco smoking, alcohol use, unhealthy diet, physical inactivity, and psychosocial stress.

In Australia, a quality improvement intervention has been developed to assist general practitioners to calculate an individual's 'absolute risk' at the point of care and to engage the patient in considering factors that might modify this risk (Peiris et al., 2009). Integrated with the healthcare provider's electronic health record, the intervention, known as 'HealthTracker', includes a real-time decision support interface using an algorithm derived from evidence-based guidelines, a patient risk communication tool including 'what if' scenarios to show benefits from risk factor improvement, as well as mechanisms for provider audit and feedback. Robust, trial-based evaluation of this electronic intervention indicates that its use is associated with a 25% relative improvement in cardiovascular risk factor screening (Peiris et al., 2015).

Successful dissemination of the intervention in a post-trial context requires understanding of how it is used in practice. How is the intervention deployed strategically in doctor-patient interaction to communicate cardiovascular risk? What are the factors that support or hinder its impact?

In this chapter we focus on a single case study drawn from a comprehensive post-trial evaluation of the intervention (Patel et al., 2014) to offer a response to these questions. Taking an interactional socio-linguistic approach (Gumperz, 1999; Roberts & Sarangi, 2005), we combine fine-grained analysis of the discourse of a transcribed, video-recorded primary care consultation in which the intervention is used with ethnographic interviews that bring the perspective of the participating patient and doctor to bear on our analysis.

Findings indicate the potential of the intervention as a valuable entry point for engaging a patient in discussion of their CVD risk. Discourse analysis brings to light the doctor's role as mediator between risk knowledge made available by HealthTracker and the patient. It uncovers the doctor's strategic, step wise deployment of HealthTracker as a means to focus the patient's attention on the cumulating factors that contribute to his overall risk. In particular, analysis reveals the doctor's skilled use of tool outputs that display risk calculations and estimations of heart age for the patient in graphic numerical and visual formulations. Yet, as Alaszewski (2010, p. 104) points out 'Individuals are not passive recipients of [risk] information ...'. Ethnographic findings from a post-consultation interview with the patient reveal how he reframes and reformulates the HealthTracker risk calculations in light of his own experience and perspective to dilute their significance.

It appears that the value of risk calculations is not fixed and immutable but subject to differing interpretations informed by different values and perspectives (Sarangi & Candlin, 2003). Risk communication that emphasises the one-way flow of knowledge from doctor to patient albeit enhanced by HealthTracker cannot take such different perspectives into account. Effective risk talk and the effective deployment of HealthTracker may rely upon the negotiation of disparate participant values and experiences by way of collaborative, co-constructed doctor-patient interaction.

The computer in the consultation

By examining an instance of electronically mediated risk communication our case study is also illustrative of the worldwide phenomenon of the computer in the primary care consultation. In Australia computerisation in general practice is all but complete with 93% of doctors using a computer on their desks for a range of functions including decision support (Pearce, Dwan, Arnold, Phillips, & Trumble, 2009). In the UK computerisation of the clinical encounter has long been in progress and the electronic patient record is now used pervasively in general practice to support patient care (Swingelhurst, Roberts, & Greenhalgh, 2011).

The presence of the computer in the general practice consultation has changed the nature of doctor-patient interaction placing new demands on the communicative expertise of doctors. The computer's presence introduces a potential third party into the previous dyadic relationship of doctor and patient opening the way for a variety of complex relational configurations and reconfigurations that would not otherwise have been possible. For example, within the newly configured 'participation framework' (Goffman, 1981) of doctor, patient and computer the doctor may act to position the computer as a 'bystander' (Goffman, 1981) to the interaction as he or she engages with the patient. Alternatively, prompted by the computer the doctor may side-line the patient to interact with the screen. The presence of the computer also provides the patient with the opportunity to disengage from what is going on as doctor and computer interact. Yet the triadic participation structure that the computer affords is of itself neither advantageous nor disadvantageous to the interpersonal relationship between doctor and patient nor to the purposeful trajectory of a consultation. The computer is an inanimate actor in the consultation. Whilst not devoid of influence it is a neutral tool that is brought into the interaction and enabled to play its part largely by the strategic work of the doctor but potentially by the actions of the patient as well. 'Integrating technology ... always requires human work to re-contextualise knowledge for different uses within complex social settings' (Greenhalgh et al., 2009 in Swingelhurst et al., 2011, p. 4).

In our case study we bring to light the discursive work that the doctor does as he strives to integrate HelathTracker purposefully and strategically into the consultation so as to persuade the patient of his CVD risk. We then go on to look beyond the immediate context of the consultation to the patient's social setting so as to access those life-world perceptions, values and attitudes that shape the patient's response to risk. Finally, in light of insights derived from analysis of the discourse of the consultation as well as from our ethnographic interview with the patient, we consider the nature of communicative expertise that is required for the effective integration of tools such as HealthTracker into the clinical encounter.

The case study

Clinical context

The patient is a 66 year-old-man who has been taking medication for high blood pressure for 40 years. He has been seeing the participating GP for two years. He is visiting the doctor today for a blood pressure check following change of medication and for comprehensive blood tests including a fasting blood sugar test following a previous high blood sugar reading.

The computer as participant

Throughout the consultation the doctor is seated in a swivel chair that allows for easy shifts in body orientation and gaze between the patient in person and 'the patient inscribed' (Robinson, 1998) in records and data on the computer screen. The patient is seated at the end of the desk and to one side of the computer. His line of vision towards the screen is unimpeded. This configuration allows for potential triadic engagement between doctor, patient and HealthTracker. Tracker is a mutually available information resource.

Space however is more complex than physical layout or physical environment. Space is socially constructed and participants in interaction 'do' or 'enact' space through their relationships with each other, with objects and with what is going on at a particular moment (Jones, 2010). Through the meanings they assign to various tools introduced into the consultation space such as blood pressure monitor, computer screen or HealthTracker, doctor and patient create and circumscribe a 'sphere of attention' (Jones, 2010). This sphere of attention may expand or contract or otherwise diversify as they shift their orientation to objects to each other and to what it is that is going on.

Thus, for a time in this consultation the doctor enacts a 'participation framework' (Goffman, 1981) or 'relational space' (Jones, 2010) that sidelines the computer screen as he engages exclusively with the patient. At other times his sphere of attention narrows to exclude the patient or expands to encompass patient and HealthTracker so that all are potential participants in the interaction. The patient too is engaged in these shifting reconfigurations of space that display his engagement or disengagement with HealthTracker and with the talk of risk that is going on.

Doctor as mediator

As the consultation gets under way, the patient raises concern about symptoms associated with withdrawal from a longstanding migraine medication. Once the doctor has reassured the patient about this matter, he turns to the tasks of taking the patient's blood pressure and extracting blood. We pick up the interaction from this point, focusing on three 'critical moments' (Candlin, 1987) to trace a risk talk trajectory that culminates in the deployment of HealthTracker to provide the patient with visual formulations of his absolute CVD risk and heart age. In this way we bring to light the doctor's discursive responses to the challenge of mediating between HealthTracker and the patient.

Consultation extract 1: Introducing risk factors

		Talk	Action	Screen information
23	D	Let's see	D swivels chair to face desk drawers at patient's side	P's electronic health record open on screen
23		I guess because the sugar was a bit high there's a very high chance that you might have diabetes so that's why the three sugar tests today yeah :	Takes tools for extracting blood from drawer as speaks	
24	P	Yep		
25	D	(inaudible)	Puts tools on desk; places hands on knees. Directs gaze towards Blood Pressure monitor but legs and torso are oriented to P	
	D	One fifty five seventy six so that's a bit high		
26	P	((nods))	P looks into middle distance; Barely perceptible nod	
27	D	What have we got you on at the moment (.) we've got the ten one sixty yeah	D turns head to glance at screen; legs, torso remain oriented towards P as fits tourniquet	
28	P	I don't know what the dosage is		

(continued)

29	D	I only give you two packs last time yeah :	
30	P	Yeah two packs	
31	D	(might have to go) to the higher strength one yeah	
		What I'm going to do today (.) is using one of the software called health Tracker (.) have a look at your cardiovascular risk factors yeah :	D swabs arm of P in preparation for extracting blood
32	P	Yeah	
33	D	Because looks like you're probably at a high risk anyway	D prepares syringe; P with head raised directs gaze towards middle distance; arm extended awaiting insertion of needle
	D	... blood pressure's a bit high you're being overweight and let's see sugar being a bit high yeah	D continues to prepare syringe. P shifts body very slightly away from D and raises eyes towards ceiling

Throughout this sequence the computer is open on the desk displaying the patient's electronic health record. Whilst the doctor glances momentarily towards the screen (27) to check the dosage of the patient's blood pressure medication his legs and torso remain directed towards the patient, communicating that his primary orientation is to the patient and the tasks at hand. The computer that houses HealthTracker remains largely outside his sphere of attention. It is positioned as a 'bystander' (Goffman, 1981) to the interaction that is going on.

Yet CVD risk and the potential for usefully deploying HealthTracker in this consultation are clearly on the doctor's mind as the following extract from his post-consultation interview confirms.

> ... I was thinking I'd done the blood pressure, I was about to take the blood and I may be thinking about 'Hey you know this fellow you know you can see he's obese, he's got blood pressure and his sugars so there you go he's the cardiovascular risk guy (.) maybe we can use the Tracker.

From turn 23 the doctor works to bring the topic of risk factors into the discourse of the consultation. As he goes about the task of preparing to extract blood, he simultaneously uses the strategy of 'thinking aloud' (Dowell, Stubbe, Scott-Dowell, Macdonald, & Dew, 2013) to share his reasoning about the likelihood that the patient has diabetes (23), and to refer to the elevated blood pressure reading (25) that might necessitate an increase in medication. Against the backdrop of these cumulating risk factors, he refers to HealthTracker as a tool for examining the patient's level of risk (31), bringing the sequence to a close with a summarising statement of known factors that already place the patient in the high risk category (33).

However this risk talk has no observable effect on the patient. Whilst he acknowledges the likelihood of a diagnosis of diabetes with an immediate and unqualified 'yep' (24), his responses to the doctor's utterances are otherwise minimal. At turn 26 he receives news of his elevated blood pressure reading with a barely perceptible nod. Further, across this sequence he directs his gaze into the middle distance as if to create a private interactional space that enables him to disengage from the interaction.

This apparent disengagement from talk of risk might be accounted for by the patient's discomfort with the process of having blood taken. As Heath (2006) has shown, patients almost invariably maintain a 'middle distance orientation' whilst subject to clinical procedures as a means to detach themselves from what is going on. Talk at such moments of detachment is unlikely to have impact as the following extract from the patient's interview suggests:

Interviewer: ... and how did you feel when he said that you're probably high risk
Patient: Oh
Interviewer: Have you been have you heard that before
Patient: I don't remember him even saying that. I think when you're facing a needle in the eye, you know (.) you tend not to tend (laughs) I'm more worried about getting jabbed in the arm than (.) you know

But it also appears that the patient is somewhat immured against risk talk. He goes on to explain that he constantly hears about his risk. The likelihood that he is at high risk is not news to him.

Patient: I mean I hear things like high risk all the time you know. Waking up in the morning's always a positive thing for me.

For the discourse analyst the patient's apparent disengagement from interaction as the doctor talks of risk represents a moment of uncertainty where the patient is difficult to read. Can his disengagement be accounted for by his need to detach himself from the process of having blood extracted as Heath (2006) would suggest? Or is he indeed hardened against risk talk and taking a fatalistic stance as ethnography implies?

Similarly, the patient's action at this particular moment may present a problem of interpretation for the participating doctor. What is the patient thinking? How can the doctor raise the stakes so that the patient is more likely to engage with the matter of risk? Will bringing HealthTracker into the interaction have some impact?

Deploying HealthTracker

As the consultation proceeds, the doctor turns his attention to HealthTracker as a means to involve the patient with the matter of risk. Following a number of turns in which the doctor completes the task of extracting the patient's blood, revisits the need for a urine test and provides the patient with a sugar drink in preparation for his upcoming blood sugar test, he initiates an interactional sequence in which, prompted by HealthTracker, he gathers data that the Tracker will use to update and modify the patient's risk profile.

Consultation extract 2: Gathering risk data

Up until this point in the consultation the computer that houses HealthTracker has been sidelined from a 'participation framework' (Goffman, 1981) encompassing doctor and patient only. But now the doctor moves to reconfigure this framework so as to bring HealthTracker into the interaction as an active and 'ratified participant' (Goffman, 1981).

		Talk	Action	Screen information
57	D	And while you're doing that *(inaudible)* might use this software to have a look at your cardiovascular risk	D swivels chair towards computer; directs gaze to computer screen P turns head slightly towards screen; sugar drink in hand	Tracker is sitting on right hand side of screen awaiting activation

		We use the Health Tracker which is on trial with the university under the torpedo study (…) when you (.)finish off that drink first	D turns head briefly towards P then returns gaze to screen	
58	P	((nods))	P nods; continues drinking from bottle; D attends to screen tracing screen prompts with left index finger P directs gaze to middle distance	
59	D	Does anyone in your family have a very high cholesterol level?	D's index finger on screen. D turns head to address patient then back to screen	Tracker prompts
60	P	Cholesterol not that I'm aware[of	P continues to drink; gaze directed at middle distance	
61	D	[not that you're aware of ok		
62	P	I mean I've got members of my family that have got heaps of other problems but not that I'm aware of cholesterol	P sustain gaze at middle distance	
63	D	Right ok and none of your family is probably (inaudible) forty five or fifty five with heart attack (.) coronary heart disease	D gazes to screen	Tracker prompts
64	P	My father died at fifty seven	D directs gaze towards P then back to screen	
65	D	Fifty seven		
66	P	With heart attack	P places empty bottle on desk	
67	D	First degree; did he have a lot of chest pain before :	D sustains gaze on screen as P responds to questions	
68	P	Yeah always had chest pain	P redirects gaze to middle distance as he answers questions	
69	D	For a few years before that=		
70	P	=yep	P sustains gaze at middle distance	
71	D	So that's probably early coronary [heart disease	D looks at screen	

(continued)

72	P	[Yep coronary that's what he had	P raises head slightly as gazes to middle distance	
73	D	Younger than sixty yeah		
74	P	Yep=		
75	D	= so that's your father=		
76	P	=yep		
77	D	So that's a tick for that	Swivels chair to face patient. Takes tape measure from drawer	Tracker prompts D to check BMI and waist

At turn 57 the doctor swivels his chair towards the computer, communicating a shift in his dominant physical orientation from the patient to the screen. In this way he signals the inclusion of HealthTracker as an active participant in the interaction. In the same turn the doctor works strategically to ratify HealthTracker, that is to invest it with the authority to act as a bona fide participant in risk talk. By deploying the inclusive institutional pronoun form 'we' he invokes an 'institutional identity' (Sarangi & Roberts, 1999) to speak as a member of the institution of general practice rather than as an individual. In this way he invests use of HealthTracker with institutional authority. In the same utterance, HealthTracker is further endorsed with reference to the university study of which it is a part.

At turn 58, the patient acknowledges this endorsement of HealthTracker with a nod. Then, as the doctor attends to the screen, tracing the Tracker checklist with his index finger, the patient redirects his gaze into the middle distance, a position that is sustained across the entire sequence. However, this shift in gaze does not appear to signify the patient's disengagement from the interaction or from the activity of risk information gathering that is going on. As Greatbatch (2006) points out the diverting of gaze by a patient whilst the doctor is engaged with computer related tasks can be seen as a means to reduce the interactional demands on the doctor. Here the patient's sustained middle distance gaze functions to free the doctor to attend to the prompts and checklist questions as they appear on the HealthTracker screen. Yet, despite their physical misalignment doctor and patient continue to occupy the same 'relational space' (Jones, 2010). As the doctor orients to the screen and the patient looks into the distance they engage with each other and with HealthTracker as they go about the task of co-constructing the risk information that Tracker requires.

From turn 59, prompted by HealthTracker the doctor initiates a question answer sequence to explore risk factors associated with the patient's family history including high cholesterol level (59) and early age heart attack (63). At turn 64 the patient states that his father died at 57. This prompts the doctor to redirect his gaze momentarily from screen to patient in an action that displays that this news is of significance to him. Then, across ensuing turns he pursues confirmation of this newly disclosed risk information with reiterated confirmation checks (65, 67, 69, 71, 73, 75) to ascertain that a family history of early coronary disease involving a first degree relative is indeed a risk factor for his patient. The patient's latched and overlapping responses to the doctor's confirmation checks (70, 72, 75) indicate his involvement in the interaction and with the risk factor topic. At this point in the consultation he is engaged in the collaborative task of building his risk profile. HealthTracker's diagnostic prompting, mediated by the doctor has enabled a collaborative conversation in which doctor and patient are mutually involved with factors that may place the latter at high risk.

In the moments that follow, and prompted by HealthTracker the doctor weighs the patient so as to calculate his body mass index (BMI) and measures his waist circumference. Doctor and patient then resume their seats as the doctor enters new and updated risk factor data into the computer so that HealthTracker might calculate the patient's absolute CVD risk.

Mobilising HealthTracker to communicate risk

The stage is now set to mobilise HealthTracker to the task of engaging the patient in recognising his current level of absolute risk as well as his risk projections.

As the consultation proceeds the doctor draws upon the combined multimodal resources of wording, gesture, gaze, pausing and silence as he works to deploy HealthTracker calculations and semiotic tools in a way that might impact on the patient. His communicative expertise in mediating between HealthTracker risk information and the patient is now on display.

Consultation extract 3: Mobilising HealthTracker

	Talk	Action	Screen information
85 D	So let's go back and have a look at this	D indicates data on screen with left hand. P shifts torso slightly and turns head to direct gaze to screen. Arms unfold and then fold again as he leans a little closer towards screen	Tracker open on screen

(continued)

	D	Ahh cardiovascular risk profile I need to add on your family history because that hasn't (.) before sixty yeah (…) and we haven't got your urine yet so when we check that yeah	D inputs data	Tracker calculates
	D	So looking at that you're (…) just refresh that (inaudible)	D presses key to refresh screen	
86	D	Your risk is sitting at about nineteen to twenty percent at the moment yeah	Indicates data on screen with left hand; turns head towards patient	Tracker calculation on screen
87	P	Ok	P sustains gaze on screen	
88	D	In terms of getting a heart attack over the next five years (pause)	P nods	
89	D	Or a stroke (pause)	D keeps index finger on data; sustains focused gaze on patient's face P slow repeated nods	
90	D	Twenty percent is about one in five (pause)	Sustains gaze towards patient's face. Left hand closes but remains at screen. P sustains gaze on screen. Nods; slight eye brow rise	
91	D	So that's (.) if you look at the colour you're in the red zone	D's hand moves across screen to indicate relationship between data and visual on screen. D looks to P and then to screen P gazes at screen	Tracker risk projection visual on screen
92	P	(nods)		
93	D	Now this is (.) not adding on the diabetes here. If you are having diabetes jump to twenty five percent easily I would think	Indicates screen with finger Directs gaze towards patient's face	
		And this is a little picture we can show you (.) your heart which is here	Indicates patient's projection on graph with left hand P sustains gaze l towards screen	Heart age graph on screen

(continued)

		compared with the general population(..) at your age	Slides finger across screen to indicate line for general population
94	P	mm	
95	D	So your heart is about (...) seventy five years old (.) even though you're only sixty five	Directs gaze towards patient's face then back to screen then back to patient. Finger remains on screen
96	P	Yeah ↑	P looks at screen
97	D	=Yeah this is taking all these [risk factors that we collected into[account	
98	P	[Yeah yeah	
99	D	Your smoking history your blood pressure and of course we haven't got the urine we haven't got the (.) I'm just going to save this (.) haven't got the urine results yet (inaudible) I'll see if I can print one out for you to look at yeah (.) I'll just print the picture instead (inaudible) cardiovascular projection	P gazes at screen
	D	So this is give you an idea about where you are at yeah :	Runs finger over screen P looks to screen
	D	What we need to monitor you here yeah :	Runs finger over screen P continues to gaze at screen

At turn 85, using his left hand to draw attention to the screen, the doctor employs the inclusive form 'Let's' to invite the patient to join him in examining the HealthTracker outputs. In response the patient reorients to the screen, shifting his torso slightly towards the computer, redirecting his gaze and leaning forward to display his attention. At the same time he unfolds and refolds his arms in a move that suggests he is settling in to take in the information that HealthTracker is about to offer. Doctor and patient are mutually engaged with Tracker.

Then, as the HealthTracker calculation appears on the screen, the doctor turns his head towards the patient, representing the calculation verbally to state that the patient has a 19–20% chance of having a heart attack or stroke over the next five years. By way of strategic pausing he fragments this statement into a series of discrete utterances that draw attention to each aspect of the message (86, 88, 89). Whilst the patient's responses are largely minimal, including the minimal acknowledgement token 'OK' and a single slight nod, the slow repeated nods (89) with which he receipts news of possible stroke suggest that he is not simply attending but absorbing and considering this information.

In the next turn (90) the doctor moves to intensify the patient's engagement by reformulating the HealthTracker calculation of a 20% risk as a one in five chance. This reframing of the patient's risk projection in more accessible terms is accompanied by the doctor's sustained gaze towards the patient's face. Gaze direction together with a further strategically placed pause invests the doctor's risk reformulation with added importance. The patient's raised eyebrow response (90) signifies that it has had some effect.

In light of these signs of the patient's receptiveness, the doctor acts to mobilise the semiotic formulations of the patient's risk that HealthTracker affords. At turn 91 he directs the patient's attention to the visual representation of his risk projection that places him in the high risk red zone. Then, as the patient nods, gaze fixed on the screen the doctor redirects his own gaze from patient to screen and back to the patient. As his gaze settles on the patient's face in a way that intensifies his message, he states that if the patient has diabetes his risk will jump to 25% (93). Whilst the patient continues to gaze at the screen, he offers no verbal or visual response.

In a final move the doctor mobilises HealthTracker's visual representation of the patient's heart age in relation to the general population. Once again he strategically deploys gesture, gaze direction, and verbal reformulation to strengthen the effect of HealthTracker's message. As turn 93 continues he uses his index finger to trace the heart age projections on the screen so as to highlight the contrast between the patient's projection and that for the general population. The patient attends but responds with a minimal 'mm'. Finally (95) as the doctor redirects his gaze from patient to screen and then back to the patient in order to emphasise his message, he brings his deployment of HealthTracker towards its completion with an upshot that summarises what this risk information means:

So your heart is about (…) seventy five years old (.) even though you're only sixty five.

This simply formulated upshot appears to have considerable impact on the patient. The patient's response 'yeah' (96), marked by a sharp rise in tone constitutes a relatively strong 'news receipt' (Heritage, 1984) that displays recognition of the significance of this news.

As the consultation draws to its close it seems that doctor and patient are in alignment with each other in their mutual understanding of the significance of the patient's high CVD risk status. Such apparent mutuality is the foundation upon which doctor and patient might go on to make shared decisions about how the patient's risk is to be managed.

But as Candlin points out (2002, p. 25) '... mutual understanding is always a shifting and temporary matter' and mutual agreement is '... an unstable state of becoming'. Alignment with the medical perspective during a consultation cannot be taken as evidence of on-going concordance. Behind the risk calculations and projections communicated to this patient in the context of the consultation lies his life-world accessed through ethnography. As finding from the post-consultation interview with the patient suggest, the meaning and value of risk calculations are not fixed and immutable but subject to reinterpretation by the patient in light of his values, attitudes, and life-world perceptions.

The patient's perspective

To what extent has HealthTracker generated knowledge of the patient's absolute cardiovascular risk been made tractable for the patient by the actions of the doctor? What value does the patient bring to this knowledge? What is its importance to him?

During the interview conducted with the patient in the days following his consultation he consistently reframes HealthTracker calculations and projections in ways that dilute their significance. For example, when asked to comment on the projection of a one in four chance of having a heart attack or stroke in the next five years he takes a positive stance to reformulate this calculation as a three in four chance that he will not experience such an event.

> ... I'm not a gambling person, but I know statistics reasonably well, and I know that, you know, okay, you've got one chance in four, but that means you've also got three chances in four that you're not going to get it, so the odds are, you know, statistically, you're alright (laughs) you know, so you've got more chances of not having a heart attack or a stroke than you have.

For the patient the upshot of this statistical information is that his chance of not having a cardiovascular event far outweighs the chance that he will. In a similar way, he takes the HealthTracker calculation of his heart age that seemed to have considerable impact during the consultation and reframes it in a positive light.

> ... you know I've only got a heart of a 75yearold and not a 95-year-old. I'm kind of a glass half-full sort of a guy, you know.

Clearly, this patient is not a passive recipient of risk information. Whilst he does not refute the risk calculations and projections that have been presented to him, the meaning that he invests in them is at odds with the medical perspective. Patient and doctor frame risk knowledge in different ways to give risk calculations different valence and such disparate perspectives are likely to affect the patient's commitment to management advice as the following interview extract suggests:

> ... it's no good saying we'll change your lifestyle ... I'm 66 years old ... I have a lifestyle, you know. I'm not an alcoholic (.) I don't over-drink (.) I don't you know I don't overeat. I'm just, just a big bloke. ... Look, I didn't walk out of there thinking, oh, I'm only going to eat salad and, you know, drink water for the rest of my life.

HealthTracker has afforded the patient the opportunity to consider those statistical calculations and projections that are indicative of his CVD risk. But it appears that contemplation of health risk as represented in statistical terms may not lead easily in a linear fashion to mutual management decisions or to determination to take action to reduce that risk. Whilst the patient appeared to defer to the authority of HealthTracker during the consultation, in the post-consultation period his life-world perceptions intervene and the impact of risk knowledge begins to decay.

Concluding comment

In this case study, discourse analysis has brought to light the doctor's skilled use of language and other semiotic means to mobilise HealthTracker to the task of informing the patient about his absolute CVD risk. Yet, as ethnographic accounts have shown, HealthTracker risk information mediated by the doctor has not been sufficient to engage the patient in acknowledging and acting upon his high risk status. Risk is not absolute but subject to different interpretations and the impact of risk knowledge on the patient is tempered by his own perceptions.

What then is the nature of communicative expertise required for the effective deployment of tools such as HealthTracker? The participating doctor in this case study adheres largely to a 'rational model of risk communication' (Alaszewski, 2010, p. 103) that characterises risk communication as the flow of risk knowledge from the knowledgeable expert, in this case HealthTracker mediated by the doctor, to the less informed patient. But such a one-way flow of information does not allow for the patient's perspective to enter the discourse of the consultation.

The effective communication of risk may require that tools such as HealthTracker be integrated into collaborative co-constructed interaction whereby the patient's perceptions and accounts can be accessed, discussed, and negotiated.

As Candlin and Candlin state (2002, p. 103) 'Expertise in the management of risk is not solely – or even primarily – a matter of knowledge but of discursive negotiation among participants' values and experiences'.

References

Alaszewski, A. (2010). Risk communication: Identifying the importance of social context. *Health, Risk & Society, 7*(2), 101–105.

Candlin, C. (1987). Explaining moments of conflict in discourse. In R. Steele & T. Threadgold (Eds.), *Language Topics: Essays in Honour of Michael Halliday*. Amsterdam: John Benjamin.

Candlin, C. (2002). Alterity, perspective and mutuality in LSP research & practice. In M. Gotti, D. Heller & M. Dossena (Eds.), *Conflict and Negotiation in Specialized Texts*. Bern: Peter Lang Verlag.

Candlin, C., & Candlin, S. (2002). Discourse, expertise, and the management of risk in health care settings. *Research on Language & Social Interaction, 35*(2), 115–137.

Dowell, A., Stubbe, M., Scott-Dowell, K., Macdonald, L., & Dew, K. (2013). Talking with the alien: Interaction with computers in the GP consultation. *Australian Journal of Primary Health, 19*, 275–282.

Goffman, E. (1981). *Forms of talk*. Oxford: Blackwell.

Greatbatch, D. (2006). Prescriptions and prescribing: Co-ordinating talk-and text-based activities. In J. Heritage & D. Maynard (Eds.), *Communication in Medical care. Interaction Between Primary care Physicians and Patients* (pp. 313–339). Cambridge: Cambridge University Press.

Greenhalgh, T., Potts, H., Wong, G., Bark, P., & Swingelhurst, D. (2009). Tensions and paradoxes in electronic patient record research: A systematic literature review using the meta-narrative method. *The Millbank Quarterly, 87*(4), 729–788.

Gumperz, J. (1999). On interactional sociolinguistic method. In S. Sarangi & C. Roberts (Eds.), *Talk, Work, and Institutional Order*. Berlin; New York: Mouton de Gruyter.

Heath, C. (2006). Body work: The collaborative production of the clinical object. In J. Heritage & D. Maynard (Eds.), *Communication in Medical care. Interactions Between Primary care Physicians and Patients* (pp. 185–213). Cambridge: Cambridge University Press.

Heritage, J. (1984). A change-of-state token and aspects of its sequential placement. In J. Atkinson & J. Heritage (Eds.), *Structures of Social Action: Studies in Conversational Analysis* (pp. 299–345). Cambridge: Cambridge University Press.

Jones, R. (2010). Cyberspace and physical space: Attention structures in computer-mediated communication. In A. Jaworski & C. Thurlow (Eds.), *Semiotic Landscapes: Language, Image, Space* (pp. 151–167). London: Continuum.

Patel, B., Patel,A., Jan, S., Usherwood, T., Harris, M., Panaretto, K., Zwar, N., Redfern, J., Jansen, J., Doust, J., Peiris, D. (2014). A multifaceted quality improvement intervention for CVD risk management in Australian primary healthcare: A protocol for a process evaluation. *Implementation Science*, 9(187), pp 1–12.

Pearce, C., Dwan, K., Arnold, M., Phillips, C., & Trumble, S. (2009). Doctor, patient and computer – A framework for the new consultation. *International Journal of Medical Informatics*, 78, 32–38.

Peiris, D., Joshi, R., Webster, R., Groenestein, P., Usherwood, T., Heeley, E., Turnbull, F., Lipman, A., Patel, A. (2009). An electronic clinical decision support tool to assist primary care providers in cardiovascular disease risk management: Development and mixed methods evaluation. *Journal of Medical Internet Research*, 11(4).

Peiris,D., Usherwood, T., Panaretto, K., Harris, M., Hunt, J., Redfern, J., Zwar, N., Colagiuri,S., Hayman,N., Lo, S., Patel, B., Lyford, M., MacMahon, S., Neal, B., Sullivan, D., Cass, A., Jackson, R., Patel, A. (2015). Effect of a computer-guided, quality improvement program for cardiovascular disease risk management in primary health care: The treatment of cardiovascular risk using electronic decision support cluster-randomized trial. *Circulation: Cardiovascular Quality and Outcomes*, 8(1), 87–95.

Roberts, C., & Sarangi, S. (2005). Theme-oriented discourse analysis of medical encounters. *Medical Education*, 39, 632–640.

Robinson, J. (1998). Getting down to business: Talk, gaze and body orientation during openings of doctor-patient consultations. *Journal of the Society for Human Communications Research*, 25(1), 97–123.

Sarangi, S., & Candlin, C. (2003). Categorization and explanation of risk: A discourse analytical perspective. *Health, Risk & Society*, 5(2), 115–124.

Sarangi, S., & Roberts, C. (1999). The dynamics of interactional and institutional orders in work-related settings. In S. Sarangi & C. Roberts (Eds.), *Talk, Work and Institutional Order* (pp. 1–57). Berlin: Mouton.

Swingelhurst, D., Roberts, C., & Greenhalgh, T. (2011). Opening up the 'black box' of the electronic patient record: A linguistic ethnographic study in general practice. *Communication & Medicine*, 8(1), 3–15.

Appendix

Table A17.1 Transcription conventions

[	A square bracket indicates the point at which a current speaker's utterance is overlapped by the talk of another.

=	Where the turns of two different speakers are connected by two equal signs, this indicates that the second followed the first with no discernable silence between them, or was 'latched' to it.
(.)	A dot in parenthesis indicates a micro-pause that is hearable but not measurable.
: :	Colons indicate the stretching or prolonging of the sound that immediately precedes them.
<u>Yes</u>:	If the letters preceding a colon are underlined, this indicates that there is a falling intonation contour; you can hear the pitch turn downwards.
Yes<u>:</u>	If the colon itself is underlined, this indicates a rising intonation contour; you can hear the pitch turn upward.
↑	An arrow indicates a strong fall or rise in pitch in accordance with the direction of the arrow.
(( ))	Double parenthesis are used to mark descriptions of events e.g. ((telephone rings)).
(word)	Words in parenthesis indicates uncertainty on the transcriber's part but represents a likely possibility.

Part VI
Regulating Risk

18
Central Banking in Risk Discourses: 'Remaking' the Economy after Crisis

Clea D. Bourne

Introduction

Global markets illustrate well Ulrich Beck's (1992) notion of the 'risk society', one marked not by more hazards, but rather a society in which risk itself has become a global discourse and set of practices; *'a gaze'* which attempts to bring the future into the present and make it calculable (Horlick-Jones, 2004: 109). As such, the risk society is increasingly preoccupied with responsibility, safety, and security generated by 'manufactured' risks emerging from new environments for which history provides little experience (Giddens, 1999). Nowhere is this truer than in financial markets, where periodic crises pose substantial risks to global and national systems. Financial crises have proved far more common than either investors or regulators expected, with 'once-in-a-century' events taking place in quick succession – from the global financial crisis which began with a 'credit crunch' in 2007; the Libor scandal which peaked in 2008; and the Eurozone crisis which emerged in 2009 – all illustrating the far-reaching consequences of spreading manufactured risks around the globe.

The aftermath of the 2007–2009 global financial crisis led to mass recriminations against those who allegedly failed to *see* the new manufactured risks inherent in financial systems. Such recriminations are symptomatic of the moral climate of the risk society, featuring a 'push-and-pull' of accusations of scaremongering on the one hand and cover-ups on the other (Giddens, 1999: 5). In the absence of a global regulator to blame for the financial crisis, the three most powerful central banks – the ECB, the Bank of England, and particularly the US Federal Reserve (Fed) – found themselves under immense public scrutiny. Central banks' role as a buffer between state and market interests had offered a unique vantage point from which to 'see' major global risks and assure market stability. Yet the combined actions and *in*action of central banks leading up to the crisis made risks less visible and less governable (Vestergaard, 2009). Specifically, central banks were accused of stoking market excesses through weak governance, inconsistent inflation targeting and 'too easy' monetary policies. The Fed was even accused of helping to contribute to the subprime mortgage crisis by producing the flawed data upon which US government policy devised home ownership loans for the financially excluded (Schwartz, 2012). Central banks were further criticised for failing to heed the pundits who *did* predict the crisis (Kirsner, 2012), for compounding the loss of trust during the early days of the crisis by denying markets liquidity, and for collectively refusing to take action as the crisis grew (Warner, 2011).

This chapter explores risk discourses by deconstructing public recriminations made during a 2008 Congressional hearing into the global financial crisis and the role of US regulators, including the nation's central bank, better-known as 'the Fed'. However, the chapter repositions the Congressional hearing as part of the Fed's own risk management of the US economy. The chapter begins by exploring risk discourses in wholesale financial markets, as well as central banks' techniques for managing attendant risks. The chapter then provides a brief background to the US Federal Reserve, before setting out the methods and materials for deconstructing US Congressional hearings as discursive events, closing with a discussion of themes arising from testimony given by Alan Greenspan, renowned economist and former Chair of the US Federal Reserve.

Central banks in risk discourses

Central banks play a distinctive role in institutionalising patterns of risk in national economies by guaranteeing overall trust in and sustaining the stability of a nation's financial system. Risk management in central banking is primarily carried out through monetary policy – the 'sharp end' of central banking – used to guarantee national money supply, set interest rates, and manage inflation (Irwin, 2013). Monetary policy encompasses a number of central bank activities, including 'open market operations' or the buying and selling of government bonds in the open market (Shafik, 2013). Central banks further manage risk by supervising, regulating and scrutinising the nation's banking system with a light-enough touch to maintain the system's vitality, while sustaining the trustworthiness of banks and financial institutions (FCIC, 2010; Woodward, 2000). Central banks also help maintain financial stability, a role straddling the first two areas of risk management, and involving additional central banking experts to identify, analyse, and communicate systemic risks. Monetary policy remains a central bank's most consistent risk-management tool and its most discursive, shaping national economies through carefully crafted 'monetary stories' (Holmes, 2014; Resche, 2004; Woodward, 2000). A central bank's monetary story is an attempt to see into the future: identifying what business conditions and inflation might be, providing careful interpretations of the economy, and communicating stability by sending advance signals or forecasts of potential risk, such as whether interest rates might move, or whether inflation is on the horizon (Holmes, 2014; Resche, 2004). Signals *must* be sent in advance so as to avoid surprising the markets, thus ruining the intended goal, namely to stabilise the economy. It is through this monetary story that an economy is 'made, remade, and unmade' (Holmes, 2014: 14). Losing control of the monetary story could mean losing control of the central bank's communicative relationship with its many publics (Holmes, 2014). Language is therefore pivotal to the central banker's job, and careful wording the 'sharp edge' of his risk-management tool. For example, 'hedging', the economist's notion for covering against risk, in central-bank-speak translates into opaque statements designed to send messages in many directions, so as not to move markets (Resche, 2004). As prudent risk managers, central bankers use language that is neutral in tone – 'non-journalistic', 'non-colourful', 'technocratic deadpan' (Holmes, 2014; Resche, 2004; Smart, 2006). A central bank's authority and trustworthiness emerges through these complex narratives. In a post-crisis era, central banks had an even more pivotal role to play. Since governments had little fiscal room to stimulate economies, central banks became more 'activist' in support of economic recovery (Shafik, 2013). Communication became the central bank's most powerful tool, as central bankers moved beyond open market operations to 'open mouth operations' (Shafik, 2013), developing a 'financial stability story' that mimicked the monetary policy story (Holmes, 2014).

Central banking activity took on mythical status in the late 1990s, when combined efforts of major central banks supposedly quelled the Asian Financial Crisis (Irwin, 2013). During this 'Age of the Central Banker' (Krugman, 1999), regulators in Western economies developed an International Financial Architecture (IFA) to share monetary policy techniques for containing the impact of adverse events such as inflation, thus 'mastering' economies (Irwin, 2013; Vestergaard, 2009). But this halcyon period of perceived low risk in Western markets proved a double-edged sword. Not only did financial markets expand and deepen; investors took *more* risks, borrowing greater amounts at cheaper rates than ever before, investing in a multitude of instruments catering to every possible risk profile, while allowing risks to spread across the globe (Irwin, 2013; Rajan, 2005).

No individual central banker was more mythologised during this period than Alan Greenspan, Chairman of the US Federal Reserve from 1987 to 2006. Widely admired for his personal ideology of 'light-touch' regulation, Greenspan was a cult figure in the business media; the man 'who held more power over the financial future than any other individual on the planet' (Irwin, 2013: 124). But only months after Greenspan demitted office in January 2006, the highly leveraged US housing market peaked, causing the value of related securities to plummet, and damaging financial institutions globally. The ensuing global financial crisis wreaked havoc on the US economy. Critiques of Greenspan (ever-present on the political left) became mainstream. In 2008, Greenspan was summoned before a Congressional hearing to defend his legacy, together with representatives of the Treasury and the Securities Exchange Commission (SEC). The Congressional hearing gave the US government a national and global platform to account for its own role in governing market mechanisms that lay at the root of the crisis. Greenspan's testimony, in particular, made headlines in all major financial news outlets. His testimony therefore provides a useful lens through which to explore how a central bank makes sense of its own risk-management techniques in the wake of one of the largest financial crises ever seen.

Risk-taking in wholesale financial markets is different from everyday financial risk. For ordinary citizens, risk-taking is a personal endeavour at the household level. Market professionals, by contrast, generally take on *other* people's risks, which they price and package up to be sold via financial products and services, thus distributing risk away from those best able to purchase safety and freedom from it (Beck, 1992; Giddens, 1999). Market professionals therefore view risk positively since it generates profit, with risk management now a specialism, where risk is segmented into various categories from credit and currency risk to market and political risk. Alongside the rational, scientific language of risk management is the contrasting language used to describe risk-*taking*, often couched in terms of exploration and bold initiatives taken at the 'frontiers of finance' (Giddens, 1999). Greenspan himself had 'waxed lyrically' about the wonders of spreading and reducing risk through financial engineering (Keegan, 2007: 8). Despite the opportunities and potential profits inherent in volatility, risk has its winners and losers. When there are enough losers, borrowing and lending becomes constrained, markets become risk averse, and 'freeze up'.

Governments view risk in quite a different way. Much of political decision-making is about managing risks which do *not* originate in the political sphere (Giddens, 1999: 5). Consequently, politicians and administrators have a 'cultural disinclination' against the exposure that comes with taking responsibility for risky decisions. Instead, governments frequently offload complex, problematic services to trustworthy third parties (Taylor & Burt, 2005: 28). States and markets both attach a 'price' to risk; but where markets sell-on risk to other parties at a profit, states shift potential blame away to third parties, who are held responsible for attendant risks of service-delivery (Taylor & Burt, 2005).

Central banks not only play an integral role in managing national economies, but also have a risk-management role in global financial markets. As 'buffer' between state and markets, central banks are subject to lobbying by market interests, intent on influencing monetary policy or deterring regulation. Meanwhile, modern central banks are typically independent of political executives, able to keep governments from interfering in markets, while enabling banks and the economic system to thrive under watchful supervision. Yet central banks are also subject to political antagonism or state interference over monetary policy (Riles, 2006). Should monetary policy prove inadequate, states can shift blame onto the central bank.

The halcyon decade following the Asian Financial Crisis rendered central banks victims of their own success. By encouraging 'light-touch regulation', central banks' own risk-management role diminished. Central bank requirements for bank capitalisation became too low, banks held too little capital during an upturn, while hoarding capital and reducing lending during a downturn (Davies, 2010; Rajan, 2005). Central banks' tendency to carry out stress-testing exercises on *individual* banks rather than the entire financial system discounted the growing influence of non-banks on financial stability (Vestergaard, 2009). Non-banks (insurers, pension funds, hedge funds) directly competed with banks by originating and securitising risks (Rajan, 2005). Once the riskiest assets lay outside of banks, either off-balance sheet or in non-bank financial institutions, central banks had diminished capacity to check the growth of systemic risks. Thus, central banks made visible the *least* vulnerable, risk-exposed financial institutions, while enabling more substantial risks to lurk in the shadows (Vestergaard, 2009: 167).

Central banks further concealed risk by promoting homogenised data from and about economies (Vestergaard, 2009). Markets increasingly relied on this data, moving away from discretionary judgments about risk toward more quantitative, market-sensitive approaches (Vestergaard, 2009). One such quantitative approach, known as Value at Risk or VaR, became the industry standard for risk disclosure, backed by the Basel Accord for international banking supervision (Davies, 2010). Quantitative tools had clear deficiencies. First, their use implied that markets had adequate self-regulation (Davies, 2010), an ill-founded assumption since quantitative tools acted as a form of trust-by-proxy between market experts, who might privately exhibit scepticism against risk-management models while publicly backing their necessity (Malsch & Gendron 2009). Second, quantitative tools created a false sense of precision by extrapolating from historical data (Martin, 2002). VaR, in particular, created artificial precision by collapsing the results of risk modelling into a single number. Reliance on such tools encouraged markets to ignore other indicators such as data on institutional compliance with standards (Vestergaard, 2009: 154). Nor could risk models calculate risks spread out of sight and off-balance sheet (Rajan, 2005). And, since everyone used the same tools, everyone moved toward or away from the same investments based on the same risk data, thus *manufacturing* more risk (Vestergaard, 2009). Finally, central banks were blamed for allowing financial institutions to believe they had an officially sanctioned safety net, a moral hazard that encouraged reckless behaviour (Warner, 2011).

About the US Federal Reserve

The US Federal Reserve System is unique among central banks, regarded as one of the most transparent in its operations. It has a presidentially-appointed Board of Governors, and a partially presidentially appointed Federal Open Market Committee (FOMC), which sets monetary policy; together with non-presidentially appointed regional Federal Reserve Banks, privately owned member banks and advisory councils (Woodward, 2000). Despite its democratic voting structure, the Fed is sometimes seen as a cloistered 'financial priesthood', aimed at 'propping up' investments of the rich (Woodward, 2000). These criticisms intensified post-crisis, with the Fed accused of presiding over the largest accumulation of manufactured risk, together with global risk contagion.

Alan Greenspan was appointed Fed Chair in 1987, and held the post for eighteen and a half years, spanning both Republican and Democratic terms in office. He had previously worked in international banking and consultancy, and was a skilful Washington insider, cautiously negotiating his central banker's role as 'buffer' between state and markets, even when the President's office attempted to influence interest rates (Woodward, 2000). Within the Fed, Greenspan orchestrated a subtle transfer of political power to the Fed Chair, often acting unilaterally on rate-setting and inflation targeting (Woodward, 2000). External, he understood that while a state's financial reputation does not begin with its central banking leadership, it almost certainly *ends* there. Consequently, Greenspan opted to 'open up' the Fed, with FOMC interest-rate decisions becoming media events. Greenspan's Congressional testimony was televised on cable channels, his statements 'combed for meaning', magnifying the importance of central bankers' decisions and focusing attention on Greenspan, who continued to testify before Congress even after the lapse of legislation requiring him to do so (Woodward, 2000). Within US borders, 'fixation' on continued economic expansion increasingly resided in monetary policy and in the central banker as economist-in-chief (Woodward, 2000). Greenspan became a means of explaining and understanding the economy; a symbol through which the United States expressed confidence in itself and in its future (Woodward, 2000: 218). Beyond US shores, Greenspan breathed life into a vision of the US economy as 'strong, the best, invincible' (Woodward, 2000: 228). Despite earning respect and admiration of many economists and political leaders, criticism of Greenspan amplified during the 2008 Congressional hearing.

Congressional hearings as discursive events

Congressional hearings are positioned here as state-produced discursive events designed to reach multiple audiences – political, business, media, and other constituencies. Many hearings are held each year at substantial cost to US taxpayers. Yet congressional hearings' contribution is disputed; they do not have much effect on legislative outcomes or on changing members' positions on issues (Diermeier & Fedderson, 2000). They are seen as everything from 'fact-finding' agencies, 'legislative courts' and 'safety valves' to mere 'propaganda channels' and theatrical performance (Diermeier & Fedderson, 2000; Huitt, 1954). Yet Congressional hearings present multiple opportunities for strategic behaviour by Committee members and witnesses alike (Diermeier & Fedderson, 2000: 52). Groups arrive at hearings with 'a ready-made frame of reference' (Huitt, 1954, 354); a list of prepared questions with expected answers resulting from extended interviews or rehearsals with witnesses, who are strategically selected to stack a hearing one way or another (Diermeier & Fedderson, 2000).

The Congressional hearing selected here interrogated 'The Financial Crisis and the Role of Federal Regulators'. Hosted by the Committee for Oversight and Government Reform, and chaired by Democrat Henry Waxman from California, the hearing took place on 23 October 2008 and was the *fourth* hearing into the 'greed and corporate excess' that triggered the global financial crisis (US Govt, 2008: 4). The first two hearings examined the collapse of Lehman Brothers and AIG, the third dissected the role of credit rating agencies. The fourth hearing completed the circle, examining the role of US regulators (US Govt, 2008: 4). Forty-one Congressional representatives heard evidence from three witnesses: Alan Greenspan, former chairman of the Federal Reserve; Christopher Cox, incumbent chairman of the Securities and Exchange Commission (SEC); and John Snow, former Secretary of the Treasury. The hearing served as an important political 'performance', an opportunity for strategic behaviour by Congress members on the eve of the 4 November general election. Committee leaders established their economic and financial expertise in the hope that associated legislation might be referred to them (Diermeier & Fedderson, 2000). The hearing was televised live by C-Span, the public affairs network, with segments transmitted globally, enabling viewers to see demonstrable anger from political representatives as the three regulators were 'called onto the carpet' for their failure to detect, manage or avert risk. Each witness fielded roughly the same number of questions. However, the analysis explores Greenspan's testimony; given the central bank's position as 'independent' buffer between state and market, the intense media and public scrutiny on the Fed, and Greenspan's symbolic role in representing the US economy (Woodward, 2000).

The hearing likewise presented an opportunity for strategic behaviour by the witnesses. Greenspan was under no obligation to appear before Congress. By 2008, he had resumed private consultancy, advising the Bank of England and institutional investors among others. He had also published a book about the years leading up to financial crisis. While the real 'heat' for the Fed's failures lay with his successor, Ben Bernanke, Greenspan was undoubtedly motivated to protect his legacy. Despite the negative media scrutiny, Congressional testimony would be a 'walk in the park' for Greenspan, whose thoughtful, unruffled demeanour made him an evasive target. He had decades of experience facing antagonistic questioning from Congress, including his defence of the Fed's actions in previous financial crises (Woodward, 2000).

The chapter's main argument however is that beyond the blame, recrimination, and individual strategic behaviour exhibited by politicians and witnesses lay a *collective* strategy by the Congressional Committee and Greenspan himself. The former Fed Chair remained an important, recognisable symbol of the US economy, and one of few influential symbols of global financial markets. In addition, Greenspan's visible habit of showing 'strain in his wrinkled forehead' as he considered Congressional questions, conveyed the image of a public representative who always appeared to be telling the truth (Woodward, 2000: 228). While he was known for opaque, convoluted language when forecasting future economic risks, on this occasion Greenspan's purpose was to account for the past and recommend solutions, an occasion when central bankers opt for simpler syntax (Holmes, 2014). It is therefore contended that, with just weeks to go before the general election, the Congressional hearing became Greenspan's centre stage. Once more, he became the 'activist' central banker (Shafik, 2013), producing a story of confidence in the US economy. Through the powerful communications medium of the US Congressional hearing, Greenspan *re-interpreted* the actions that led to the financial crisis, laying the groundwork for stability going forward. In so doing, he remained true to the primary role of central banker as risk manager, laying the groundwork for economic stability by 'making' and 'remaking' the economy.

Analysis of findings

The Congressional hearing is analysed as both written and oral genre. The Committee chairman and all three witnesses opened by presenting written testimony, while some questions were minuted, with witnesses asked to submit a written response for the record at a later date. The three witnesses, Greenspan, Snow, and Cox, fielded more than 80 questions in all; some questions were posed to all three witnesses, some targeted individuals. Member contributions were limited to ten minutes each. The hearing lasted for nearly four hours; from 10:00am to 1:55pm with short breaks. The entire 68-page transcript was analysed, highlighting material connected with the Federal Reserve. The material was extracted and condensed in order to explore Greenspan's attempt to 'remake' the US economy. The findings are organised and presented in four themes. The first over-arching theme examines post-2008 risk prospects for the US economy. The remaining themes explore arguments and counter-arguments concerning risk management in financial markets.

'The Global financial crisis was a once-in-a-century event'

The hearing was opened by Committee Chairman, Henry Waxman, who excoriated all three witnesses for 'the regulatory decisions they made' and 'failed to make' (US Govt, 2008: 5). Waxman questioned the notion that 'free, competitive markets are by far the unrivalled way to organize economies', arguing that regulators 'became enablers rather than enforcers', and that the US economy had paid the price (US Govt, 2008: 18). Specifically, Chairman Waxman accused Greenspan of allowing his libertarian ideology to trump governance. Several Committee members would return to Greenspan's ideological 'flaws' throughout the proceedings. Greenspan responded in various ways, reverting to his idiosyncratic, convoluted speaking style to modify yet justify his personal beliefs:

Mr Greenspan: To exist, you need an ideology. The question is, whether it ... is accurate or not ... yes, I found a flaw, I don't know how significant or permanent it is, but I have been very distressed by that fact. [I] found a flaw in the model that ... defines how the world works, so to speak.

Chairman Waxman: In other words, you found that your view of the world, your ideology, was not right, it was not working.

Mr. Greenspan: That's precisely the reason I was shocked, because I had been going for 40 years or more with very considerable evidence that it was working exceptionally well. (US Govt, 2008, 18)

Defending his libertarian ideology, Greenspan maintained that the financial crisis 'turned out to be much broader' than he imagined (US Govt, 2008: 9). He repeated his longstanding argument that as new markets developed in Asia and elsewhere, global structure had changed and with it, established economic models for defining the global economy. Changing global trends had curtailed central banks' efforts to 'make' economies, resulting in a 'major decline in real long-term interest rates globally' (US Govt, 2008: 64). Consequently, central banks 'lost control of ... the longer end of the market' (US Govt, 2008: 64). When they tried to raise interest rates (e.g. to prevent a housing bubble), long-term rates did not move at all. Despite this, Greenspan insisted that the 'credit tsunami' which prompted the financial crisis had been 'almost surely a once-in-a-century phenomenon' (US Govt, 2008: 60). By claiming that a global crisis would not recur, Greenspan attempted to 'remake' the US economy and with it, global financial markets.

'The markets are chastened, risky products have disappeared never to return'

From the start of the hearing, Greenspan quickly dispatched blame for the financial crisis to 'securitizers, banks, credit rating agencies and risk management models' (US Govt, 2008: 30). The Committee's response to this was mixed. Some found Greenspan disingenuous for ignoring the Fed's role in promoting 'adjustable rate mortgages that fuelled the subprime market' and the explosion in public and private debt (US Govt, 2008: 31), crystallised in an observation by the Committee Chair:

Chairman Waxman: ... you said in your statement that ... the whole intellectual edifice of modern risk management collapsed ... Now that sounds to me like you are saying that those who trusted the market to regulate itself ... made a serious mistake.

Mr. Greenspan: Well, I think that's true of some products, but not all ...

Chairman Waxman: Well, where did you make a mistake then?

Mr. Greenspan: ... in presuming that the self-interest of organizations, specifically banks and others [meant] they were best capable of protecting their own shareholders ... (US Govt, 2008: 17)

Greenspan launched a four-point defence of the Fed's role vis-à-vis market abuse. First, he declared that the crisis had been precipitated by 'a failure to properly price such risky assets', compounded by the fact that even 'the most sophisticated investors' in the world had 'wrongly viewed' subprime mortgages 'as a steal' (US Govt, 2008: 10). Greenspan further insisted that he *had* provided advance warning of 'dire consequences' of 'underpricing of risk' as early as 2005 (US Govt, 2008: 9). Defending his 'light-touch' regulatory approach, Greenspan asserted that many risky products had disappeared, never to return; that the markets had punished their own and would be 'chastened' in future. He conceded one change: that securitisers be required to have 'skin in the game' henceforth, retaining 'a meaningful part of the securities they issue' (US Govt, 2008: 11). But he emphasised that the global crisis would pass, and America would 'reemerge with a far sounder financial system' (US Govt, 2008: 11).

However, the Committee disputed the notion that markets were either chastened or sufficiently punished. Democratic Congressman Kucinich of Ohio argued that it was the American people not the markets who were 'getting punished', with millions of Americans losing their homes (US Govt, 2008: 30). Republican Congressman Mica of Florida protested that markets had been inadequately disciplined, that taxpayers wanted 'someone held accountable' and for 'people to go to jail' (US Govt, 2008: 35). Democratic Congressman Cummings of Maryland lamented that constituents were 'losing their jobs', 'losing their investments', 'unable to get student loans', while businesses were going under (US Govt, 2008: 27). Democratic Congressman Sarbanes of Maryland chastised Greenspan for failing to acknowledge the true flaw in his libertarian ideology; that while markets might claim to punish their own, the financial crisis had instead injured many 'innocent bystanders' (US Govt, 2008: 59).

'Light-touch regulation is determined by Congress not the central bank'

The Committee repeatedly interrogated Greenspan and his co-witnesses about the need to strengthen financial regulation. Unlike his counterparts, Greenspan argued that the regulation required to prevent a once-in-a-century phenomenon would be so onerous as to 'suppress the growth rate in the economy' (US Govt, 2008: 60). Greenspan also maintained that politicians, not central bankers, dictated the nature of central banking regulation; arguing that as a public servant he had been appointed by the executive, who in turn devised the rules governing the US economy:

Mr Greenspan: I took an oath of office when I became Federal Reserve chairman ... to uphold the laws of the land passed by the Congress, not my own predilections.

I think you will find that my history is that I voted for virtually every regulatory action that the Federal Reserve board moved forward on ... because I perceived that was the will of the Congress. (US Govt, 2008: 18)

Democratic Congressman Tierney of Massachusetts challenged Greenspan on his attempt to shift accountability to the political executive, arguing that the Homeownership Equity Protection of 1994 gave the Fed a 'clear directive' to 'prohibit acts or practices in connection with refinancing of mortgage loans found to be associated with abusive lending practices ...' (US Govt, 2008: 37). In addition to grilling Greenspan on the Fed's poor governance, both Republican and Democrat Committee members descended into political rivalry, each blaming the other for watering down existing regulation or failing to introduce appropriate new regulation (US Govt, 2008).

'It remains impossible for regulators to see all future risks'

A major point of interest for Committee members was how to address weaknesses in the regulatory framework so as to make risks more 'visible' in future. Unlike his co-witnesses, Greenspan continually deflected these questions. He protested that the 'extraordinarily complex global economy' was 'very difficult to forecast in any considerable detail', and that even the Fed's unrivalled cadre of economists could not 'see events that far in advance' (US Govt, 2008: 43).

Mr Greenspan: If we are right 60% of the time in forecasting, we're doing exceptionally well ...

We at the Federal Reserve had a much better record forecasting than the private sector, but we were wrong quite a good deal of the time ... forecasting ... never gets to the point where it's 100% accurate. (US Govt, 2008: 34)

Greenspan argued that weak regulation was often based on forecasting whether products might go bad or whether the market cycle would turn. He declared that no one should 'expect perfection in any area where forecasting is required' (US Govt, 2008: 43), and that regulation based on forecasting could never be sound. Yet the Committee was adamant on the need to strengthen risk-management tools, as articulated by Republican Congressman Issa of California:

Mr Issa: ... the doomsday scenario we now live with undoubtedly could have been modeled but wasn't ... by any of the agencies of government and delivered to Congress ... If that modeling is available today, please tell me. Otherwise tell me, do you think we should be investing in that? (US Govt, 2008: 50)

Greenspan responded that the 'vast risk management and pricing system' governing global markets had collapsed because of a fatal flaw (US Govt, 2008: 10). The risk pricing system only modelled two stages of the economic cycle – the 'euphoria stage' and the 'fear stage' (US Govt, 2008: 10) but failed to construct a third model to identify which of the other two was about to happen. Greenspan argued that even if such a risk-management system could be devised it could never prevent a financial crisis since financial crises must, of necessity, be unanticipated. If they were anticipated, they would only 'be arbitraged away' (US Govt, 2008: 51).

Discussion and conclusion: 'Re-setting' ... not remaking the economy

The 2008 US Congressional hearing into the global financial crisis demonstrates the concerns of the contemporary 'risk society', in which political and market interests collide, entangling discourses of trust with discourses of risk. The hearing further highlights the integral role played by states at the highest levels of global financial markets. Despite the opprobrium levelled at central banks for contributing to financial risk contagion, in a post-crisis world, state priorities were economic stability and recovery as a means of managing *political* risk. Accordingly, through the Congressional hearings, multiple political voices – in government and opposition – managed political risk by shifting blame everywhere and yet nowhere; from guilty financial institutions to regulators who were 'asleep at the wheel'. Beyond the need to shift blame away from politicians, for Congress, the greatest risk posed by the crisis was to the US itself, as the pre-eminent global economy. To this end, Congress briefly restored Alan Greenspan's role as central banker, 'symbol' of the US economy and chief activist for economic recovery. Using the Congressional hearing as platform, Greenspan affirmed a 'financial stability story'. His testimony gained intense global media coverage, as central bank watchers searched for meaning in his words, treating Greenspan once again as the 'tuning fork' of the US economy (Woodward, 2000).

It has been argued that central banks play a distinctive role in institutionalising patterns of risk in national economies by guaranteeing overall trust in and sustaining the stability of a nation's financial system. Congressional hearings are just one discursive event through which US central bankers make, unmake or remake an economy (Holmes, 2014). The 2008 Congressional hearing provided a powerful communications tool through which Greenspan could fashion a 'financial stability story' for the US economy in the wake of the global financial crisis. Symbolically acting as central bank 'buffer' between state and markets, Greenspan used the Congressional hearing to diverge from the tougher stance adopted by the Treasury and the SEC. Whereas these agencies were keen on increasing regulation, Greenspan made the case for inertia rather than action; contending that markets would be motivated to price risk properly in future before spreading it more widely. Yet, while insisting that good regulation could be based on forecasting, Greenspan failed to say what 'good' regulation might look like.

The chapter concludes by acknowledging a singular, damaging contribution financial markets now make to the risk society. Financial risk-spreading is a consequence of a 'steady diet of distrust' by professional risk-takers (Keegan, 2007) who increasingly sidestep traditional trust mechanisms, substituting risk-management tools as trust-by-proxy. Central banks operate at the intersection of trust and risk discourses in financial markets, becoming increasingly conflicted in their support for professional risk-spreading. This conflict highlights shifting power relations in financial markets, as the interests of short-term financial speculators diverge from those of long-term investors and governments. Whereas states seek economic stability in order to facilitate growth, and long-term investors seek similar stability in order to protect wealth, economic stability is of little interest to powerful speculators (hedge funds, high frequency traders, etc.) who increasingly dominate global markets. Speculators seek volatility, and indeed actively seek out 'crisis' to gain profit at the expense of citizens who must live with the havoc wrought by such volatility. Greenspan publicly distanced himself from the rise of such harmful risk practices, despite overseeing their expansion during 18-plus years at the helm of the world's most powerful central bank. While Greenspan's public testimony on 23 October 2008 may have restored some confidence in the US economy and in the US Federal Reserve, it soon became clear that financial markets were anything but chastened. On 24 October 2008, a whistleblower called the New York Fed to report widespread manipulation of Libor, the interest-rate setting mechanism.

References

Beck, U. (1992). *Risk society*. Sage Publications.
Diermeier, D. & Fedderson, T.J. (2000). Information and congressional hearings. *American Journal of Political Science*, 44 (1), 51–65.
Davies, H. (2010). *The financial crisis: Who is to blame?* London: Polity Press.
FCIC (2010). Evidence from Alan Greenspan. Financial crisis inquiry commission. 7 April. Retrieved from: http://fcic-static.law.stanford.edu/cdn_media/fcic-docs/2010-04-07%20Alan%20Greenspan%20Written%20Testimony.pdf.
Giddens, A. (1999). Risk and responsibility. *The Modern Law Review*, 62 (1), 1–10.
Holmes, D.R. (2014). *Economy of words: Communicative imperatives in central banks*. University of Chicago Press.
Horlick-Jones, T. (2004). Experts in risk? ... do they exist? *Health, Risk & Society*, 6 (2), 107–114.
Huitt, R.K. (1954). The Congressional Committee: a case study. *American Political Science Review*, 48 (2), 340–365.
Irwin, N. (2013). *The alchemists: Inside the secret world of central bankers*. London: Headline Publishing.
Keegan, W. (2007). The terrible consequences of the financial world's diet of distrust. *The Observer*, 16 December, 8.

Kirsner, D. (2012). Trust and the global financial crisis. In S. Long & B. Sievers (Eds). *Towards a socioanalysis of money, finance and capitalism*, pp. 278–291. Abingdon: Routledge.

Krugman, P. (1999). *The liquidity trap*. Retrieved from: http://web.mit.edu/krugman/www/trioshrt.html.

Malsch, B. and Gendron, Y. (2009). Mythical representation of trust in auditors and the presentation of social order in the financial community. Critical Perspectives on Accounting, 20, 735–750.

Martin, R. (2002). *Financialization of daily life*, Temple University Press.

Rajan, R.G. (2005). *Has financial development made the world riskier?* Economic Policy Symposium, Jackson Hole, Federal Reserve Bank of Kansas City, August, 313–369.

Resche, C. (2004). Investigating 'Greenspanese'. *Discourse & Society*, 15 (6), 723–744.

Riles, A. (2006). Real time: Unwinding technocratic and anthropological knowledge. In M.S. Fisher & G. Downey (Eds), *Frontiers of capital*, pp. 86–107. Duke University Press.

Schwartz, H.S. (2012). Anti-oedipal dynamics in the sub-prime loan debacle. In S. Long & B. Sievers (Eds), *Towards a socioanalysis of money, finance and capitalism*, pp. 321–334. Abingdon: Routledge.

Shafik, N. (2013). Communication, engagement and effective economic reform: The IMF experience. *Chartered Institute of Public Relations Annual Maggie Nally Lecture*, Houses of Parliament, 30 July.

Smart, G. (2006). Writing the economy: activity, genre and technology in the world of banking, London: Equinox Publishing.

Taylor, J. & Burt, E. (2005). Managing trust, generating risk. *Information Policy*, 10, 25–35.

US Govt (2008). The financial crisis and the role of federal regulators. *Hearing before the committee on oversight and government reform*. House of Representatives, 110th Congress, 2nd session, 23 October, US Govt Printing Office, Serial No. 110–209.

Vestergaard, J. (2009). Discipline in the global economy, New York: Routledge.

Warner, J. (2011). The governor – Ready to rule the economy? *Sunday Telegraph*, 29 May, 6.

Woodward, B. (2000). *Maestro: Greenspan's Fed and the American boom*. New York: Simon & Schuster.

19
Projecting a Definition of Risk Situation: Travel Advice and the Prudential Traveller

Arthur S. Firkins and Christopher N. Candlin

Introduction

The projected situation as it relates to risk has not been widely addressed, yet it remains in our view a key problematic in the communication of risk. In this chapter we focus on how risk situation is projected in the field of international travel. The right to travel is increasingly hampered by a heightened concern over risk. In the global risk environment the 'prudential traveller' is the 'informed traveller' and a core aspect of this information process is the issuing of a particular type of risk communication account to travellers, known as travel advisories. Travel advisories are key risk communication technologies, through which a risk communicator renders visible the risk situation evident in particular countries to 'the traveller'. Travel advisories essentially enable governments to assert some form of control or influence over the destinations their citizens may travel to and as such are motivated in their design to influence the traveller's decision-making processes. The issue we consider in this chapter is how travel advisories linguistically direct the traveller to evaluate the risks involved in travelling to a particular country. Our discussion proceeds from an analysis of a small corpus of Travel Advice texts taken from the 'Smart Travellers Website' (Australian Department of Foreign Affairs and Trade).

Travel advisories are constructed to influence action without appearing to be overtly directive and so the way they frame country risk is not only crucial for the well-being of the traveller, but is equally important for the public accountability of the institution responsible for providing the advice (Oded, 2007). The institutional framing of risk through travel advisories is therefore an example of the forensic orientation of a culture to hold the 'institutional agents', who are the definers of risk, accountable (Douglas, 1990). Institutionally, risks are defined by what Beck refers to as the 'relations of definition' which are the panoply of institutions and agencies involved in the uncovering and the subsequent communication of risk, of which government agencies are a part of (Beck, 1996).

Our reason for analysing travel advisory texts is that they exemplify the role of rhetoric in the communication of a risk situation, and hence how a situation of risk is linguistically projected to an audience. To date, the historical basis of risk communication in which rhetoric has played a significant role in moving the public and individual opinion has not been well examined (Choi, 2001). In this chapter we argue that, far from being competing paradigms, rhetoric and statistics are in fact historical bed-fellows when it comes to communicating risk. Rhetoric and statistics have been used simultaneously by risk communicators to frame a risk situation and interchangeably to effect actions, changes in behaviour, or even to retrospectively justify policy that may have already been put into operation (Douglas, 1990).

Moreover, travel advisory texts manifest many of the social phenomena that concern risk communication theorists and practitioners. They are focused towards the individualisation of risk which is shifting the responsibility for risk management to the individual person (Beck, 1992, 2006). They focus on the perceived prevalence of manufactured and ecological risk which could present some form of harm to the traveller (Giddens, 1990, 1991). In addition, they are also examples of what governmentality theorists of risk refer to as 'control by distance', and encode rhetorically the paternal voice of the government projecting influence over the places a citizen may visit (Foucault,1991; Rose & Miller, 1992; Rose, 1993). Furthermore, travel advisory texts frame risk in a 'bounded space', because the risk situation which is rendered visible within the advice is defined within a confined bounded geographical space, and although some risks may transcend national boundaries, such as exposure to terrorism, the framing of situation within a travel advisory account is generally confined to a specific country or region. Each of these phenomena can be found linguistically realised in how the communicator anchors the account of country risk to the environing surrounds while simultaneously aligning the assessment with the intuitive judgment of the traveller, using modality and evoking layers of affect. We will elaborate these points further on in the chapter.

Travel advisory accounts are highly reflexive of the situations they construct (Garfinkel, 1964, 1970). Moreover, evoking Goffman's dramaturge metaphor (1959), both 'backstage' and 'front stage' rendering processes are of equal importance in producing a travel advisory account. Although considerably more resources may have gone into the backstage analysis of the various separate pieces of information which are recontextualised into the account, the front stage projection of the risk situation though the written travel advisory text itself is equally consequential to a range of audiences and carries considerable face threatening risks for the communicator. In other words the risk of projecting an incorrect definition of the risk situation to the audience may expose both the communicator and the traveller to certain hazards. In the case of the traveller it could result in a threat to safety and security, leading to injury or possible death and in the case of the communicator the consequences could be reflexive criticism or public sanctioning, leading to possible prosecution.

Constructing a travel advisory text is therefore an important institutional performance in Goffman's (1959) terms that warrants close attention as its communicative purpose is not simply about ensuring the traveller has an enjoyable time overseas, but is also a core part of a government's accountability to protect its citizens. In essence to construct a travel advisory text is to engage in risk work of a particular nature; that is to 'control' without appearing to 'control' and to shift the sole risk decision-making responsibility onto the traveller (Ewald, 1991).

The traveller as institutional member

Travelling has become an increasingly popular, but risky activity. The traveller is a particular type of member, who by circumstance, belongs to the institution of 'overseas travel'. The traveller, by the nature of the activity of buying an airline ticket, going on a business trip, or purchasing a holiday in another geographical location acquires institutional membership, and is potentially placing himself or herself in a situation of danger and may by this action be deemed to be a 'risk taker'. The membership of the institution, 'traveller', is further nourished through a particular type of discourse in which overseas travel is equated with positive risk, as a symbol of personal freedom and promoted as a way of nurturing peace and prosperity, broadening the mind, understanding cultures, and developing as an individual (Bianchi, 2007). Therefore, overseas travel as an activity has also become a critical part of building self-identity and a way of projecting social status (see de Botton, 2003; 2005).

The notion of freedom of travel arguably serves as the orienting disposition for the risk communicator who constructs a travel advisory text (Dake, 1991). In Australia, as in other western countries, travel is promoted as something Australians should do. The freedom to travel and to make decisions as to where to travel is not simply tied up with liberal views of freedom (Urry, 2000, 2002) but is also linked with the economic systems which underwrite these views. In short, we agree with O'Byrne who states 'freedom to travel is literally freedom of trade' (2001:409). In other words the orientation against which the risk communicator formulates the travel advice text is not only *travel broadens the mind* but also *travel at your own risk,* two metaphors which in a sense encompass what it means to be a prudential traveller and represent the two poles of the space within which the risk communicator formulates an account of the risk situation at hand (Candlin, 2003, 2006, Candlin & Maley, 1997). The resulting tension between these two communicative purposes represents a 'critical moment' in the communication of risk, which for the texts we have examined is manifest in the tension between compulsion and persuasion (Candlin, 2001).

In sum, then, the idea of membership is central to our conceptualisation of a risk communicator (Garfinkel, 1967). Risk communication is principally about communicating; who is at risk from what and in which situations. Hence we believe in a discussion of risk communication there also needs to be a sense of the agency of the risk communicator. In our opinion a risk communicator can be an institutional or an individual agent, virtual or real, present or distant. Therefore a risk communicator can be physically present but also temporally or geographically removed from the immediate audience. Societal members are both communicators and interpreters of risk (Goffman, 1983).

Travel advice texts

A key aspect of the advisory process in Australia is the posting of a particular type of advisory text known as 'Travel Advice', a specific type of written text available to travellers through internet web sites or via press releases and sometimes provided by travel agents or government departments in a pamphlet format. Travel Advice as issued by different countries and also by transnational organisations such as the World Health Organisation is so similar that it represents a specific genre of risk communication (Bhatia, 2002).

The genre similarities evident in the construction of Travel Advice across the English speaking world and their focus on similar sub-areas of risk arguably evidences a 'community of fate' (Beck, 2003) as countries, such as Canada, the United Kingdom, the United States, New Zealand, Australia and South Africa focus attention on similar global, regional, and country risks within their respective advice. Generically, the construction of Travel Advice follows the same method and is written in a similar style in the United States, the United Kingdom, Canada, and New Zealand (Flood Inquiry, 2004:55), making them an identifiable tool of risk communication.

Travel Advice texts follow a standard format worldwide. Each Travel Advice text is typically five to six pages in length and begins with an overall rating of risk and clearly states when it was last up-dated and briefly how it has been changed. A brief overview of the risk situation is provided in the summary section with a more detailed framing of the risk situation in the body of the account. The main part of the Travel Advice text is divided into the sub-heading of significant risk. Each sub-heading, includes such themes as 'Safety and Security', 'General Health', 'Crime', 'Local Travel', 'Airline safety' 'Consular Assistance', 'Insurance', 'Transport' and 'Natural Disasters', 'Local Laws', 'Money and Valuables' and 'Wild Life'. Hence these are the major themes of risk against which the communicator formulates the account and are selective rather than all encompassing.

The selective nature of the areas of risk which Travel Advice tends to focus on lends support to Douglas (1990, 1992) who claims that a culture, at the end of the day, selects some risks, while ignoring others and therefore risk situation as framed within Travel Advice is a 'culturally determined' categorisation selected from among an array of possible framings of 'risk situation'. In other words, Travel Advice demonstrates a tendency towards a focus on health and terrorism, but in countries such as Pakistan and Zimbabwe cheap cigarettes are widely available and smoking is not framed as a risk for travellers. Arguably, more Australians die of cigarette related illnesses than terrorism or malaria and there is no warning in respect to smoking evident in any of the Travel Advice texts in our corpus.

Travel Advice is an active text in the sense that it is a fluid account of situation which should change as the risk situation within each country changes and hence should be regularly up-dated. The assessment of risk is calibrated along a 'four-descriptor risk assessment scale' with the lowest pole being 'Exercise Normal Safety Precautions' proceeding to Exercise a High Degree of Caution, to 'Reconsider Your Need to Travel', with the highest pole indicated with the warning 'Do not Travel'. These four indicators remain consistent across all of the Travel Advice posted on the website, so the risk communicator must align their judgment of the risk situation with one of these four descriptors (Firkins & Smith, 2002). Each such risk descriptor is realised through a Modal

Adjunct, which is 'institutionally crystallised' as an 'indicator' of risk. Each descriptor denotes the risk communicator's judgment of the level of harm a particular situation could potentially pose to the travellers' safety and the hazards that they might encounter.

In similarity with the framing of risk situation in other contexts where institutionally ratified risk assessment is a core activity, the framing of risk must take account of the 'ritual constraints' and consequently the account must be squeezed into one of these four ratified descriptors (Goffman, 1981). Hence there is an institutionally imposed constraint on the communicator's judgment of risk (Firkins & Smith, 2002; Firkins & Candlin, 2006).The choice of which descriptor is appropriate to signal the level of risk evident in the country of concern and the one ultimately projected to the traveller rests with the risk communicator who frames the situation and needs to ultimately align with what is institutionally deemed to be acceptable or unacceptable risk.

The projected situation

Risk situations are rendered visible through discourses and this framing of a risk situation is part of the multiple genres of risk visualisation (Eppler & Aeshimann, 2009). Projecting a risk situation is a complex process of multiple framing and reframing of information, reports, and narrative. The risk communicator not only has to make sense of all of the relevant pieces of often disparate 'intelligence', they must then collate it together a produce a credible account of the risk situation at hand and then convey it to an audience which is more than likely geographically and temporally removed. Projection of the situation at hand is therefore underpinned by particular feature of grammar which the communicator selects to frame and project the account of risk.

We now turn our attention to consider the division of grammatical labour involved in linguistically projecting the risk situation within our corpus of Travel Advice texts and show how each of the phenomena we have discussed in the introduction is consequential to the construction of each account. Halliday (1985) identifies three main forms of projection available in the grammar of English and hence three central resources available to the communicator with which to project the situation. These are linguistic categories of projection: as a Quote, as a Report. or as a Fact. All three types are evident in our corpus (Halliday, 1985:251) and in the way the risk communicator projects the situation.

The situation can be projected by the communicator through the resource of Quote (Halliday, 1985:228) which is realised primarily through a Verbal Process which in turn can project a Locution i.e. the exact words of the agent being quoted, as in ['The official said the risk is very high'] or as an Idea i.e. a projection of the thoughts of the participant being quoted, as in [He thought 'I will be very careful next time']. Interestingly, there is minimal projection through Quotes evident in our corpus, and it is only used where the words of a trusted expert source appears to enhance the projection of the situation, exemplified here where the risk communicator uses the words of the World Health Organisation projected by the Verbal Process 'declared' [¹On 5 May 2014, the World Health Organisation declared the recent international spread of wild poliovirus a 'public health emergency of international concern']. The almost complete absence of direct quotes in our corpus supports Beck (2002) who points out modernity is characterised by a lack of trust in the institutions which are supposed to communicate risk. Hence there is tendency against directly quoting, and thereby identifying, institutionally affiliated expert sources within our corpus.

Instead Travel Advice typically directly projects the rhetorical voice of what we have called the 'risk communicator'. Hence the risk situation as framed is primarily projected through the other two forms of linguistic projection. The communicator makes frequent use of Report to project the risk situation (Halliday, 1985:230). Of key importance in risk communication is the 'reliability' and 'credibility' of the source of information. Not only does the analyst need to determine the accuracy of the intelligence, but the risk communicator also needs rhetorically to 'proclaim' the event to be highly warrantable or otherwise distance themselves from the credibility of the information. Hence using Report can be a way of assigning credibility through projection as the judgment of harm can be assigned to the communicator directly signalled by such elements as 'we', 'our information suggests', 'reports we have obtained support claim'. An example of this can be seen in Extract 1 frame 2 and is signalled by the Verbal Process 'recommendations' and cued by the grammatical element 'that' and extended by a Finite Modal Operator such as 'may affect', 'should ensure', or 'should avoid'.

Alternatively projection of risk situation through Report is frequently used by the risk communicator to align the source of the information with specific participants. For example [The Australian Government has recommended that Australians in Syria depart immediately by commercial means while it is possible to do so] projects an assertion 'depart immediately' with the report from the 'Australian Government', cued by the Verbal Process 'has recommended'. Hence projection through Report can be via a locution, realised by a Verbal Process or through an Idea, realised through a Mental Process such as 'considers' in the following example [The Australian Government considers that paying ransoms increases the risk of further kidnappings].

In this same way the communicator uses Report to rhetorically project particular types of voice such as 'postulation', for example [We continue to receive a stream of reporting indicating that terrorists are planning attacks against any venues frequented by foreigners.] or a 'hypothetical', [There has been some evidence suggesting that recent attacks in Libya have deliberately targeted foreign workers] an 'actuality' for example 'long standing policy' in [The Australian Government's longstanding policy is that it does not make payments or concessions to kidnappers], or a 'certainty' for example 'are planning' in [Information we have indicates that militants are planning attacks against hotels frequented by foreigners in Karachi over the Christmas period] or as a 'possibility' for example 'may be planning' in [The US Government warned that unspecified terrorists may be planning to target Western-affiliated international schools in Islamabad at an unknown date and time]. Extract 1 projects the risk situation in a series of Reports through locution, 'We continue to receive reports', 'issued temporary recommendations', 'the UK Embassy announced', or 'the US Government advised':

Extract 1
[[1]We continue to receive reports that indicate terrorists are planning further attacks and expect attacks to be ongoing for the next several weeks]. [[2]The Libyan authorities have issued temporary recommendations that may affect travel to the country and will affect travel to certain regions].[[3]On the 4th August 2014, the UK Embassy announced that it will no longer provide consular assistance to its nationals in Libya]. [[4]On 26 July, the US Government advised that staff at the US Embassy in Tripoli have been evacuated]. [[5]Since October 2014, a number of foreigners have been attacked in public places by individuals who may have been inspired by anti-western sentiment].

Fact is the most frequently used form of projection found in Travel Advice. A Fact is a projection with no process of saying or thinking which projects it, hence no participant doing the projecting (Halliday, 1985:243). Therefore a Fact is a way of projecting a situation impersonally and is realised linguistically through a Relational process, typically signalled by 'there' as in 'there is' [There is a low incidence of crime in Khartoum, however there are high levels of violent crime in Darfur], projects the situation as a proof. or 'there have been' for example [There have been a number of recent mass casualty attacks on sites and ceremonies associated with religious pilgrimages], projects the situation as a 'case' or 'there was' [There was a significant attack in Damascus involving the use of chemical weapons] and 'there remains' [There remains a very high threat of kidnapping in Iraq], projects the situation as a 'chance' and [There is a need to take care of your personal security], projects the situation as a 'need'. In Extract 2 (below) the country risk situation is projected through a cascade of three projections as Fact, signalled by 'has deteriorated', 'are now active', 'has conducted' or 'there is'.

Extract 2
[¹The security situation in the country **has deteriorated** significantly. Armed opposition groups **are now active** in many parts of Iraq, including in Iraqi Kurdistan]. [²The US-led coalition, including Australia, **has conducted** targeted airstrikes against militants in Iraq]. [³With the escalating conflict, **there is** an increased threat to foreigners throughout Iraq, particularly journalists and NGO workers].

Although we have isolated the three types of projection for the purposes of discussion, in actuality our corpus evidences that the risk communicators uses all three types of projection in order to project the risk situation to the traveller. Framing the risk situation is also a process of rendering visible an account through actively aligning to the audience while simultaneously anchoring it to the various events, participants, locations and agents in the country of concern the communicator wants to focus on.

Reference is to specific participants, events, places, and times is dominant within our corpus and anchors the account of risk to the environing surrounds and give 'empirical credibility' to the claims being made by the risk communicator (Snow & Benford, 1988; Snow et al., 1986). Travel advisories anchor the account of risk through reference to specific hazardous agents which might exist in the situation such as 'terrorists' (Extract 2) 'armed opposition groups', participants such as 'Libyan authorities', 'UK Embassy', 'US Government', 'staff at the US Embassy in Tripoli' (Extract 1) and 'The US-led coalition', 'journalists and NGO workers' and 'foreigners' (Extract 2). Reference is made to specific events such as 'security situation' and 'targeted airstrikes' (Extract 2) as well as to the situational scope encompassing locations such as 'travel to the country' and 'travel to certain regions' (Extract 1) or specific reference 'including in Iraqi Kurdistan' , 'throughout Iraq' (Extract 2). Thus reference serves to anchor the risk situation to the specific bounded space the communicator is intending to highlight.

The Travel Advice accounts in our corpus also simultaneously align the risk situation with the traveller's judgment by evoking multiple levels of affect which amplify the abnormal elements within the situation, such as 'harmful wildlife', 'indiscriminate crime', 'further attacks' , 'unexpected attacks' to up-scale or down-scale the hazards which may be evident in a particular country (Slovic, 2004, 2006, 2007). The communicator further amplifies affect through the use of intensifiers such as 'strongly', 'under no circumstance', 'significant', 'heightened', and 'extremely', found in our corpus and serve to amplify the sense of danger and foster avoidance of the location. For example 'marked increase' serves to intensify increase [In 2014, there has been a marked increase in the number of reported kidnappings of NGO workers and journalists]. Intensity is upscaled through negative circumstances, such as 'threat of kidnapping', 'violent civil unrest', has been evacuated, 'targeted airstrikes', 'armed conflict', and violent crime all build a sense of risk through affect. Hence intensifiers function to up-scale or down-scale how the institutional member judges seriousness of possible harm or threats to safety. For example Extract 2 emphasises the risk through the intensifiers 'deteriorated significantly' and 'increased threat'. Some choices turn the volume up for example 'extremely important', 'sharply increasing', and 'dangerously close' and some choices turn it down such as 'fairly urgent', 'largely dealt with', 'basically contained', and 'steeply declining'.

The communicator projects attitude towards the risk situation by sharpening the judgment of the risk, using for example 'out of control', 'deadly virus', 'unsafe environment', or obscuring it from view, for example, in no 'evidence to suggest' or 'affects only a small number'. In our corpus, such intensifiers are dispersed throughout the projected situation and can attenuate the risk situation through the chaining and the repetition of particular attitudinal lexis, be it antonyms or synonyms, thereby creating 'directional track' in the projection of affect (Goffman, 1974). This is seen in Extract 3 where the risk communicator frames the risk situation in Sudan as a series of five projected Fact frames. The sense of a high risk situation is built through the chaining of intensifiers, violent civil unrest, threat of terrorist attack, 'armed conflict' and through repetition of 'threat of kidnapping' (Halliday & Hasan, 1985:83–94).

Extract 3

[[1]We recommend you reconsider your need to travel to Khartoum due to the possibility of **violent civil unrest**, the **threat of terrorist** attack and the **threat of kidnapping**]. [²We strongly advise you not to travel outside of Khartoum due to **the possibility of armed conflict, the threat of terrorist attack, the incidence of violent crime and threat of kidnapping**]. [³Australians in Sudan should have personal security measures in place and contingency plans to depart Sudan if necessary]. [⁴There is a **threat of kidnapping** throughout Sudan. Foreigners, including NGO workers, could be targeted.' [⁵You should avoid **demonstrations, protests and large crowds** throughout Sudan, including in Khartoum, as **violence** could occur with little warning and terrorist may be planning attacks].

Our corpus also foregrounds the resources of moduality and modality to project the communicator's judgment of the risk situation (Firkins, 2012). Modality refers to 'the area of meaning that lies between yes and no, the immediate ground between positive and negative polarity' (Halliday, 1985:335). This linguistic grading of certainty (Simpson, 1993:46–85) allows the institutional member to 'entertain' possibility (Martin & White 2005), through the use of 'may be planning' and 'possibility of violent civil unrest' (Extract 3), but leaves sufficient space for the traveller to decide if they fly or not to the country in question for example 'should' in the [It should always be kept in a safe place], and could in [Terrorist tactics could include bombings or smaller-scale attacks, such as drive-by shootings and kidnapping]. Modality in our corpus is realised through a Modal Adjunct for example 'appropriate personal security measures', 'may turn violent', or 'may prompt demonstrations' or alternatively via a Finite Modal Operator for example 'you should ensure', 'you should avoid', 'maybe targeting', 'does not make', 'Should you choose to remain', 'should pay close attention', or 'may sometimes restrict'

The resource of Modality and Modulation allows the communicator to traverse the pragmatic space between freedom and safety by presenting the proposition as one of a range of possible actions the traveller can take. Hence the traveller can follow the advice or choose to ignore it. The situation is framed in degrees of certainty, with information of complete certainty being 'substantiated' or 'determined', in essence proclaimed by the institutional member and information of less certainty framed as 'alleged', 'disputed', or 'implied' and therefore 'entertained' by the communicator as modulation or through a modulation.

Hence, the situation can be projected by the communicator as a degree of Probability for example as cued by 'possibility' as in the example [We strongly advise you not to travel outside of Khartoum, due to the possibility of armed conflict] and 'some evidence' in [There has been some evidence that recent attacks in Libya have deliberately targeted foreign workers]. The most frequent form of modality projected in our corpus is degree of Usuality, as cued by 'frequency' and 'usually' in [The frequency and intensity of localised, sporadic, and usually spontaneous civil disturbances has increased significantly in recent years over economic issues] and as cued by 'typically' in [Attacks typically target Libyan Government or security interests and increasingly].

Further, Inclination is most frequently projected as a mental process as cued by 'consider' [In planning your activities, consider the kind of places known to be terrorist targets and the level of security provided at venues] and 'reconsider' as in the following example [We advise you to reconsider your need to travel to Pakistan at this time due to the very high threat of terrorist attack and volatile security situation]. Finally, Obligation projects what the traveller can and cannot do as is frequently cued in our corpus by 'allowed to' or 'required to' expressions such as 'allowed to' and 'able to' as in [We understand, however, that Australian citizens may be allowed to enter Lebanon if they are able to prove they have an appointment with the Australian Embassy in Beirut].

At the higher end of the risk scale, the communicator attempts to deter the traveller completely through a discourse strategy we have elsewhere termed 'rhetorical panic' (Firkins & Candlin, 2011; Firkins, 2012) which is the purposeful up-scaling of the discourse in such a manner that makes the hazards 'stand out', ultimately stigmatising the location as an undesirable place to visit (Turner, Gregory, Brooks et al., 2008). For example travel advisories typically up-scale the 'high dread consequences' of exposure to such things as terrorism and epidemics (Giddens, 1990). Even Travel advisory texts for fairly tame and risk benign countries such as Canada and New Zealand amplify certain risks, so as to avoid any form of liability on the part of the government, should some misadventure befall the traveller.

Finding the correct balance between 'advising' and 'directing', in other words aligning risk to one of the four institutionally provided indicators which Travel Advice must align to, can pose difficulties for institutionally accountable risk communicators. A mistake in framing the situation may result in the shifting of blame for a potential disaster onto the organisation, which framed the assessment. We agree with Goffman's observation that 'failure to regulate the information acquired by the audience involves a possible disruption to the projected situation' (Goffman, 1959:77). For instance, on the day of the bombing in Bali (12 October, 2002), the Travel Advice issued by the Australian

Department of Foreign Affairs and Trade (DFAT) and posted on their website which was active and in effect for the 12 October, simply 'cautioned' Australian tourists to 'maintain a high level of personal security awareness', while emphasising that tourist services were functioning 'normally' across the country, 'including Bali', a major resort and holiday destination (see Extract 4 below). The framing of risk, the language selected as well as the way the risk situation was projected to travellers on that day were all reflexively examined through inquiry process and media attention on the tragedy.

Extract 4 is taken from the summary section of the Travel Advice posted on the day of the Bali Bombing (Flood Inquiry, 2004) linguistically down-scales the presenting risks in a number of ways. The advice singles out 'terrorist activity' as being the concern and identifies what geographical areas the traveller should avoid in Indonesia. However Bali is specifically cited through the dynamic Modal Adjunct 'operating normally', suggesting normality of situation, which could be read by a traveller as meaning 'the absence of risk'. Moreover the risk communicator aligned the situation at the second lowest risk level on the department's own risk scale.

Extract 4 Travel Advice Indonesia Posting 20 September 2002: Effective on the Day of the Bali Bombing (DFAT)
[In view of the ongoing risk of terrorist activity in the region, Australians in Indonesia should maintain a high level of personal security awareness.] [Australians should avoid travel to west Timor (outside Kupang), Maluku, North Malunku and Aceh. Australians in Papua (Irian Jaya) and North Sulawesi should exercise caution and seek current information from the Australian Embassy prior to travel. Australians in Poso, the middle of central Sulawesi should avoid inter-provincial and inter-city bus travel and exercise caution following recent attacks on passenger buses. Tourist services elsewhere in Indonesia are operating normally, including Bali. (Flood, 2004)

In the view of the Flood Report (2004) the advice provided to travellers on the day of the bombing unduly attenuated the actual risk to travellers. In other words, the institutional member who framed the account dampened the advice through the verbal phrase 'operating normally' and as a result it was subsequently argued tourists would not even suspect such an event could occur (Flood, 2004). In contrast, the Travel Advice posted for 13 October 2002 the day after, amplified considerably the risk by advising the need to 'defer travel' and 'advance departure';

in other words don't go and if you are already there leave (Refer to Extract 5, below).

> Extract 5 Consular Services – Travel Advice Indonesia Posting 13 October 2002: The day after the Bali Bombing (DFAT)
> Extract 5 Australians are advised to defer all travel to Bali. Australian visitors in Bali should consider advancing their departures on available flights. In the meantime they should remain in their hotels, avoid public places where possible and call home to advise families of their well-being. (Flood, 2004)

Given the importance of Bali as a holiday destination the risk communicator appears to have overly attenuated the risk situation. In the view of the Australian Senate Inquiry (2004) the advice provided to travellers on 12 October 2002 not only unduly attenuated the actual risk to travellers, but also was not 'up-dated' to reflect the change of situation from 20 September to 12 October, an oversight given the availability of information to the Australian intelligence agencies (Flood, 2004; Senate Inquiry, 2004).

The government was held to be liable for how it institutionally framed the risk situation within the Travel Advice. In other words, there was in fact a serious disruption in the projection of the risk situation by the risk communicator. The Modal Adjunct 'operating normally' (Extract 4) was identified to be responsible for the deaths of 202 people, including 88 Australians. The use of the noun 'normal' and the process 'caution' arguably unduly down scaled and arguably dampened the traveller's risk perception. In other words the risk communicator framed the advice tending towards 'freedom' over 'duty' (Extract 4).

Conclusion

Travel Advice texts anchor the situation to ecological and manufactured risks while simultaneously aligning the situation to the traveller's concerns as to what risks they may need to be aware of. They are examples of risk technologies which aim to control behaviour from a distance. Our analysis highlights how a risk situation can be projected through discourse alone. This is achieved though the linguistic resources of reference and modality, while at the same time creating levels of affect by up-scaling and down scaling the assessment of the situation through the use of intensifiers. Travel Advice texts are found to project the risk situation at hand primarily through Report and Fact frames.

Within our corpus a tension can be identified between the risk communicators' attempts to 'advise' action without appearing to be 'overly directive' as to what the traveller should or should not do against the 'institutional responsibility' to clearly communicate the risks involved in travelling to the country in question. Hence the risk communicator frames the risk situation within a tension between 'compulsion and persuasion', in which the grammatical resources of modality and modulisation are emphasised. Travel Advice therefore highlights dangers but nevertheless leaves its management to the individual agency of the prudent traveller.

References

Australian Senate Committee Inquiry (2004). *Bali 2002: Security Threats to Australians in South East Asia*. The Commonwealth of Australia.

Beck, U. (1992). *Risk Society, Towards a New Modernity*. London: Sage.

Beck, U. (1996). World risk society as cosmopolitan society: Ecological questions in a framework of manufactured uncertainties. *Theory, Culture & Society*, 13, (4): 1–32.

Beck, U. (2002) On world risk society. *Legos*, 1, (4), 1–18.

Beck, U. (2003). Towards a new critical theory with a cosmopolitan intent. *Constellations*, 10, (14): 453–468.

Beck, U. (2006). Living in the world risk society. *Economy & Society*, 35, (3): 329–345.

Bhatia, V. K. (2002). Applied genre analysis: A multi-perspective model. *Ibérica*, 4, 3–19.

Bianchi, R. (2006). Tourism and the globalisation of fear: Analysing the politics of risk and (in) security in global travel. *Tourism and Hospitality Research*, 7, 64–74.

Candlin, C.N. (2003). Issues arising when the professional workplace is the site of applied linguistic research. *Applied Linguistics*, 24, (3): 386–394.

Candlin, C.N. (2006). Accounting for interdiscursivity: Challenges to professional expertise. In M. Gotti & D.S. Giannoni (eds.), *New Trends in Specialised Discourse Analysis*. Bern: Peter Lang.

Candlin, C.N. & Maley, Y. (1997). Intertextuality and interdiscursivity in the discourse of alternative dispute resolution. In B.L. Gunnarsson, P. Linell & B. Nordberg (eds.), *The Construction of Professional Discourse*. London: Longman. pp. 201–223.

Candlin, C.N. (2001). Medical discourse as professional and institutional action: challenges to teaching and researching languages for special purposes. In M. Bax, & J. W. Zwart (eds.), *Reflections on Language and Language Learning*. In Honour of Arthur Van Essen. Amsterdam: John Benjamins. pp. 185–208

Choi, T.Y. (2001). Writing the Victoria city: Discourses of risk, connection, and inevitability. *Victorian Studies*, 43, (4): 561–581.

Dake, K. (1991) Orienting dispositions in the perception of risk. *Journal of Cross-Cultural Psychology* 22, (1): 61–82.

De Bottom, A. (2003). *The Art of Travel*. United Kingdom: Penguin Books.

De Bottom, A. (2005). *Status Anxiety*. United Kingdom: Penguin Books.

Department of Foreign Affairs and Trade (2015) SmartTraveller The Australian Commonwealth Government http://smartraveller.gov.au

Douglas, M. (1990). Risk as a forensic resource. *Deadalas*, 119, (4): 1–16.
Douglas, M. (1992). *Risk and Blame. Essays in Cultural Theory*. London, New York: Routledge.
Eppler, M.J. & Aeshimann, M. (2009) A systematic framework for risk visualization in risk management and risk communication. *Risk Management*, 11, (2): 67–89.
Ewald, F. (1991). Insurance and risk. In G. Burchell, C. Gordon & P. Miller (eds.), *The Foucault Effect: Studies in Governmentaility*. London, Harvester Wheatsheaf. pp.197–210.
Firkins, A.S. (2012) Discourse and the Framing of Risk. Unpublished PhD Thesis. Sydney, Australia : Macquarie University.
Firkins, A. & Candlin, C.N. (2006). Framing the child at risk. *Health, Risk & Society*, 8, (3): 73–291.
Firkins, A.S. & Candlin, C.N. (2011). 'She is not coping': Risk assessment and claims of deficit in social work. In C. Candlin & J. Crichton (eds.), *Discourses of Deficit*. Basingstoke, Hants: Palgrave. pp.81–98.
Firkins, A. & Smith, S. (2002). Judgement as a resource in child protection practice. In C. Candlin (ed.), *Research and Practice in Professional Discourse*. Hong Kong: City University of Hong Kong Press. pp.309–332.
Foucault, M. (1991). Governmentality. In G. Burchell, C. Gordon & P. Miller (eds.), *The Foucault Effect: Studies in Governmentality*. Chicago: University Of Chicago Press. pp.119–150.
Flood, P. (2004). *Australian Government Inquiry into Australian Intelligence Agencies*. Commonwealth of Australia. Retrieved on 20th June 2008 from http://www.fas.org/irp/world/australia/flood.pdf
Garfinkel, H. (1964). Studies of the routine grounds of everyday activities, *Social Problems*, 11, (3): 225–250.
Garfinkel, H. (1967). *Studies in Ethnomethodology*. Cambridge: Polity Press.
Garfinkel, H., & Sacks, H. (1970). On formal structures of practical actions. In C. McKinney & E. A. Tiryakian (eds.), *Theoretical sociology: Perspectives and developments*. New York: Appleton-Century-Crofts. pp. 338–366.
Giddens, A. (1990). *The Consequences of Modernity*. Cambridge: Polity Press.
Giddens, A. (1991). *Modernity and Self-Identity: Self and Society in the Late Modern Age*. Cambridge: Polity Press.
Goffman (1959) *The Presentation of Self in Everyday Life*. London: Penguin Books.
Goffman, E. (1974). *Frame Analysis*. New York: Harper & Row.
Goffman, E. (1981). *Forms of Talk*. Philadelphia: Pennsylvania, University of Philadelphia Press.
Halliday, M.A.K. & Hasan, R. (1985). *Language, Context, and text: Aspects of Language in a Social-Semiotic Perspective*. Victoria: Deakin University Press.
Martin, J.R. & White, P.R.R. (2005). *The Language of Evaluation: Appraisal in English*. New York: Palgrave Macmillan.
Oded, L. (2007). The responsibility to responsibilize: Foreign offices and the issuing of travel warnings. *International Political Sociology*, 1, (3): 203–221.
Rose, N. (1993). Government, authority and expertise in advanced liberalism. *Economy & Society*, 22, (3): 283–299.
Rose, N. & Miller, P. (1992). Political power beyond the state: Problematics of government. *British Journal of Sociology*, 43, (24): 174–205.

Simpson, P. (1993). *Language, Ideology and Point of View*. London & New York: Routledge.
Slovic, P. (2004). What's fear got to do with it? Its affect we need to worry about. *Missouri Law Review*, 69: 971–990.
Slovic, P. (2006). Risk perception and affect. *Current Directions in Psychological Science*, 15, (6): 322–325.
Slovic, P. (2007). Affect, reason and mere hunches. *Journal of Law, Economics and Policy*, 4, (1): 191–211.
Turner, N.J., Gregory, R., Brooks, C., Failing, L. & Satterfield, T. (2008). From invisibility to transparency: identifying the implications. *Ecology and Society*, 13, (2): 7. [online] URL: http://www.ecologyandsociety.org/vol13/iss2/art7/
Urry, J. (2000). *Sociology Beyond Societies: Mobility's for the 21st Century*. London: Routledge.
Urry, J. (2002). *The Tourist Gaze: Leisure and Travel in Contemporary Societies*. London: Sage.

20
Suicide Candy: Tracing the Discourse Itineraries of Food Risk

Rodney H. Jones

Introduction

In a YouTube video entitled 'Suicide Candy' a poster named Sir Sebastian displays a bag of White Rabbit Creamy Candy and declares dramatically:

> I'll be the first one to admit I've eaten a lot of weird candy treats in my day. Scorpion lollypops. Mexican maggots. But never, NEVER, have I ever been in so much danger than I am with the creamy candy by White Rabbit, which has been known to actually POISON people!

Suddenly, two title slides flash up on the screen:

- Banned in 2007 for Formaldehyde contamination.
- Banned AGAIN in 2008 for Melamine contamination.

> 'Truthfully', the narrator continues:
> It doesn't really look all that dangerous. It's a regular Chinese wrapping with a little white rabbit on there. But then again, I have no idea what a white rabbit means in Asia. So it could as well just mean poison!
>
> (http://youtu.be/oRm1uG7ie2k)

Despite the humour of this video, it highlights many important aspects of contemporary discourses of food risk: the prevalence of 'food scandals', the increasing sense that even the most innocuous seeming foodstuffs many contain 'hidden dangers' that it has become more difficult to interpret the information we are given about our food. It also highlights the unpredictable way consumers sometimes respond to information about food risks, in this case the narrator of this video appropriating the supposedly tainted candy to stage a mock suicide online.

In many ways, food risk is a very special kind of risk. It is a risk that we cannot avoid, since we must eat for our survival (Kjærnes, 2006). At the same time, we are increasingly loosing trust in the food we eat, partly because of the plethora of food scandals reported in the media, involving, for example, 'Mad Cow disease', salmonella in eggs, and horsemeat tainted hamburgers. Such incidents have deteriorated our trust not just in food producers and retailers, but also in the scientists and public officials that are supposed to keep us safe.

Nowhere is this more apparent than in China, where food scandals have been so widespread as to be regarded as major threat to the nation's standing in the world (Huang, 2012). Recent scandals involve noodles tainted with calligraphy ink, fake eggs, 'sewer oil' recycled for cooking oil, soy sauce made out of human hair, and milk and milk based products (like the candy referred to above) adulterated with the industrial chemical melamine (Ross, 2012; Yan, 2012). In many of these scandals not just factory workers and farmers, but also managers, entrepreneurs, and government officials have been implicated, contributing, in the words of Yan (2012: 705) 'to a rapid decline of social trust … that has far-reaching social and political ramifications'.

These issues, of course, are not unique to China. In other contexts as well, inaction or obfuscation by government and industry spokespeople, and sometimes disproportionate responses by consumers, have resulted in similar breakdowns of trust. The classic example is the British BSE scandal in which mixed messages from the government about beef infected with bovine spongiform encephalopathy caused long-term damage not just to the British beef industry but also to the public's confidence in the government (Washer, 2006).

Even when information about food risk is communicated in a clear and transparent fashion, however, the kinds of messages that finally reach consumers and how they actually apply these messages to decisions about buying and consuming food are notoriously unpredictable.

This is because communication about food is almost never a matter of straightforward transmission of information. Instead, every act of choosing what to eat occurs at the *nexus* of multiple 'itineraries' of discourse (Scollon, 2008) that form complex connections among people, objects, texts, and social practices.

Risk communication and discourse itineraries

A problem with many models of risk communication is that they assume that information about risk travels along rather unproblematic pathways from experts to the media and then to the public, who apply it in a more or less rational fashion (Beacco et al., 2012). When partial or 'inaccurate' messages reach the public, or when people apply the information in unexpected ways, this is usually blamed on media 'distortions' (Suhardja, 2009) or the failure of people to understand the information. This 'deficit model' of risk communication (Kjærnes, 2006) proposes that all we need to do is improve the quality of information reaching the public, and educate the public so they can apply that information better.

What these models ignore, first, is the fact that people rarely respond to risks based on a single stream of 'authoritative' information, but rather act on the basis of a complex mixture of many, sometimes contradictory, 'voices' coming from such sources as the mainstream media, medical professionals, the internet, friends, and family members. Second is that fact that decisions about food are rarely purely 'intellectual decisions', but are instead forms of *social action* which are performed with other social actors in the context of complex sets of social practices. Many of the choices we make about food are not made consciously, but rather have their source in long histories of habitual practices that have become part of the cultures of our families or our communities, practices which circulate in the form of stories, jokes, prohibitions, and prejudices that become sedimented in our 'historical bodies' (Nishida, 1959). Finally, this model fails to account for the ways discourse about risk is 'resemiotised' (Iedema, 2001) as it flows 'back and forth between academic experts, regulatory practitioners, interest groups, and the general public' (Leiss, 1996: 86) taking on 'diffuse intertextual forms' (Beacco et al. 2012: 280) along the way. What is needed, then, is a model that can trace these flows of discourse across texts, actions, practices, people, institutions, and objects, across time and space, and across multiple modes and materialities.

One approach that offers the potential to do this is *mediated discourse analysis* (Norris & Jones, 2005; Scollon, 2001). Mediated discourse analysis is centrally concerned with the relationship between discourse and action, in particular, what role discourse plays in making some actions more possible than others. One of the key challenges to understanding this is the fact that often discourse that plays a key role in amplifying or constraining a social action may be far removed from that action. A law prohibiting the manufacturer of a product from making a particular package claim (that the product, say, helps prevent heart disease) is not present in the supermarket when a customer chooses that product from the shelves. The law is manifested in the *absence* of the claim. The law has been *resmiotised* into the action of the manufacturer removing the claim, resulting in a new piece of discourse. This trajectory of discourse and action intersects, at that moment when the customer is deciding whether or not to choose the product, with the customer's own trajectory of discourse and action associated with the product, beginning perhaps with a blood test that showed high levels of cholesterol which was resemiotised into a decision to eat 'heart healthy' foods, which was later transformed into the action of choosing this product because of the original claim, an action that may have solidified into the habitual practice, until the moment in the supermarket when she notices that the discourse upon which this practice is based is no longer present. Of course, these are not the only trajectories of discourse and action that are circulating through this 'site of engagement' (Scollon, 2001). Others include the trajectory of ordering, stocking, and shelving the product on the part of the supermarket, the trajectory of manufacturing the product, and even trajectories of scientific research about what foods reduce the risk of heart disease.

Scollon (2008) calls these trajectories 'discourse itineraries'. The central task of mediated discourse analysis, he says, is to map these 'itineraries of relationships among text, action and the material world' using an analytical strategy known as *nexus analysis* (Scollon, 2001; Scollon & Scollon, 2004). Nexus analysis always begins with a particular 'site of engagement', such as the moment a customer decides against purchasing a particular product in the supermarket. Any site of engagement is considered a *nexus* where particular people, objects, texts, and social practices come together, each with their own "histories" – the trajectories along which they traveled to reach this moment. The analytical utility of the concept of discourse itineraries, as opposed to, say, notions of "flows of information" (see for example Beacco et al., 2012) is that it highlights how discourse might enter into a particular site of engagement in many possible forms – in the form of a 'text' such as a story or a rumour, or as a habitual practice such as the practice of only buying vegetables from a certain place, or even as an object such as a particular kind of kitchen appliance. The goal of nexus analysis is twofold: first, to understand what made this moment possible, to trace the histories of these people, objects, texts, and social practices and try to understand the conditions that led to them coming together at this particular site of engagement; and second, to understand the consequences of this nexus on human agency. To what extent does this particular momentary configuration of people, objects, texts, and social practices 'open up a window' (Scollon 2001:4) within which certain social actions are possible?

Two cautions are in order here. The first concerns the possibility of an exhaustive account of the many itineraries converging at any given moment. Even with careful ethnographic investigation, such an account is unlikely, especially since many of histories that lead up to a particular text being the way it is or a particular practice being performed the way it is are buried deep in the past and invisible to those using these texts or engaging in these practices. Such an exhaustive account, however, is not really the main goal. A more important goal is to understand something about the different processes of transformation discourse undergoes as it travels along these itineraries, processes that Scollon (2008) calls *processes of resemiotisation*. By observing these processes the analyst can uncover the basic mechanisms by which discourse about food risk gets transformed into concrete social actions and habitual social practices as it travels along multiple, intersecting discourse itineraries.

The second caution regards human agency. In one sense, nexus analysis presents a challenge to rationalist models of risk that imagine that individuals freely take actions based on the logical assessment of information. Instead it suggests that agency is *distributed* among multiple people, objects, texts, and social practices. This should not, however, be taken as a denial of the importance of 'individual choice'. The nexus of discourse itineraries that converge on a site of engagement is not determinative of what actions social actors will take: rather it provides an environment for the negotiation of risk among these various agentive entities (see also Latour, 2007).

The site of engagement

The starting point for this analysis is a stretch of conversation collected during a research study on the discourse of food in Hong Kong.[1] As part of the study, a research assistant conducted in-depth fieldwork with five families, spending a week with each family, observing meals, trips to the market, and casual conversations around food. The excerpt is part of a conversation involving a middle-aged British woman who has lived in Hong Kong for over 20 years (whom I will call Janet), her seven-year-old daughter, and a friend, who was a guest for dinner. It occurred while the woman and her daughter were preparing the meal (Excerpt 20.1). I will use this conversation as a *reference point* from which to trace the discourse itineraries that converged at this moment of talk about food

1	Mother	White Rabbit y'know th very <u>fa</u>mous [milk] sweets::?
2	Friend	[yeah]
3		(0.4) mmm hmm=
4	Mother	=uh there was a big <u>scan</u>^dal wuzit last year? (0.6)
5		they had ehhh:: RAT poison in in them? Or wuzit
6		what was in th[em?
7	Daughter	[yeah yeah [it's] rat poison=
8	Mother	[that's] = rat poison
9		to to (0.1) y'know uh sort of uh by::product <u>of</u> it [to uhum]
10	Friend	[mm hmm]
11	Mother	to make them whi::te (0.2) and to sort of give them elas[ticity?
12	Friend	[mm hmm
13	Mother	(0.3) so <u>those</u> definitely go straight in the bin if they come back
14		[with them]
15	Daughter	[yeah so::] it was like Hallo[<u>ween</u>
16	Mother	[I took my:: (0.1) I took my <u>friend</u>
17		my friend teaches in a public eh <u>pri</u>vate school in England and
18		many of the students are from Hong Kong (0.1) and China=
19	Friend	=really?
20	Mother	en she said (.) I told her and she said (.) oh eh (.) he (.) Chinese
21		New Year of course they eh that's <u>ba</u>sically what the students [had
22	Friend	[mm hmm
23	Mother	they would share out these White R[a b b i t s]
24	Friend	[hm hm hm]
25	Mother	just be <u>care</u>^ful=
26	Daughter	=and [em
27	Mother	[coz they were banned in the Phillipines
28		which is why I was sending Farah the article first [of all] actually
29	Friend	[oh wow]
30	Mother	it just makes you shocked (.) cos [these sweets] are <u>really</u> popular
31	Daughter	[I rememba ah: ah
32	Friend	[yeah
33	Mother	here^
34		(1.6)
35	Daughter	I just rememba:: (.) [It was like (.3)] hhh I rememba it was
36	Mother	[It was a learning experience]
37	Daughter	like Halloween and em (0.2) me n mama jus went shopping and
38		got a <u>whole</u> bag of them (0.1) hhh n then (???) showed us the
39		<u>arti</u>[cle
40	Mother	[ha=
41	Friends	=eh [heh] [heh]
42	Daughter	[I jus go like <u>in</u> the bin=
43	Friend	eh he <u>he</u>:
44	Daughter	[in the bin]
45	Mother	[n when you brought them] in your trick or treat we just
46	Daughter	[yeah]
47	Mother	took them out=
48	Daughter	= we just took all them out and (0.8) yeah (.) and (.)
49		it's just (0.2) I used to always like (.) <u>love</u> them and <u>eat</u> them (0.4)
50		But (1.0) they were like <u>nice</u> and (0.8) why would you add that
51		into something like that when (0.8) h. children enjoy eating it (.)
52		and it's just really no point so=
53	Friend	=hmmm

Excerpt 20.1

risk, itineraries that involve not just the history of Janet and her family, but also itineraries involving the food she is discussing, White Rabbit Candy, its involvement in various food scandals and how those scandals were responded to by government agencies, reported in the press and discussed in public. Therefore, along with this conversation, I will also use as data other interviews with Janet and her daughter, interviews with volunteers from Hong Kong and China about their experiences with White Rabbit Candy, a collection of news reports and material from blogs and internet forums.

It is useful to point out at the outset that there are some inaccuracies in the information that Janet is presenting as the basis for her decisions about this product. One inaccuracy is the statement that 'rat poison' was found in White Rabbit Candy. Instead, the dangerous substance usually associated with the candy is melamine, an industrial chemical. This inaccuracy might be something of a concern for those who take an 'information deficit' approach to risk communication or worry about how ordinary people often 'distort' information about food risks. From a mediated discourse perspective, however, our main concern is whether or not this 'distortion' has actually changed the action Janet takes (would she be more comfortable eating an industrial chemical?), as well as whether or not 'rat poison' has any particular utility in terms of what she is trying to do in this conversation

The questions I will attempt to answer, then, have less to do with whether or not Janet's characterisation of the health risks involved in eating this candy are accurate or whether or not the family's risk reduction strategy is effective, and more to do with what this conversation tells us about the *processes of resemiotisation* that various texts and actions related to White Rabbit Candy have undergone on their pathways to this moment in Janet's kitchen.

The *processes of resemiotisation* I will focus on are:

Certification/authorisation: processes through which an authoritative person or institution issues a statement about the safety or danger of a particular product, or through which a particular piece of information or practice is verified based on its association with a recognised authority (e.g. 'They were banned in the Philippines ...').

Iconisation/stigmatisation: when a particular brand or kind of food comes to be regarded as a symbol of either quality/safety or lack of quality/danger (e.g. 'Y' know the very famous milk sweets?).

Metonymisation: The process of attributing the safety or danger of a product to an ingredient or part of the product, or of attributing the safety or danger of a product to a larger class of products that it belongs to (e.g. *'It just goes to show that you can't trust ANYTHING from China anymore'*). Related to this is the process of substitution, when one attribute of safety or danger is substituted for another (e.g. *'They had RAT poison in them.'*)

Narrativisation: When certain aspects of quality/safety or danger or certain practices associated with risk are transformed into stories that circulate through social groups.

Practice: when concrete social actions are submerged into the historical bodies of those who perform them, becoming 'practices' that are not just repeated but can also themselves be narrativised, authorised, and iconised (e.g. 'in the bin!').

Suicide Candy

Food scandals involve the delegitimation of some food product that was previously regarded as safe. This process of delegitimation depends crucially on some previous process of *legitimation*. In the transcript above, what makes the scandal involving White Rabbit Candy particularly salient is the popularity of the brand, which, as Janet remarks, formed an important part of celebrating Chinese New Year for students studying at the school her friend works at. So one important set of discourse itineraries that circulate through this conversation are what might be called *itineraries of legitimation* in which White Rabbit Candy has come to be associated with certain expectations of quality.

These itineraries can be traced back to 1943 when the ABC candy factory in Shanghai began to manufacture a milk candy named Mickey Mouse Candy. In the anti-Western political climate after the Communist victory, the candy was renamed White Rabbit, and quickly became a staple at Chinese New Year gatherings. With the reform and opening up of China beginning in the 1980s, the brand began extending its reach beyond China, first to other Asian countries and then to Europe, North America, and Australia.

The rapid *inconisation* of White Rabbit Candy came as the result of the intersection of the company's itineraries of manufacturing and marketing with a number of other discourse itineraries involving the rise of China as a political and economic power. For example, in 1959, the Chinese government distributed the candies as official souvenirs of the Tenth Anniversary of the People's Republic, and in 1972 President Nixon was presented with a package on the occasion of his historic visit to China. These actions constituted forms of *certification*, the recognition by authorities of the candy's quality and its status as a symbol of

Chinese economic progress. They also demonstrate how the intersection of discourse itineraries can affect the future trajectories of these itineraries. That moment, for example, when President Nixon was handed a bag of White Rabbit Candy not only affected the trajectory of US China relations, but also the trajectory of White Rabbit as a famous brand. These implicit acts of certification were made explicit in the early 2000s when the Ministry of Commerce conferred upon White Rabbit Candy the designation of 'time honoured brand' (中华老字号), given to brands which the government deems symbolic of Chinese quality ingenuity.

Another important itinerary White Rabbit Candy intersected within the reform and opening-up period was the growth of the concept of 'quality' (*suzhi*) as an important cultural category. Here I am not referring just to the 'quality' of the product, but to a broader discourse that developed as China embraced modernisation, urbanisation and capitalism. The term subsumes notions of 'civilisation', 'taste', 'etiquette', the 'quality' of consumer products, the quality of consumers themselves, and perhaps most importantly, the 'quality of the population' both bodily and spiritually (Anagnost, 2004; Jones, 2010). Interestingly, the primary intersection between White Rabbit Candy and this powerful cultural discourse itinerary occurred due to its association with milk. In the late eighties, milk began to be regarded in China as a symbol of modernity, westernization, good nutrition, and, most of all, 'quality'. Between 1980 and 2007, milk consumption in China increased from one million metric tons to 35 million metric tons (Ross, 2012). The *iconisation* of milk was also helped along by various *certifying* statements from government officials, such as Premier Wen Jiabao 2006 declaration: 'My dream is for every Chinese person especially children to be able to drink 500 ml of milk a day' ('Drink milk...' 2006).

White Rabbit Candy benefited from this association of milk with the discourse of 'quality' through the process of *metonymisation*, by which the candy came to be seen in terms of one of its key ingredients: whole milk powder. The company, in fact, encouraged this association though an advertising campaign in the mid-1980s that claimed that seven White Rabbit candies is equivalent to a glass of milk ('七颗大白兔奶糖等于一杯牛奶'), leading some at that time to adopt the practice of melting the candy in hot water and drinking it. Ironically it was this entanglement of the manufacturing and marketing itineraries of White Rabbit Candy with those of the Chinese dairy industry that was responsible for the scandal the candy became embroiled in in 2008, and which, led Janet and her daughter (two years later) to relegate the candy to the 'bin'. Also ironic is the fact that it was the

Chinese dairy industry's entanglement with the discourse of 'quality' that led to the massive demand for the milk that created the conditions for its adulteration (see below).

Of course none of these processes could have resulted in the *iconisation* of White Rabbit Candy had these itineraries not intersected with the itineraries of the everyday lives of consumers – itineraries of parenting, of celebrating festivals, and of rewarding students in school – where the meaning of the brand underwent further transformations through the processes of *narrativisation* and *practice*. Nearly all of the Chinese informants I talked to related fond childhood memories associated with White Rabbit Candy, often associated with Chinese New Year. One 26-year-old informant recalled, 'my family bought it a lot during Chinese New Year. I used to always get it with my *Lai See* (red envelopes containing money)'. An older informant (aged 61), who had been a student in the 1980s, said, 'It was the best milk candy back then ... sometimes we would dissolve it in our dormitories to make milk.' Another, aged 22 related a story of how her mother used the candy to make medicine more palatable

> I was always sick and had Chinese medicine almost every day. Mother would put a White Rabbit Candy in a bowl of black Chinese medicine and say, 'good girl, after finishing this the candy is yours.' Then I'd hold my tears and finish the medicine. If I finished the medicine all at once, mother would give me the White Rabbit Candy as a reward.

Although in the conversation in Janet's kitchen there is no mention of Richard Nixon or the 'discourse of quality', or the practice of students melting pieces of White Rabbit Candy in their dormitories to make milk, in a sense all of these itineraries of discourse and practice have converged to make this conversation possible. Their convergence is what makes Janet and her daughter regard the candy as 'famous', what makes them buy a bag of it in preparation for Halloween, and, of course, what heightens their horror when they find that this presumably reliable brand of candy may be, in fact, poison.

Tangled as the itinerates contributing to the legitimation or *iconisation* of White Rabbit Candy may appear to be, those associated with its *stigmatisation* are equally complex, implicating the brand in at least three major scandals.

The first scandal that affected White Rabbit Candy had nothing directly to do with the candy, but ended up tarnishing the name of Chinese exports in general and pre-figuring a problem with the Chinese food supply that was later to impact the candy. In 2007 several brands of pet food containing ingredients imported from China were recalled in the United States, Europe, and South Africa after dogs and cats started appearing at veterinary clinics with renal failure. At first it was believed that the pet food contained Aminopterin, a kind of rat poison, but it was later found to contain melamine, an industrial chemical that had been added to wheat and rice gluten to increase its apparent protein levels. This scandal is important for several reasons, the most important being that it was the same widespread use of melamine by Chinese farmers and food manufacturers that led to the 2008 milk scandal that directly affected White Rabbit Candy. It was also the beginning of a growing international distrust of Chinese food products (Cai et al., 2009). Finally, the initial suspicions that the contaminant was rat poison, and the circulation of that rumour in the media, may explain why Janet identifies rat poison as the adulterant in White Rabbit Candy in the conversation above. In fact, it is not unusual to find in media reports and internet blog post about the later milk scandal melamine described as a kind of rat poison (see for example 'Watch out for candy...', 2008); it is not.

The second scandal that likely affected Janet's judgment of White Rabbit Candy involved allegations in the summer of 2007 by the Philippines Bureau of Food and Drugs that White Rabbit Candy contained formalin. Independent tests by both a Swiss company and the Singapore government failed to find any contamination, and the manufacturer of White Rabbit claimed that the likely culprit was counterfeit White Rabbit Candy. The intersection of this scandal with Janet's conversation might be responsible for her mention of the Philippines as the country that had banned the candy, though this particular scandal did not result in any conclusive findings regarding the danger of White Rabbit Candy.

The most damaging scandal in which the candy was implicated was the 2008 Chinese milk scandal, which began when children in China started developing kidney stones at an alarming rate. The cause was found to be the contamination of milk powder with melamine. At least 300,000 Chinese children were affected, and at least six died (Ross, 2012). Although the government and dairy industry initially reported that the contamination was limited to supplies from small companies, it soon became apparent that some of the country's largest and most respected brands, including products from *Sanlu*, the nation's largest dairy company, had been adulterated. The revelations precipitated moves to ban Chinese milk and items containing milk (like White Rabbit Candy) throughout the world. Inside of China the scandal resulted in the firing and jailing of public officials and the execution of at least two people (Yang, 2013).

It would be impossible to untangle all of the different itineraries of discourse and action involved in this scandal. Some have suggested, for example, that government authorities may have suppressed information about the tainted milk to protect the country's image in the run up to China's hosting of the 2008 Olympics (Reporters Without Borders, 2008). As I said above, another itinerary, the one that led milk to be associated with 'quality' in the 1980s and 1990s, likely created the conditions for its adulteration: under the pressure of increased consumer demand, dairy companies and the small farmers who supplied them took to watering down their milk and then 'fortifying' it with melamine to make it appear to be higher in protein (Kuehn, 2009). Other itineraries of legitimation that might have contributed to the scandal include the one that resulted in a *certification* of the *Sanlu* company whereby it was granted a quality inspection and quarantine waiver. This is one reason why products from the company went untested for so long.

Although milk powder is one of the main ingredients in White Rabbit Candy, the extent to which the product was actually contaminated is far from clear. Many retailers began removing the candy from their shelves before any tests had shown melamine contamination, simply because of the product's strong association with milk (Kelleher, 2008; 'Chinese sweets axed...', 2008). On September 24, the Hong Kong Centre for Food Safety released the results of tests showing that White Rabbit Candy had been found to contain melamine. Similar findings were released in Australia, Singapore, and New Zealand, but did not in all cases result in recalls due to the belief that the amounts of melamine posed only a minor health risk. A news report in Singapore (Neo and Tan, 2008) estimated that 'a 60kg adult ... would have to eat more than 47 White Rabbit sweets ... every day over a lifetime to exceed the tolerable threshold,' (an interesting contrast to the company's previous claim that seven White Rabbit Candies are the equivalent of one glass of milk). Nevertheless, most international distributors recalled the candy, and the company stopped exporting it for several months until 2009 when it resumed the export of candy, claiming that now it was made only with milk from Australia and New Zealand.

The apparent safety of the White Rabbit Candy which was back on the market at the time of this conversation in Janet's kitchen attests to the difficulty food manufacturers have in recovering from the stigmatising effects of food scandals, effects that are driven by processes of *metonymisation, narrativisation,* and *practice.*

Metonymisation, as I said before, is a process by which a food product either comes to be associated with one ingredient in the product, or it becomes defined by its membership of a broader class of food products. This process is obvious along many of the itineraries associated with White Rabbit Candy, including regulatory and retail itineraries in which the candy was singled out for testing or even removed from the shelves without testing because of its association with milk. Ironically, this was an association that the manufacturer strongly encouraged before the scandal, and was partly responsible for the candy's iconic status. A similar process can be seen in the Candy's stigmatisation because of its membership of a larger class of products, specifically products from China (a form of *metonymisation* we can refer to as *generalisation*). Such reactions are common in the data I collected both from informants and from the internet. One writer on the Candy Blog, for example, commented:

> I couldn't believe it when I saw that there is melamine in White Rabbits! I used to love these candies when I was a kid. I guess it just goes to show that you can't trust ANYTHING from China anymore.

Similarly, when asked if they ate White Rabbit Candy, several Hong Kong informants remarked that they tried to avoid eating foods imported from China altogether. One Hong Kong student said:

> I try my best to avoid eating anything from China, partly because of all the bad news about the food in China, and partly because I'm in Hong Kong, and we have lots of good alternatives here.

When these Hong Kong students make such statements, it is difficult not to infer an intersection between the history of White Rabbit Candy and the Chinese melamine scandal and the itineraries of recent political developments in Hong Kong involving both concerns about political development of the former British colony, and growing discontent about Chinese tourists who have taken to buying large quantities of milk powder in Hong Kong for resale on the mainland, which has led to higher prices and shortages of milk powder in the territory.

This process of generalisation is also evident in Janet's household. In interviews she repeatedly told us that she avoids buying food from China, and, speaking about another product – potato chips – her daughter told an interviewer:

> I wouldn't get those y'know the Chinese ones because I don't really feel comfortable with them.

From the point of view of families and individuals, though, perhaps the most potent process contributing to the stigmatisation of food products is *narrativisation*, the process by which perceived dangers are transformed into 'horror stories', which circulate through social groups. Above we saw how important personal narratives were in creating and sustaining brand loyalty to White Rabbit Candy. In the conversation in Janet's kitchen we can see how narratives can be equally powerful tools in stigmatising products and sustaining that stigmatisation over time. Narratives themselves, however, also form at the nexus of multiple discourse itineraries, often combining elements from actual events, hearsay, media reports, and 'cultural storylines' (myths, folktales, etc.).

The transcript above contains two narratives, both of them portraying moments of 'narrow escape' from the dangers of eating White Rabbit Candy. The first involves Janet warning a friend of hers who teaches overseas Chinese students in the UK to 'be careful' and sending her an 'article' voicing her concerns, taking on herself the role of issuing a kind of 'food safety warning' not unlike those issued by the governments that banned the sale of White Rabbit Candy.

A much more interesting story, however, is told by her daughter, involving how the family had bought a bag of White Rabbit Candy to give out at Halloween, only to be alerted to the danger of the candy (via what was presumably the same 'article'), and immediately responding by throwing the candy 'in the bin'. This is followed by the mother and daughter co-narrating a related account of how, when the daughter returned with her own trick or treat candy, they search through it for White Rabbit Candy and 'took them out'. What is so interesting about this story is the way it intersects with the itineraries of narratives about poisoned Halloween candy which have been circulating in the media and among parents for decades, stories which have little basis in fact, but occasionally intersect with the itineraries of actual product poisonings (such as the 1982 Tylenol poisonings — see Jones 2013), and actual cases of the adulteration of candy (like the 2008 contamination of White Rabbit Candy by melamine). In her book, *Candy: A century of panic and pleasure*, Samira Kawash (2013) discusses the persistence of tales of poisoned Halloween candy, tracing them back to the mid-seventies when changing economic and social norms (racial integration, gender equality) precipitated an increase in societal fears about the safety of children and the danger of strangers. By linking the itinerary of the adulteration of White Rabbit Candy with the persistent cultural storyline of the Halloween poisoner (whose favourite weapon is, of course, 'RAT poison'), Janet and her daughter have created a discursive artifact that is likely to sustain the family's stigmatisation of White Rabbit Candy for some time to come.

Of course, the most powerful process of stigmatisation occurs when discourse is transformed first into concrete social actions (such as deciding not to eat a particular product) and then into more durable social practices. The practice of avoiding White Rabbit Candy depends on the intersection of the candy's own scandal-marred history with itineraries of social practices that began in this family long before the safety of White Rabbit Candy was called into question. One such practice is Janet's practice of paying attention to media reports about food safety. On several occasions during the study she referred to 'articles' she had read which had led her to 'ban' certain foods from her home. In a group interview with participants from other families, for example, she said:

> There's one item on that table that's absolutely <u>banned</u> from my house that ... that's the Burger King crisps because I read ... I read an article that said they found something **ter**rible in them ...

Practices are chains of actions that are not just repeatedly performed, but also which come to take on 'a life of their own', becoming 'objects' that can be referred to with labels or 'catch phrases'. Clearly the practice of policing the food items her children bring home and relegating unacceptable items to 'the bin' has become, for Janet and her family, a recognisable practice. In fact, the phrase 'in the bin' was one Janet used often when talking about food safety. She used the term, for example, in an earlier interview when discussing whether she let her children drink soda pop ('If they dare to bring home a can of Coke it goes right in the bin'), and it is also a phrase she uses at the beginning of the transcript of this interaction ('so those definitely go straight in the bin if they come back with them').

Social practices are often the end result of processes of *certification/ authorisation*, *inconisation/stigmatisation*, *metonymisation*, and *narrativisation*, and they can also be sustained by these processes. Janet and her daughter's story about throwing the White Rabbit Candy 'in the bin', told and retold to friends and other family members, helps to sustain the practice of throwing 'dangerous' food 'in the bin' among the members of this family. In fact, by telling this story, the daughter demonstrates the extent to which she has been socialised into this practice, and the gesture she makes while uttering this phrase (see Figure 20.1) shows the extent to which the practice has come to be submerged into her historical body. In fact, one can imagine this gesture regularly

Figure 20.1 'In the bin'

accompanying this phrase when it is uttered in this family as a kind of icon of this social practice. So just as White Rabbit Candy has become *stigmatised*, the *practice* of avoiding it has become *iconised*.

Conclusion

It is not hard to see that the complex intersection of discourse itineraries that result in Janet and her daughter throwing White Rabbit Candy 'in the bin' does not much resemble the model of rational decision making that many theories of health communication promote. That is not to say, however, that Janet's practices of managing food risk are not 'rational'. Rather they follow a logic much more complex than that imagined by health promoters and regulatory bodies, one which involves negotiating the complex interaction of discourses, practices, objects, and people that come together at different sites of engagement. One might object that maybe Janet is being too harsh on White Rabbit Candy, that the business about rat poison is a bit over the top, and after all, as far as we know, White Rabbit Candy manufactured at the time of this conversation was melamine free. But such observations tell us less about the appropriateness of Janet's response and more about the challenge food companies face when they also have to negotiate the complex intertwining of multiple discourse itineraries characterised by the same processes of *inconisation* and *stigmatisation*.

It would also be a mistake to assume that processes of stigmatisation like those catalogued here lead naturally to people avoiding certain foods. The process of generalisation observed above, for example, which resulted in some participants avoiding food from China, for other participants led to a more resigned attitude. One student from Mainland China, for example, declared: 'Most of the Chinese food is kind of poisonous, and I've been exposed to it since birth, so I'm immune to it.' Similarly a poster on the Candy Blog remarked: 'Face it, what DOESN'T cause cancer these days ...' Finally, there are even cases where *stigmatisation* becomes the very reason to eat a certain food, as in the parody video I discussed at the beginning of this chapter.

The complex and contradictory discourses that coalesce around food scandals provide analysts with unique opportunities to explore how social practices around food risk develop. What this analysis highlights is how difficult it is to posit any direct relationship between 'official warnings' or media reports and people's actual behaviour, since that behaviour always takes place at the intersection of multiple itineraries of discourse and action. At the same time, it also shows how individuals operate as active agents at these sites of engagement, appropriating and mixing discourses from different itineraries and subjecting them to their own processes of *iconisation/stigmatisation, authorisation/certification, metonymisation* and *narrativisation*, and how these processes lead to the transformation of these discourses into durable practices that become part of their unconscious and habituated behaviour around food risk.

Note

1. This research was made possible by a grant from the General Research Fund of the Hong Kong Research Grants Council Grant # CityU 144110.

References

Anagonst, A. (2004). The corporeal politics of quality (Suzhi). *Public Culture, 16*(2), 189–208.
Beacco, J.-C., Claudel, C., Doury, M., Petit, G., & Reboul-Touré, S. (2002). Science in media and social discourse: new channels of communication, new linguistic forms. *Discourse Studies, 4*(3), 277–300.
Cai, P., Ting, L. P., & Pang, A. (2009). Managing a nation's image during crisis: A study of the Chinese government's image repair efforts in the 'Made in China' controversy. *Public Relations Review, 35*(3), 213–218.
'Chinese sweets axed in milk scare' (September 24, 2008). *Daily Express*. Retrieved March 5, 2015, from http://www.express.co.uk/news/uk/62961/Chinese-sweets-axed-in-milk-scare.
'Drink milk everyday to benefit your health' (May 27, 2006). Xinhua. Accessed March 3, 2015 from http://www.xinhuanet.com/xhft/20060527/wzsl.htm (in Chinese).
Huang, Y. (August 17, 2012). China's corrupt food chain. *The New York Times*. Retrieved January 2, 2015 from http://www.nytimes.com/2012/08/18/opinion/chinas-corrupt-food-chain.html.
Iedema, R. (2001). Resemiotization. *Semiotica, 137*(1–4), 23–39.
Jones, R. H. (2007). Imagined comrades and imaginary protections: Identity, community and sexual risk among men who have sex with men in China. *Journal of Homosexuality, 53*(3), 83–115.
Jones, R. H. (2013). *Health and risk communication: An applied linguistic perspective*. London: Routledge.
Kelleher, S. (September 25, 2008). 'China's milk scandal has Seattle-area stores pulling candy, drinks'. *The Seattle Times*. Retrieved March 5, 2015, from http://www.seattletimes.com/seattle-news/chinas-milk-scandal-has-seattle-area-stores-pulling-candy-drinks/.
Kjærnes, U. (2006). Trust and distrust: Cognitive decisions or social relations? *Journal of Risk Research, 9*(8), 911–932.
Kuehn, B.M. (2009). Melamine scandals highlight hazards of increasingly globalized food chain. *JAMA, 301*(5), 473–475. doi:10.1001/jama.2009.35
Latour, B. (2007). *Reassembling the social: An introduction to actor-network-theory*. Oxford; New York: Oxford University Press.

Leiss, W. (1996). Three phases in the evolution of risk communication practice. *Annals of the American Academy of Political and Social Science, 545,* 85–94.

Nishida, K. (1959). *Intelligibility and the philosophy of nothingness.* Tokyo: Maruzen Co. Ltd.

Neo, C. C., & Tan H. L. (September 22, 2008). Crying over spilt milk. *Today Online.* Retrieved on January 15, 2015 from https://docs.google.com/viewer?url=http%3A%2F%2Fwww.nuh.com.sg%2Fwbn%2Fslot%2Fu1753%2FPatients%2520and%2520Visitors%2FMedia%2520Articles%2FSep%252008%2F24th%2520TODAY.pdf

Norris, S., & Jones, R. H. (2005). *Discourse in action: Introducing mediated discourse analysis.* London: Routledge.

Reporters Without Borders (October 2, 2008). Open letter to Margaret Chan, WHO director, about the contaminated milk powder scandal, Retrieved March 26, 2013 from http://en.rsf.org/china-open-letter-to-margaret-chan-who-02-10-2008,28791.html.

Ross, K. (2012). Faking it: Food quality in China' *IJAPS, 8,* 33–54.

Scollon, R. (2001). *Mediated discourse: The nexus of practice.* London: Routledge.

Scollon, R., & Scollon, S. W. (2004). *Nexus analysis: Discourse and the emerging internet.* London: Routledge.

Scollon, R. (2008). Discourse itineraries: Nine processes of resemiotization. In V. K. Bahtia, J. Flowerdew & R. H. Jones (Eds.), *Advances in discourse studies* (pp. 233–244). London: Routledge.

Suhardja, I. (2009). *The Discourse of 'distortion' and health and medical news reports: A genre analysis perspective.* Unpublished PhD. dissertation, Edinburgh: University of Edinburgh.

Washer, P. (2006). Representations of mad cow disease. *Social Science & Medicine, 62*(2), 457–466.

'Watch out for candy made in China this Halloween' (October, 2008). *Harmonix Forum.* Retrieved February 25, 2015 from http://forums.harmonixmusic.com/discussion/92764/watch-out-for-candy-made-in-china-this-halloween.

Yan, Y. (2012). Food safety and social risk in contemporary China. *The Journal of Asian Studies, 71*(3), 705–729.

Yang, G. (2013). Contesting food safety in the Chinese media: Between hegemony and counter-hegemony. *The China Quarterly, 214,* 337–355.

Index

9/11 86–87
  Commission 86

absolute risk
  of a cardiovascular event 285
abstracts 212
account 310, 315, 324
accountability 6, 11, 57, 68, 78, 172, 318
  government 318, 325
  public 323
  mechanisms 172
accountable 334, 323
accounts 56
actuarial justice 88–91
  exclusion 89
  resistance to 88
  sanctions 88–89
  sentencing 88–90
actuarial risk 145
adaptation 21, 217, 223
adolescent transition 109
adolescent transition and unemployment 109
aged care 52
alignment 193, 299
  as evidence of concordance 299
analyst's paradox 3
anthropology 171, 172
appraisal 10, 248
appropriate level of protection 237
Asian Financial Crisis aftermath 311
Asian Financial Crisis quelled 309
assessment 1, 4, 7, 27, 47, 52, 67, 72, 108, 124, 128, 215, 324, 327
attribution 6, 18, 23, 26, 28, 99, 245, 249
audiences publiques 198

Bayesian probability 211
Beck, Ulrich 1, 2, 8, 85, 99, 149, 156, 172, 173, 175, 182, 184, 245, 307, 324

biographical narratives 157, 160
biosecurity 1, 5, 11, 229, 234, 238
  Australian context 229–232
  discourse 232, 235
  risks 234–237
  communication 235
  advice 237, 239
blame 6, 24, 26, 28, 30, 49, 99, 115, 76, 78, 127, 149, 155, 263, 273, 307, 311, 314, 320, 324,
  allocation 172, 180
  avoiding 49
  shifted to third parties 311, 317, 318, 320
body orientation 288
bystander to interaction 287, 290

caesarean section 72–73, 79
Canada 189
cardiovascular risk communication of 286
case study 190
categorisation 5, 19, 54, 67, 106, 251, 327
CCTV 929
central banks
  'open mouth' operations 309
  activism 315
  and monetary policy 309
  buffer role 307, 311
  central-bank-speak 309
  inaction 308
  managing national economies 311
  mythical status 309
  open market operations 308
  reputation 313
  risk management role 308–309, 311
certification authorisation 346, 347, 348, 351, 353
child protection 7, 139–140
  child protection meeting 140
  child protection enquiry 143

childbirth 68–70, 79–80
  natural 70, 80
  uncertainty 79–80
children's mental health issues
  depression 267
choice 5, 7, 69–72, 75–77, 79–81, 328, 332, 342
  ethical 213
  freedom of 71
  individual 144, 344
  informed 69, 71, 75
  maternal 86
  rational 121
CLAWS4 247
climate sensitivity 209
coding 212
communicative expertise
  nature of 63, 64, 287, 295, 301
  of doctors 287, 295
communities 41, 90, 99, 105, 108, 176, 178, 180–184, 199, 212, 278, 342
  expert 219
  indigenous 8, 178
  local 171, 179, 189
  working class 112
  local conflict 177
community of fate 326
community relations 176, 178, 181
  community relations management systems 171
computer
  as participant 288
  in the consultation 286, 287
concordance 299
conflict cost of 177, 178
congressional hearing
  2008 hearing into financial crisis 310, 314
  and 'risk society' 320
  as discursive event 313
  as genre 315
  strategic behaviour 314
  transcript 315
consent
  free, prior and informed 179, 183–185
  informed 69–72, 77, 179

contingency plans 175, 333
corporate culture 176, 182, 184
Corporate Social Responsibility 171, 178, 180
corpus 245, 250, 258, 259, 261, 323, 328, 333
  corpus annotation 247
  corpus-based study 245
credibility 29, 172, 173, 197, 202, 279, 329
  empirical 331
  source 197
credit crunch 307
crime prevention 91–99
  police transformation 92–93
  situational 91–92
crisis 178, 180, 308, 321
  aged care 52
  financial 8, 307, 310, 314, 318
  global 317
  in confidence 203
  management 179
  psycho-social 200
  situation 178, 180
  subprime 308
Critical Discourse Analysis 5, 234
critical moments 3, 278
cross-cultural communication 8, 178
crucial sites 1–5, 11
cultural bias 182

danger 1, 43, 73, 79, 94, 98, 114, 138, 158, 325, 352
  hidden 340
dangerisation 108
dangerous 138, 145, 148, 172, 215, 340
  behaviours 37, 41, 94, 134
  climate 215
  food 354
  substance 346
debate 100, 132, 156, 164, 189, 196, 198, 200, 209, 220
  climate change 215, 226
  policy 225, 235
  public 197
  risk 164, 189, 196, 202
  social science 212

decision making 1, 7, 78, 88, 90, 138, 140, 141, 145, 148, 183, 190, 194, 198, 204, 213, 221, 311, 323, 325
delegitimisation 347
deliberation facilitated 197
democracy 196, 200
dialectical discourse 190, 193, 195, 199, 203
dialogic space 248, 249, 256, 257, 259, 262
disclosure 3, 6, 17, 20, 24, 26, 32, 37–38, 43, 46
  conversations 30, 31
  training 29
discourse 2, 4, 6, 31, 67, 87, 108, 122, 127, 131, 139, .141, 142, 147, 150, 156, 159, 162, 167, 171, 192, 229, 246, 250, 265, 270, 301, 325, 328
  anthropological 171
  choice 70
  medicalisation 269
  safety 95
  social performance 184
  tipping point 219
  mediated 342, 353, 346
  post risk 88
  trust 320
  workplace 3
discourse analysis 5, 67–69, 229, 232, 273, 296
  ethnographic 67, 68
  methods 68, 69
discourse itineraries 341, 342, 343, 344
discourse of quality 348, 349
discourse strategy 324
discourse trajectory 343
discourses
  competing, conflicting 144
  midwifery 69–74
  national 69–71
discursive action 40
discursive approach 245
discursive event state-produced 313
discursive work of doctor 287
distribution 5
distrust institutional 195

diversity linguistic and cultural 51
Douglas, Mary 172, 180, 181, 327

ecological risk 324
ecological modernisation narratives of 157
ecological validity 4
economics 212, 217, 223
economy making/remaking 111, 218, 230, 313, 315, 320
ecospeak 86, 160, 203, 285, 290,
electronic health record 285, 290
elicitation 86, 160
emplotting 57
empowerment 69–72, 143, 182, 199
energy development 189, 198
engagement
  civic 192
engagement 3, 6, 9, 11, 115, 122, 155, 232, 248, 250
environmental risk
  awareness of 158, 160
  discursively grounded empirical approach to 157, 163
  knowledges of 158, 165
  locally situated 157
  perceptions 155 &156, 158–9, 162, 164, 165, 166
  psychosocial approach to 159, 167
  social construction of 156, 157, 163
  study of 157, 162
ethnography 5, 67, 292, 299
extractive industries 171, 176, 182

families 31, 36, 104, 138, 146, 138, 342, 353
feminist STS 163–4
fields 2, 115, 161, 169, 232, 234
  prevention 91
  youth justice 126
  academic 156
  conceptual 165
  interdisciplinary 220
  international travel 323
financial markets
  attitude to risk 310
  expansion of 310
  hazards of 307
  stability of 308

Index 361

financial stability
    as story 320
    maintaining 308
    stress-testing 311
fines 97–98
floods 218
focal theme 11
focus groups 157, 162, 164
food risk 340, 341, 344, 346
food safety 245, 246, 247, 254, 255, 261, 263, 264, 265,
forensic mental health 39
formulations
    visual 286, 289
frame
    country risk 323, 328
    risk situation 324, 327, 328, 329, 332, 333, 334, 336
    perception 215, 219
frames 8, 30, 55, 115, 209, 212, 222, 324, 332
framework 93, 144, 164, 190, 210, 225, 230
    conceptual 163
    intervention 122
    meta-analytic 105
    open disclosure 21
    practice 104
    preventative 88
    regulatory 5, 319
    scientific 233

gaze
    direction 298
    diverting of 294
    into middle distance 291
    redirecting of 294, 295, 297, 298
    shifts in 294
    sustained 294, 298
Giddens, Anthony 1, 56, 171, 172
global
    risk 52, 155, 167, 184, 229, 308, 314, 317, 323
global financial crisis
    aftermath 307
    congressional hearing 314, 316, 317
    prevention 319
Good Lives Model 147

governance 5, 85, 92, 99, 157, 189, 220, 221,
government
    Australian 329.330
governmentality 157, 324
governments
    attitude to risk 311, 324, 336
    contrast with markets 321
    financial reputation of 313
Greenspan, Alan 310, 312, 313, 314, 315, 317, 318, 320
grievance mechanisms 183

hard news 245, 247, 263
healthcare 53
helping families 147–149
Historical body 342, 347, 354
human rights 144, 146, 148, 171, 177, 183, 184, 281

iatrogenic 73–74, 78–79
iconisation 346, 348, 349
identities
    localised 157, 160, 325
identity
    discredited/discreditable 37
import risk analysis 230
incident disclosure
    apology 21, 23
    blame 24, 26, 28, 30
    clinical risk 30
    ethical risk 18
    financial risk reduction 18
    legal reform 18
    legal risk 18, 19, 20, 21, 24, 29, 30, 31, 32
    models 18
    openness 18, 31, 32
    patient expectations 18, 23, 24, 25, 26, 32
    policy development 18, 21
    practice 18, 20
    procedures 18, 20
    psychological risk 18, 19
    reputational risk 18
    research 18
    responsibility 18, 19, 20, 23, 24, 25, 26, 27, 28, 29, 30, 31, 32
    responsiveness 18

restorative justice 18, 32
training 18, 29
inclusive pronoun form
 use of 294
indigenous peoples 178, 179, 183, 185
individualisation 111, 122, 127, 156, 324
individualisation of responsibility
 for risk 156–157, 159, 160
institutional
 framing 323
 member 325
 performance 325
 practice 264, 294
 ratified 328
insurance 218
interaction
 active participant in 294
 bystander to 287, 290
 collaborative and
  co-constructed 286, 301
 disengagement from 292, 294
 ratified participant in 292
 triadic engagement in 288
intercultural communication 53
International Financial Architecture
 (IFA) and central banks 309
interpenetrating contexts 4
interpretive risk research 155, 157, 158, 163, 165, 166, 167
interventionism 122, 130
IPCC 215

John Ruggie 177
joint problematisation 9, 234
judgement 42, 53, 62
 intuitive 324
 probability 144

late modern society 156, 163–4, 166
legitimacy 81, 105, 173
 legitimate decisions 138, 141
legitimisation 347, 349
life-world
 of patient 299
 perceptions 287, 300
likelihood 28, 79, 124, 133, 171, 174, 192, 210, 221
local methods 54

manual handling 58
mass preventive justice 94–100
 and individual justice 99
 drink driving 94–95
 on the spot fines 97
 resistance to 97–99
 safety cameras 95–96
 speeding 94–96
Massachusetts 189
media 5, 11, 22, 93, 98, 121, 139, 148, 157, 178, 190, 199, 204, 245, 265, 275, 310, 335, 356
Mediated Discourse Analysis 342
mediation 5, 196, 202, 246
medicalisation 72–75, 79–81, 269
Megan's Law 90
megaproject 177
mental health disclosure 46
mental health in children
 Attention Deficit Disorder 267
 Autism Spectrum Disorder 267
 Bipolar Disorder 267
metonymisation 346, 348, 351, 352, 354
midwifery 68–81
migrants 52
modality 324, 333
modulation 333
monetary policy 308
 influencing 311
 risk management tool 309, 320
monetary story
 and central banks 309
moral 37, 42, 57, 68, 76–78, 80–81, 88, 99, 106, 127, 139, 150, 160, 165, 182, 307, 312
moral panics
 social worker as 'folk devil' 149
motivational relevancies 10
multi-perspective approach 3
multiple perspectives 40
mutuality 299

narrative context 57
narrative identity 57
narrative intervention 57
narrativisation 347, 349

Index

natural gas
  liquified 190
negotiation 5, 64, 110, 161, 180, 213, 220, 238, 245, 250, 264, 286, 301, 344
neo-conservative
  correctionalism 121, 122, 123
neo-liberal responsibilisation 121
new penology 88
news receipt 299
Nexus Analysis 341, 343, 340
nexus of practices 245
non-technical risks 176, 177
normality 70, 73, 77, 79–81
nuclear energy 173

orienting disposition 325
othering 106
  normalisation 142

panic
  rhetorical 324
participation 92, 122, 146, 155, 199
  civic 190, 196, 204
  Consumer culture 111
  public 177, 194, 198, 199
  social 37
participation framework 287, 288, 292
part-of-speech 247, 252
perception 219
perspective
  alignment with 299
  of patient 299
police and policing 92–93, 94–97
policy discourses 150
political mobilization 182
positivism 108
poverty 106, 112, 278
  social exclusion 112
power 5, 67, 69, 78, 80, 141, 148, 162, 203, 234, 264, 310, 321
powerlessness 113, 150
practical relevance 6, 54, 58
practice 1, 7, 11, 28, 40, 57, 80, 90, 139, 147, 165, 170, 263, 341, 349, 351
  best 90, 285
  biosecurity 234
  care 32
  corporate 181

disclosure 17
discursive 30
everyday 68
evidence based 123, 130
global 104
interpretive 163
lending 318
medical 281
models 122
precautionary 144
professional 1, 10, 63, 81, 150, 246
risk assessment 128
safe work 51, 58, 61
site of 9
social 341
practice frameworks 104, 147
precaution 86, 90, 221, 327
precautionary
  approach 218, 142
  measures 192, 209
  principle 210, 218
pregnant women 68–69, 71–72, 74–81
preparedness 85
preventive justice 88–91
print media
  Australian 245, 247
  Chinese 245, 247
privacy management 37, 38, 47
probability and impact matrix 174, 184
probation 90
profession 69–71, 73, 75–78, 80–81
professional
  language 233
project management 171, 173, 174, 176
project risk 173, 175, 177
projection 5, 299, 325
  affect 332
  climate 211, 223
  fact 331
  future 209, 213
  idea 329
  quote 328
  risk 295, 298
  situation 328
protest 171, 178, 181, 182, 184

psychosocial bias   123, 126, 128, 129
public discourse
  factors of good   193
  formalisation in   196
  grounded rationality in   197
  mediation in   202
  moderation in   197
  radical difference in   194
  reciprocity in   195
  reflexivity in   193
public irrationality   197
public meetings   197
public testimony

qualitative   22, 124, 128, 157, 161,
    162, 175, 210, 220, 233, 271
  psychometric   128, 133, 156, 159,
    162, 182, 213, 220, 312
quarantine   231, 351
Québec   189

radicalisation   138, 139, 145–146
reductionism   121, 122, 124, 127,
    134
reflexivity   10, 157, 158, 160, 163,
    167, 193
regulation   5
  absence of global regulator   307
  good vs bad   320
  light-touch   310, 311, 317, 318
  regulator
reification   275
  reification   of risk   140, 146
reputational
  damage   71, 77, 80, 82
  fears   18
  risk   18
resemiotisation   342
resilience   87
resistance   88, 97–99,
responsibility
  acceptance of   23, 29,
  acknowledgement of   25, 28, 29, 31
  attribution to others   23, 25, 26, 30
  complexity   27, 28, 30, 31, 32
  dispersed risk   27, 28
  guilt   30
  proactive   24
  refusal to attribute   23, 24, 30

reluctance to discuss   18, 19
retrospective   24
unable to determine   23, 27
revoice   59
rhetoric   40, 72, 139, 149, 203, 281,
    324
rhetorical
  activity   2,
  devices   39
  construction of risk   194
risk   1, 68, 70–81, 245, 254, 261, 262,
    265
  acceptability   233
  activity   235
  risk amplification of   332, 334
  analysis   230
  and responsibility   245
  anxiety   150
  appetite   176
  appropriate levels of   234
  as absolute   236
  assessment   210, 233
  risk assessment of   324, 327, 328,
    334, 326
  attenuation   332, 335, 336
  calculation of   285, 286, 299, 300
  clinical   30
  communication   211, 233, 246,
    257, 262
  communication of   171, 183
  complexity of   30
  conceptualisation of   156, 157,
    158, 165
  consequences   238
  cultures   173, 182
  decision making   325, 326
  definitions   233
  discourses   107, 245, 246, 308, 320
  ethical   18
  factorisation   104, 105
  factors   104
  financial   18
  formula   232
  frames   209
  hegemony of   103, 106
  ideologies in   233
  incalculability of   172, 175, 184
  influence on policy   104
  interpreter   326

risk – *continued*
  legal   18, 19, 20, 21, 24, 29, 30, 31, 32
  limitations of the risk factor model   104
  management   173–176, 178, 181–184
  management plans   175
  management systems   182
  manufactured   307, 312, 324
  mitigation   180, 184, 184
  non-technical   176, 177
  patterns of   308, 320
  perception   234, 239
  perception of   172, 173
  pricing   311, 317, 319
  primary   30, 31
  probabilities   234
  professional perspectives   107
  projection   295, 298
  psychological   18.19
  psychosocial   64
  quantification   211
  reformulation of   298
  religions   173. 182
  makers   258, 260, 264
  society   307, 320
  spreading   321
  social   51, 178, 179, 181, 183, 184
  social construction of   192
  society   1, 246
  systemic   309, 311
  technologies   323, 336
  tolerance   176, 181
  visibility of   308, 311, 319
  social amplification of   156, 160, 203
  sociocultural theory of   156, 157, 165
risk communication   2, 4, 7, 9, 11, 138, 140, 149, 166, 182, 189, 190, 192, 194, 195, 203, 232-, 233, 239, 246, 257, 262, 296, 323, 326, 327, 329, 342
risk communicator   9, 323, 326, 329, 332, 336
risk assessment   7, 8, 11, 42, 90, 107, 122, 124, 128, 132, 133, 156, 163, 171, 182, 190, 192, 198. 201, 209, 211, 220, 231, 233, 237, 325
  Asset   124, 125, 128, 129
  AssetPlus   122, 132, 133
  inclusive   194
  technological   194
risk averse   311, 141, 148
risk calculations
  meaning and value of   5, 299
  risk calculations reframing of   286 299
risk communication
  climate   219
  electronic mediated   286
  cross-cultural   178
  deficit models   342, 346
  electronically mediated   286
  functional strategies   8, 204
  patient   285
  rational model of   301
  schools   200
  tools   327
  trajectory of   289
risk discourse   2, 106, 108, 122, 127, 142, 146, 149, 158, 166, 232, 308, 321
Risk Factor Prevention Paradigm   122, 125, 128, 131
Risk Factor Research   121, 123, 126, 129, 132, 134
risk factors   8, 90, 104, 115, 123, 127, 133, 144, 148, 238, 290, 294
risk framing   157, 165, 167, 225
risk management   48, 73, 123, 133, 140, 171, 175, 178, 181, 185, 194, 217, 221, 237, 239, 268, 308, 311, 324
  flood   221
  methodology   124
  models   317
  policies   239
  practitioners   176
  tools   319, 321
risk perception   6, 11, 162, 164, 166, 172, 195, 197, 208, 336
  amplification   160
  culturally embedded   166
  environmental   155, 165
  knowledge   165

studies 165
public, expert 156, 158, 160
risk situation
 aligning 9, 324, 331, 334, 336
 anchoring 331, 332
 attenuated 336
 judgement 333
 framing 324, 327, 328, 331, 335
Risk Society 1, 8, 87, 93, 115, 121, 145, 149, 158, 159, 233, 145, 158, 233, 246, 270, 307, 320
risk-taking 310
risk thinking 139, 140, 143, 146, 150
risk-needs 90–91
risks
 sociotechnical 155, 158, 162, 165, 166

safety 7, 9, 26, 51, 54, 58, 61, 68, 70–73, 75, 79–80, 98, 140, 143, 148, 325, 328, 346, 351, 353
 communication 51, 58, 60, 62
 coordinator 21, 26,
 discourse 95
 information 56, 58
 patient 19, 32
 practices 58
 practitioners 103
 procedures 51, 58
 product 11
 regulations 55, 202
 road 96
 training 60
 workplace 52
salience 173
Sarah's Law 90
Scaled Approach 122, 131, 133
Science and Technology Studies (STS) 155, 161, 163–4, 166, 210
security 86–89, 92, 94, 191, 327, 331, 333
 consciousness 93
 home 92, 93
 information 93
 international 2, 9
 technology 93
severity 174
Signs of Safety Approach 147
site of engagement 6, 11, 243, 245, 343, 344

siting conflicts 189
social arena concept of risk debates 196
social constructionism 139, 142
social impact assessment 171, 182–184
social licence to operate 183, 184
social sensitivity 37, 38, 41, 42, 43
social work 2, 7.40, 48.138.140, 143, 146
 English system 138, 140,
socio-cultural change 155, 163, 167
sociological imagination 142, 143
space 6, 36, 45, 112, 113, 169, 221, 250, 288, 326, 333
 private interactional 291
 reconfiguration of 288
 relational 288, 294
 bounded 324, 326, 332
 perception 161
 pragmatic 333
 social 113
speculative pre-emption 87
sphere of attention 288, 290
stakeholder
 management 175
stakeholders 17, 31, 175, 182, 183, 185, 191.196, 200, 231, 235, 238
stance 3, 23, 30, 39, 61, 246, 249, 250, 320
 marker 249, 250
stereotyping 53, 164
stigma 6, 36, 110
stigmatisation 108, 334, 346, 348, 352, 357
stigmatised 109, 127, 356
strengths based practice 143, 147
subjectivity 128, 249, 278, 281, 161, 167, 220, 225, 246
 epistemic 164
 as constructed in risk discourse 156, 158, 164, 165, 167
 subjectivity powerful moments of 161
surveillance 74, 81, 89, 90, 140, 146, 284
 monitoring 141,
 state 139, 144
sustainability 155, 167, 173, 195

tacit models 9
technologies of governance 157
terrorism 85, 87, 138, 139, 232, 324, 327, 334
  risk of 86
travel
  advice 323, 326
  advisories 323, 324, 328, 334
  freedom 326, 327, 328, 332
traveller
  prudential 323, 326
triadic
  engagement 288
  participation structure 287

uncertainty 8, 19, 20, 42, 79–80, 87, 132, 139, 146, 156, 158, 159, 160, 164, 166, 208, 210
  frames 209, 211, 213, 218, 220, 224, 246, 274

  management 160
  quantification 210
USAS 248

war gaming 86
war room 179
W-Matrix 248, 250, 252

youth crime
  as structured social action 113
  car crime 111
  causation 104, 107
  delinquent solutions 109
  entrepreneurship 111
  impact of structure 105, 106
  reputation 110
  seductions of crime 113
  selling stolen goods 111
  status 112
  subjective meanings 105
Youth Offending Teams 124, 125, 134